Mathematical Analysis and Proof

The direction in which education starts a man will determine [his] future life.

Plato, *The Republic, Book IV* [427-347 B.C.]

Mathematics possesses not only truth, but supreme beauty—a beauty cold and austere, like that of sculpture, and capable of stern perfection, such as only great art can show.

Bertrand Russell in *the Principles of Mathematics* [1872-1970]

Talking of education, "People have now a-days" (said he) "got a strange opinion that every thing should be taught by lectures. Now, I cannot see that lectures can do so much good as reading the books from which the lectures are taken. I know nothing that can be best taught by lectures, expect where experiments are to be shewn. You may teach chymestry by lectures. — You might teach making of shoes by lectures!"

James Boswell: *Life of Samuel Johnson, 1766* [1709-1784]

ABOUT OUR AUTHOR

David S.G. Stirling is Senior Lecturer in Mathematics at the University of Reading, where he has been since 1972. Born in 1947, he was educated at the High School of Glasgow, the University of Glasgow (BSc) and the University of Edinburgh (PhD). His PhD thesis was on *Banach algebras*. Subsequently, his research interests have been in integral equations, in particular on methods of approximating solutions to certain types. He has published papers in the *Glasgow Mathematics Journal*, the *Journal of the London Mathematical Society*, the *IMA Journal of Numerical Analysis*, and the *Proceedings of the Edinburgh Mathematical Society*.

As well as earlier editions of the predecessor book *Mathematical Analysis*, published by Ellis Horwood, he has co-authored with David Porter the book *Integral Equations*, published by Cambridge University Press in 1990.

Throughout his career at Reading David Stirling has been involved in the teaching of analysis, at all levels of the undergraduate curriculum, including course design and development, as well as a range of other topics. Other responsibilities have been admissions tutor, course adviser, careers liaison, Secretary of the British Mathematical Colloquium in 1993 and Hon. President of Reading Association of University Teachers (1993-95). He represents the London Mathematical Society on the Undergraduate Mathematics Teaching Conference Committee.

Apart from his mathematical interests David Stirling is interested in classical music, opera and railways, in particular railway signalling, on which he has written many papers, and is an Associate of the Institution of Railway Signal Engineers. He also designed the Scottish Railway Exhibition at Bo-ness in the Forth Valley.

Mathematical Analysis and Proof

David S.G. Stirling
Senior Lecturer in Mathematics
University of Reading

Albion Publishing
Chichester

First published in 1997 by
ALBION PUBLISHING LIMITED
International Publishers
Coll House, Westergate, Chichester, West Sussex, PO20 6QL England

British Library Cataloguing in Publication Data
A catalogue record of this book is available from the British Library

ISBN 1-898563-36-5

Printed in Great Britain by Hartnolls, Bodmin, Cornwall

Table of Contents

Table of Contents

Author's Preface

Analysis tackles the issues which were fudged in the development of the calculus. With the trend away from formal proof in school, it may not be evident to students beginning higher education that there is a problem to be attended to here. Indeed, most school leavers have seen virtually none of the ideas of proof and do not necessarily accept that it is a vital part of mathematics. This book's forerunner, *Mathematical Analysis: A Fundamental and Straightforward Approach*, Ellis Horwood, 1987, was written in acknowledgement that most students then had this background. Since that time there has been further substantial change in the mathematical background of those entering higher education, and the present book addresses the change.

The book begins with much material on proof, the logic involved in proof and the associated techniques. To some extent this requires facing issues which used to be taken for granted: manipulation of symbols, inequalities and the like. Since the ideas of proof require to have some vehicle to carry them and through which they can be illustrated, the early chapters take the opportunity of practising manipulative skills to an extent rather greater than necessary for the proof alone. The ideas of proof are of little use without the means of carrying them out.

The main aim of the book is to present the accepted core material of analysis in such a way that the development appears fairly natural to the reader. Some of the ideas (such as the sum of an infinite series) are first introduced slightly less formally and later treated formally. This is intended to ease the reader's passage, but care has been taken to avoid inconsistencies or confusion. One of the difficulties of teaching analysis is that it is some time before we reach results which a newcomer would consider both worth knowing and not achievable by other less demanding means. A detailed discussion of the real number system, which is necessarily technical, is postponed until other matters have highlighted the need for it, while I have tried to maximise the number of results whose value can be appreciated from a standpoint other than that of the analyst, so that the subject is not seen as merely self-serving. This approach, while it could not be sustained throughout a degree course, seems to be correct for the start of a subject. The technical jargon of analysis cannot sensibly be avoided but it can be reduced and I have taken the view that a definition is not worth the sacrifice of memory unless it is used often.

The principal difference between this book and many others is that attention is devoted not only to giving proofs but to indicating how one might construct these proofs, a rather different process from appreciating the final product. The completeness of the real number system is assumed in the form of Dedekind's axiom of continuity, because this is more plausible than some of its immediate consequences.

Logically, one can study analysis with no knowledge of calculus, but it would be rather pointless to do so, and I have tacitly relied on calculus for some of the motivation. This is particularly true of Chapter 14, on functions of several variables, where the experience of grappling with the problems which arise in practice is a necessary supplement to the theory.

The book contains many problems for the reader to solve, designed to illustrate the main points or to force attention onto the subtler ones. Tackling these problems is an essential part of reading the book although the starred problems may be regarded as optional, being more difficult or more peripheral than the others.

I should like to thank my colleagues at Reading, especially Leslie Bunce, David Porter and David White, for comments and useful conversations over the years and the students who have been subjected to courses based on this material. Others who have influenced my views of analysis and how it should be taught have been some of the participants in the Undergraduate Mathematics Teaching Conference, Johnston Anderson, Keith Austin and Allan Norcliffe in particular.

I am grateful to Rosemary Pellew for typing the manuscript and coping with both my handwriting and the demands of setting mathematical symbols in type. Ellis Horwood and his staff at Albion have been very helpful with their editorial and production skills, and for the co-operative way in which they proceed.

Reading,April 1997   David Stirling

1

Setting the Scene

"If a man will begin with certainties, he shall end in doubts; but if he will be content to begin with doubts, he shall end in certainties."   Francis Bacon.

1.1 Introduction

We have all seen mathematical formulae like

$$1 + 2 + 3 + \ldots + n = \tfrac{1}{2} n(n+1)$$

or $\sin(x + y) = \sin x \cos y + \cos x \sin y$.

What do these actually mean? - and why should we believe them?

In both cases, we mean more than we have said. The first statement has n in it, which we have not explained, but it is understood that n is some positive whole number. If n is a specific positive integer (using the more imposing word integer for a whole number), say 4, then it is easy to check that $1 + 2 + 3 + 4$ and $\tfrac{1}{2} 4(4+1)$ both equal 10. However we would usually interpret the statement not as being true for one particular value of n but for all positive integers. Obviously we can test this for as many different values of n as we like but, however many we test, there will remain lots of integers for which the formula has not been tested. Mathematics gives us reasons to believe that the formula is true also for these untested values and that there will be no surprises. The old proverb that "the exception proves the rule" is not part of the mathematical folklore!

The second statement involves two complicated functions, sin and cos, and we shall gloss over the detail of what these mean. The formula here claims that if we choose x and y to be two numbers, which do not have to be integers, then the formula holds for these values. Again there is an argument, more complicated in this case, why we should believe this. This formula, however, holds for a wider range of values of the "variable" x (and a second "variable" y) in that x can be any real number, not just a whole number. We really ought to distinguish the two by saying for which values of the "variable" the formula is supposed to hold. Of course, human nature tends to brush these things aside as "obvious" (although what is obvious today may not be so obvious tomorrow!). The statements should really have been something like "for all positive integers n, $1 + 2 + \ldots + n = \tfrac{1}{2} n(n+1)$" and "for all real numbers x and y, $\sin(x + y) = \sin x \cos y + \cos x \sin y$".

The important point, however, is that in both cases above we believe the result to be true not only for those values of the variables which we have tested but for others as yet untried. This is one of the features of mathematics: it "predicts" results before we have tested them. In the first example above we are confident that even if we substitute some huge new number for n the result will still be correct. The basis for this confidence is that we have proved the result (or, if you have not already done that, because you soon will). That is, we show that, given certain other results which we accept, the statement we have made must be true.

If we wish to establish results like these, and more interesting examples in due course, we need to start somewhere. That is, we need to begin with a body of information which we accept, perhaps provisionally, as true and deduce what follows from this. Exactly where we start is a matter of choice, for we can always go back later and show how the assumptions we have made can all be deduced from other more fundamental assumptions. One approach, investigated about the beginning of the twentieth century, is to try to deduce the whole of mathematics from pure logic. This turns out to be abstract and difficult, and it takes a great deal of work to reach the ordinary realms of mathematics; it also introduces some philosophical problems. The starting point we shall take in this book is to accept the number systems we know and deduce things from there. These assumptions can be replaced by more elementary ones, but we shall leave that to courses on logic and the foundations of mathematics.

We are going to start with the number systems we know, so we had better state what these are.

1.2 The Common Number Systems

The **natural numbers** (or positive integers), $1, 2, 3, \ldots$ are the fundamental numbers by which we count. In due course we shall have to list their properties, but for the moment let us just notice that there is a smallest (or first) natural number and that each one has a succeeding one. We can add and multiply two natural numbers and the result is in each case another natural number; these operations obey various rules which, again, we can specify in detail later when we need to. The natural numbers are rich enough to be interesting - for example, some can be factorised into the product of two smaller numbers and others cannot - but they have some limitations for everyday calculations: for example, we cannot solve the equation $x + 3 = 2$ within the framework of the natural numbers.

The **integers** consist of the whole numbers, positive or negative or zero, that is, $0, 1, -1, 2, -2, 3, \ldots$.

The **rational numbers** are those numbers which are fractions (or *ratios* of two integers), that is, numbers of the form a/b where a and b are integers and $b \neq 0$. These numbers clearly have the advantage over the integers in that we can always solve the equation $bx = a$ within the system, if a and b are rational numbers and $b \neq 0$. However we need a little caution with rational numbers, for they can be written in many ways in the form a/b (so that $\frac{2}{3} = \frac{4}{6} = \frac{-12}{-18}$), that is, the numerator and denominator are not uniquely fixed once we know the number. If necessary we could specify that a and b have no common factor greater than 1 and that b is positive, to produce a "standard" way of representing a rational number. It may take a moment's thought to convince yourself that this is possible. Notice also that in passing to the rational numbers we have lost the idea of there being, in any natural way, a "next" such number after a particular one. For if x and y are two rational numbers, then there are rational numbers between them, for example $\frac{x+y}{2}$; this would be impossible if y were the "next" rational number greater than x.

The last number system we shall consider in this book is the system of **real**

numbers. These are intended to allow us to describe the lengths of lines in geometrical figures, including such things as the circumference of a circle. Pictorially we can think of this as being the numbers required to describe all of the distances from zero of points on a straight line, the positive distances being to the right of 0. It is not obvious that we need numbers which are not rational to do this, but we do, as we shall see later. The numbers $\sqrt{2}$, π and e (the base of natural logarithms) are real numbers which are not rational; such numbers are called **irrational**.

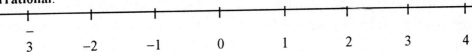

Example 1.1 Let us accept the number systems above as given, and get down to some work. Take the formula

$$1 + 2 + 3 + \ldots + n = \tfrac{1}{2}n(n+1) \ ,$$

which we would like to show is true for all natural numbers n. (Actually, what we want to do is prove that, with the given assumptions about the numbers, the formula has to be true.)

Let n be a natural number, and let $S = 1 + 2 + \ldots + n$.

Then
$$1 + 2 + \ldots + n = S$$
and
$$n + (n-1) + \ldots + 1 = S \ ,$$
so that by adding corresponding terms we have

$$\underbrace{(n+1) + (n+1) + \ldots + (n+1)}_{n \text{ terms}} = 2S \ .$$

Therefore $2S = n(n+1)$ so $S = \tfrac{1}{2}n(n+1)$. Since we assumed of n only that it is a natural number, this all holds for all such n.

The result is proved to be true, but notice that we have used some properties, such as the property that the order of the numbers added up does not affect the sum. $\square$

Example 1.2 Some other problems lead to rather different outcomes. Consider the three equations

$$\sqrt{(x+3)} = \sqrt{(1-x)} + \sqrt{(1+x)} \tag{1}$$
$$\sqrt{(x+3)} = \sqrt{(1-x)} - \sqrt{(1+x)} \tag{2}$$
$$\sqrt{(x+3)} = \sqrt{(1+x)} - \sqrt{(1-x)} \tag{3}.$$

In each case we wish to find all the real numbers x which satisfy the equation. As usual the symbol $\sqrt{\ }$ denotes the non-negative square root.

Let us begin with equation (1)

$$\sqrt{(x+3)} = \sqrt{(1-x)} + \sqrt{(1+x)} \ .$$

The square roots are a nuisance so we square both sides, giving

$$x + 3 = 1 - x + 2\sqrt{(1-x)}\sqrt{(1+x)} + 1 + x \ ,$$

which we rearrange to give

$$x + 1 = 2\sqrt{(1-x)}\sqrt{(1+x)} \ .$$

Then squaring again removes the square roots and we have

$$x^2 + 2x + 1 = 4(1-x)(1+x) = 4 - 4x^2 .$$

The rest is easy, for we obtain in turn

$$5x^2 + 2x - 3 = 0 ,$$
$$(5x - 3)(x + 1) = 0 ,$$
$$x = \tfrac{3}{5} \text{ or } x = -1 .$$

We conclude that the solutions of equation (1) are $x = \tfrac{3}{5}$ and $x = -1$. We can check these if we wish, for putting $x = \tfrac{3}{5}$ into the left hand side of (1) gives $\sqrt{\tfrac{18}{5}} = \tfrac{3\sqrt{2}}{\sqrt{5}}$, while substituting $x = \tfrac{3}{5}$ on the right hand side gives $\sqrt{\tfrac{2}{5}} + \sqrt{\tfrac{8}{5}} = \sqrt{\tfrac{2}{5}} + 2\sqrt{\tfrac{2}{5}} = \tfrac{3\sqrt{2}}{\sqrt{5}}$; $x = \tfrac{3}{5}$ is a solution. It is simpler to check that $x = -1$ is also a solution.

Now let us tackle equation (2). Squaring the original equation gives

$$x + 3 = 1 - x - 2\sqrt{(1-x)}\sqrt{(1+x)} + 1 + x ,$$

which rearranges to $\qquad x + 1 = -2\sqrt{(1-x)}\sqrt{(1+x)} .$

Squaring gives $\qquad x^2 + 2x + 1 = 4 - 4x^2 ,$

as before, and again we solve this to give $x = \tfrac{3}{5}$ or $x = -1$. Now substituting $x = \tfrac{3}{5}$ in the left hand side of (2) gives $\tfrac{3\sqrt{2}}{\sqrt{5}}$, while substituting into the right hand side gives $\sqrt{\tfrac{2}{5}} - 2\sqrt{\tfrac{2}{5}} = -\tfrac{\sqrt{2}}{\sqrt{5}}$. The two sides are not equal, so $x = \tfrac{3}{5}$ is not a solution. However, $\sqrt{(-1+3)} = \sqrt{2}$ and $\sqrt{(1-(-1))} - \sqrt{(1+(-1))} = \sqrt{2}$, and therefore $x = -1$ is a solution. In this case we have found one true solution and one spurious "solution".

Equation (3), tackled the same way gives

$$x + 3 = 1 + x - 2\sqrt{(1-x)}\sqrt{(1+x)} + 1 - x ,$$

and again we follow this through to obtain $x = \tfrac{3}{5}$ or $x = -1$. In this case neither number is a solution. (Check that!) $\square$

This is rather disturbing, for the processes above lead to false "solutions", which we need to check. Obviously we need to know when this checking process is essential and when it can be omitted, which means we need to know why these spurious solutions appear. Have we made some mistake? It certainly does not seem so.

Example 1.3 Let us consider another situation. It is well known that if a, b and c are real numbers, with $a \neq 0$, then the quadratic equation

$$ax^2 + bx + c = 0$$

has solutions $\qquad x = \dfrac{-b \pm \sqrt{\left(b^2 - 4ac\right)}}{2a} .$

In the case where $b^2 \geq 4ac$ these are real numbers, while if $b^2 < 4ac$ there are no real numbers satisfying the equation. We shall not be interested in solutions which are not real numbers.

Solving a cubic equation is considerably more complex. The general equation

$ax^3 + bx^2 + cx + d = 0$, with $a \neq 0$, has the same solutions as $x^3 + (b/a)x^2 + (c/a)x + d/a = 0$ and by substituting $y = x - b/(3a)$ we can obtain an equation for y in which the coefficient of y^2 is zero. We need to solve equations of the form

$$y^3 + py + q = 0 \tag{4}$$

where p and q are known real numbers. (They have rather messy expressions in terms of a, b, c and d.)

There is a trick for doing this, which has been known since the early sixteenth century, although its inventor, Ferro, kept it secret. We find two real numbers s and t satisfying

$$\left.\begin{array}{c} s - t = -q \\ \\ st = \left(\frac{p}{3}\right)^3 \end{array}\right\} \tag{5}$$

and then set $y = s^{\frac{1}{3}} - t^{\frac{1}{3}}$. Here $s^{\frac{1}{3}}$ means the number z satisfying $z^3 = s$. By (modern) algebra we can check that

$$
\begin{aligned}
y^3 + py + q &= s - 3s^{\frac{2}{3}}t^{\frac{1}{3}} + 3s^{\frac{1}{3}}t^{\frac{2}{3}} - t + p(s^{\frac{1}{3}} - t^{\frac{1}{3}}) + q \\
&= s - t - 3s^{\frac{1}{3}}t^{\frac{1}{3}}(s^{\frac{1}{3}} - t^{\frac{1}{3}}) + 3s^{\frac{1}{3}}t^{\frac{1}{3}}(s^{\frac{1}{3}} - t^{\frac{1}{3}}) + q \\
&= 0
\end{aligned}
$$

and so all we have to do is solve (5). But (5) can be reduced to a quadratic

$$0 = s^2 - st + qs = s^2 + qs - \left(\frac{p}{3}\right)^3$$

giving

$$s = -\frac{q}{2} \pm \frac{1}{2}\sqrt{\left(q^2 + 4\left(\frac{p}{3}\right)^3\right)}.$$

This yields real numbers only in the case where $4p^3 + 27q^2 \geq 0$.

Try this method on the equation

$$y^3 - 3y + 2 = 0,$$

which factorises into $(y-1)(y-1)(y+2) = 0$. We have $q^2 + 4\left(\frac{p}{3}\right)^3 = 4 + 4(-1)^3 = 0$, giving $s = -\frac{q}{2} = -1$. Then from (5) $t = 1$ and $y = -1 - 1 = -2$. The method does give a solution, but notice that it gives only one of the solutions. □

This is a phenomenon we need to be aware of: there are solution processes which give only some of the possible solutions of the problem, but not all. This is the complementary case to the one earlier where we obtained extra spurious "solutions"; this time there are some missing. The technique for solving cubic equations is no longer of much interest, but we shall meet solution processes which yield only some of the solutions of a problem later. The immediate question is why we have only some solutions, and whether we can detect when this is going to happen.

2

Logic and Deduction

"Contrariwise", continued Tweedledee, "if it was so, it might be; and if it were so, it would be; but as it isn't, it ain't. That's logic."   Lewis Carroll.

2.1 Introduction
We said in Chapter 1 that if we start with the equation
$$\sqrt{(x+3)} = \sqrt{(1+x)} - \sqrt{(1-x)}$$
then an apparently correct argument leads us to show that $x = \frac{3}{5}$ or $x = -1$, yet neither of these numbers satisfies the equation. The process produces spurious solutions. We also saw the converse phenomenon with the cubic equation: that process produces only correct solutions, but it does not find all of them. In both cases we have to do additional work to modify the list of solutions we first obtain. What on earth is happening here?

Let us begin with the equation
$$\sqrt{(x+3)} = \sqrt{(1+x)} - \sqrt{(1-x)} \ .$$
We square both sides of this, to produce
$$x + 3 = 1 + x - 2\sqrt{(1+x)}\sqrt{(1-x)} + 1 - x$$
and carry on from there. Now what we really mean at this step is that if $\sqrt{(x+3)} = \sqrt{(1+x)} - \sqrt{(1-x)}$ then the square of the left hand side equals the square of the right hand side. From that second step we notice that if it is true then
$$x + 1 = -2\sqrt{(1+x)}\sqrt{(1-x)}$$
and, omitting some detail, then $x = \frac{3}{5}$ or $x = -1$. That is, what we have actually shown is that *if* x satisfies the original equation *then* $x = \frac{3}{5}$ or $x = -1$. Nobody has yet said that x *does* satisfy the original equation, but we have obtained some useful information. If x satisfies the equation then it has to be one of two numbers, so we have cut down the range of potential solutions from the whole collection of real numbers to just two, and we can check individually whether or not these two potential solutions do satisfy the equation. The checking process establishes whether or not the statements "$x = \frac{3}{5}$ is a solution" or "$x = -1$ is a solution" are true. This could be put slightly more formally by saying that we check whether or not it is true to say that "if $x = \frac{3}{5}$ then $\sqrt{(x+3)} = \sqrt{(1+x)} - \sqrt{(1-x)}$".

The checking process may be tedious and it is obviously important to know when we need to use it. Moreover, if we consider the example of the cubic equation, we see that we also need to know what to check. For the cubic equation $y^3 + py + q = 0$, we find s and t satisfying $s - t = -q$ and $st = (p/3)^3$ and then set $y = s^{\frac{1}{3}} - t^{\frac{1}{3}}$. The logic here is that if we can find s and t and construct y, then that y satisfies the equation $y^3 + py + q = 0$. That is, the y we find will be a solution of the cubic, but the process is tantamount to saying that a number of a certain kind is a solution; it does not tell us whether some numbers of a different kind are also solutions.

The point to notice here is that in both cases what we have actually shown to be true is a statement of the form "If A then B" where A and B are statements themselves. This is actually the heart of mathematics, in that the subject proceeds by noticing that if one thing holds then necessarily some other statement has to be true. The statement "if A then B" is often expressed as "A implies B" and put in symbols as $A \Rightarrow B$. What this tells us is that the two statements A and B are related in such a way that should A happen to be true then B must also be true. It does not indicate whether or not A and B themselves are true, nor does it give us any useful information about B in the case where A happens to be false. From the two statements $A \Rightarrow B$ and $B \Rightarrow C$ we can deduce that $A \Rightarrow C$.

Let us return to the equation $\sqrt{(x + 3)} = \sqrt{(1 + x)} - \sqrt{(1 - x)}$. Then, using implication signs, we have

$$\sqrt{(x+3)} = \sqrt{(1+x)} - \sqrt{(1-x)} \quad \Rightarrow \quad x+3 = 1+x - 2\sqrt{(1+x)}\sqrt{(1-x)} + 1 - x$$
$$\Rightarrow \quad x+1 = -2\sqrt{(1+x)}\sqrt{(1-x)}$$
$$\Rightarrow \quad x^2 + 2x + 1 = 4 - 4x^2$$
$$\Rightarrow \quad 5x^2 + 2x - 3 = 0$$
$$\Rightarrow \quad (5x - 3)(x + 1) = 0$$
$$\Rightarrow \quad 5x - 3 = 0 \quad \text{or} \quad x + 1 = 0$$

where we mean here that each statement implies the one immediately following it, so we conclude that

$$\sqrt{(x+3)} = \sqrt{(1+x)} - \sqrt{(1-x)} \Rightarrow x = \tfrac{3}{5} \quad \text{or} \quad x = -1.$$

This is not exactly what we want, but it is a great step in the right direction, since we have now shown that at most two numbers are candidates for solutions of the equation $\sqrt{(x + 3)} = \sqrt{(1 + x)} - \sqrt{(1 - x)}$. We now need to check the truth of the opposite implications, e.g. $x = -1 \Rightarrow \sqrt{(x + 3)} = \sqrt{(1 + x)} - \sqrt{(1 - x)}$, which happens to be false. Also $x = \tfrac{3}{5}$ does not imply that $\sqrt{(x + 3)} = \sqrt{(1 + x)} - \sqrt{(1 - x)}$ so that neither of the candidates is a solution. We conclude that the equation has no real solutions.

2.2 Implication

There are, then, two possible implications connecting the statements A and B: "$A \Rightarrow B$" and "$B \Rightarrow A$". For a given pair of statements A and B it may happen that one of these implications is true but not the other, or they may both be true, or neither may be true. The most satisfactory case, which occurs frequently, is when both are true. For example, if A is "$x = y + 5$" and B is "$x - y = 5$" then $A \Rightarrow B$ (if A is true, subtract y from both sides and we see that B is true), and $B \Rightarrow A$ (if B is true, add y to both sides and we see that A is true). Since we can write $B \Rightarrow A$ just as neatly as $A \Leftarrow B$ (and we take that to be the meaning of $\Leftarrow$) we shall write $A \Leftrightarrow B$ to mean "$A \Rightarrow B$ and $B \Rightarrow A$". This means that if A is true then B is true, and if B is true then A is true, so that either A and B are both true or they are both false. This is the most useful logical connection, although in practice it is usually easier to establish "$A \Rightarrow B$" and "$B \Rightarrow A$" separately. "$A \Leftrightarrow B$" is usually pronounced "A (is true) if and only if B (is true)".

It is important to distinguish between the three logical connectives $\Leftrightarrow$, $\Rightarrow$ and $\Leftarrow$, particularly since everyday speech often blurs the distinction. If, in ordinary life, I say "I will give you this book if you pay me a pound", then in addition to the statement just made there is an unspoken presumption that I will not give you the book if you do not give me the pound. Everyday human behaviour may lead us to that conclusion, and it may be correct, but it is not what was actually said.

In our arguments above we used the statement

$$\sqrt{(x+3)} = \sqrt{(1+x)} - \sqrt{(1-x)} \Rightarrow x+3 = 1+x - 2\sqrt{(1+x)}\sqrt{(1-x)}+1-x \ .$$

This is as much as we can say immediately: if two numbers are equal then their squares are equal. The converse is false in general since the squares of two numbers may well be equal without the two numbers being equal; for example $2^2 = (-2)^2$. We cannot say that the right hand side above implies the left hand side without additional information.

Consider a simpler problem. We wish to solve $x^2 + 2 = 2x + 1$.

$$\begin{aligned} x^2 + 2 = 2x + 1 &\Leftrightarrow x^2 - 2x + 1 = 0 \\ &\Leftrightarrow (x-1)^2 = 0 \\ &\Leftrightarrow x - 1 = 0 \\ &\Leftrightarrow x = 1 \ . \end{aligned}$$

We deduce from that that $x^2 + 2 = 2x + 1 \Leftrightarrow x = 1$, that is, if x satisfies the orginal equation $x^2 + 2 = 2x + 1$ then $x = 1$ and if $x = 1$ then x satisfies the equation. In this case a separate checking step is unnecessary, as the working above has shown that x satisfies the equation if and only if $x = 1$. Therefore, to see whether a separate checking step is necessary we need to notice whether all the connections are "$\Leftrightarrow$" or whether some of them are "one way" connections.

2.3 Is This All Necessary – or Worthwhile?

This rather fussy examination of the logical connections and implications may seem to be splitting hairs. At a practical level, for simple problems, we can find the right answer without it, so is it all necessary? If we could always produce arguments where the connections were $\Leftrightarrow$ (even if we did not notice that) then perhaps we could be said to be being fussy. However, problems soon arise where we have to be content with partial information and patiently assemble that to give what we want. To take a specific example, suppose we had some good reason to find all the solutions of the equation

$$x^{17} + e^x + 1 = 0 \ .$$

This is not going to be easy if we wish exact solutions and it soon becomes plausible that we are not going to be able to express the solutions exactly in terms of known quantities. If we believe this (and we might be pleasantly surprised if it later turns out to be wrong) then we have to ask ourselves what we *could* do. We could deduce properties that all the solutions have, assuming there are any solutions: for example, that all solutions are negative and lie between -2 and -1. (Forget the detail of how we might do this, it does not matter.) We might find a value of x for which $x^{17} + e^x + 1$ is very close to zero. Does it follow that x is close to an actual solution of the equation? If so, how close? Now all of these things can be done, and in this case a graphical argument is convincing, but the point is that we are assembling scraps of information and we have to be absolutely clear about the status of these "scraps". Looking further ahead, we may wish to solve problems where there are several variables and graphical arguments are no help. If we can gain some intuition in how to use the logic on its own, then that may be useful in the more complicated situation when other techniques fail.

When all is said and done, however, logical deduction is what much of mathematics is about. In relatively simple problems it may be possible to sidestep the logic or to pass glibly over it, but the business of a mathematician is to deduce that from certain mathematical assumptions (perhaps that a number or a function satisfies a certain equation) some other consequences follow.

Mathematical argument is based on putting together logical statements to form a convincing demonstration of what we wish to show. To do that we may have to demonstrate that one statement implies another, so we need some techniques for doing this.

Example 2.1 Show that $\sqrt{(1+x)} = 1 + \sqrt{(1-x)} \Rightarrow x^2 = \frac{3}{4}$.

(To be strictly correct, we ought to say that x is to be a real number, which is tacitly assumed here. Whether or not one does this here is a matter of taste, for one can overdo correctness to the extent of obscuring the main point!)

Solution: Suppose that $\sqrt{(1 + x)} = 1 + \sqrt{(1 - x)}$

Then	$1 + x = 1 + 2\sqrt{(1 - x)} + 1 - x$	(squaring).
Therefore	$2x - 1 = 2\sqrt{(1 - x)}$	(rearranging)
and	$4x^2 - 4x + 1 = 4 - 4x$	(squaring again).
Therefore	$4x^2 = 3$	
and so	$x^2 = \frac{3}{4}$.	

From the above it is clear that if, as we supposed, $\sqrt{(1 + x)} = 1 + \sqrt{(1 - x)}$ then it follows that $x^2 = \frac{3}{4}$. A concluding statement is not really necessary, and we have justified most of the steps with the reasons in brackets. In this sort of deduction we can save some labour by using the symbol $\therefore$ for "therefore" but the main feature of this style is that having made the initial supposition, all the subsequent statements are plain, unconditional ones like "$4x^2 = 3$". This sort of layout is well-suited to complicated arguments where one line may be true because of a combination of earlier information. $\square$

We could have proved the result in the last example by using implication signs:

$\sqrt{(1+x)} = 1 + \sqrt{(1-x)}$	$\Rightarrow$	$1 + x = 1 + 2\sqrt{(1-x)} + 1 - x$ (squaring)
	$\Rightarrow$	$2x - 1 = 2\sqrt{(1-x)}$ (rearranging)
	$\Rightarrow$	$4x^2 - 4x + 1 = 4 - 4x$ (squaring again)

$$\Rightarrow \qquad 4x^2 = 3$$
$$\Rightarrow \qquad x^2 = \tfrac{3}{4} \ .$$

Again the layout makes it clear that the first statement implies the last. The second form has the advantage that it is easy to check that the connections are all compatible; from $A \Rightarrow B$ and $B \Rightarrow C$ we can deduce that $A \Rightarrow C$ but if we know $A \Rightarrow B$ and $B \Leftarrow C$ then we do not have enough information to deduce any connection between A and C.

The second form has the feature that we may be able to see that each of the connections is actually $\Leftrightarrow$. (Even if this is true the reasons for one implication may be different from the opposite implication.) Its drawback is that each line has to be a consequence purely of the preceeding line, which may not be convenient for a complicated deduction.

Example 2.2 Show that if x is an odd integer then x^2 leaves a remainder of 1 on division by 8.

Solution: Suppose that x is an odd integer. Then, for some integer n, $x = 2n + 1$.

Therefore
$$x^2 \ = \ (2n + 1)^2 \ = \ 4n^2 + 4n + 1$$
$$= \ 4n(n + 1) + 1 \ .$$

Now $n(n + 1)$ is a multiple of 2, because either n is or $n + 1$ is, and so $4n(n + 1)$ is a multiple of 8. It follows that x^2 is of the form of (a multiple of 8) + 1, so it leaves a remainder of 1 on division by 8. $\square$

The argument just given is not very subtle, but notice that if we wrote it using $\Rightarrow$ signs, the step from $x^2 = 4n(n + 1) + 1$ to x^2 having a remainder of 1 on division by 8 would need a line or two of justification as an aside. It is a matter of personal preference whether this is simpler or more complicated to understand than the one given. The proof given is a little fortunate in that we began with the information that x was an odd integer and were able to calculate something about remainders on division by 8. It is more tedious to carry out in this example, but if we wish information about remainders on division by 8 then we may find it best to start with that sort of information. That is, we could have started by assuming that x was of the form $8n + r$ for integers n and r, with r one of the possible remainders on division by 8, that is 0, 1, ..., 7. This is longer than the proof above, but it is more routine in some ways. One of the skills we have to keep at our disposal is to notice – and keep noticing – a variety of ways in which we might deduce what we wish, and then choose the one which seems simplest. This choice will, of course, occasionally be mistaken, in which case we revert to the method which looks to be the next best.

2.4 Using the Right Words

Tweedledum (or was it Tweedledee?) said that when he used a word it meant exactly what he wanted it to mean. This is fine for him, but not of too much use if no-one else knows the intended meaning. Mathematicians tend to move some way along the same road, in that we use words to mean what we want them to mean - but we announce this in advance by making a **definition**, and then sticking to that definition. This may be fairly informal or, particularly if we will need to refer to it later, it may be put formally and highlighted by the heading "Definition".

The first definition we shall make is to say what we, meaning mathematicians, mean by "or". When we say "$x = 0$ or $y = 0$" we shall mean that one or other *or perhaps both* of the statements "$x = 0$" and "$y = 0$" is true. If we ever wish to exclude the possibility of both statements being true we shall say so explicitly. For example, one of the properties of real numbers to do with the order can be stated as saying that "exactly one of the three statements $x < y$, $x = y$, $x > y$ is true", meaning that one, but not more than one, is true. Had we merely said "$x < y$ or $x = y$ or $x > y$" then we would not be disallowing the possibility of more than one of these statements being true.

As a matter of course we construct complicated statements by piecing together simpler ones and we need to be able to deduce the complicated ones from information about the simpler ones and vice versa. In most cases this is just a matter of noticing, or in complicated examples proving, which implies which.

Example 2.3 In each case below decide on the most appropriate logical connection ($\Rightarrow$, $\Leftarrow$, $\Leftrightarrow$ or none) between the statements A and B. x and y denote unknown real numbers and we wish the connections to be true for all real x and y.

(i) A: $xy = 0$ B: $x = 0$ or $y = 0$

(ii) A: $xy = 0$ B: $x = 0$ and $y = 0$

(iii) A: $xy = 0$ B: $x \geq 0$ and $y \geq 0$

(iv) A: $xy = 0$ B: $x^2 + y^2 \geq 1$.

Solution: In (i) if $xy = 0$ then one or other (or perhaps both) of x and y is zero, so $A \Rightarrow B$. Also $B \Rightarrow A$ since if $x = 0$ or $y = 0$ then $xy = 0$. In (i) then the correct connection is $\Leftrightarrow$. In (ii) it is certainly true that if $x = 0$ and $y = 0$ then $xy = 0$ so $A \Leftarrow B$. However if $xy = 0$ then $x = 0$ or $y = 0$ but we need not have both $x = 0$ and $y = 0$. (To convince yourself of that beyond all doubt, notice that if $x = 0$ and $y = 1$ A is true but B is false.) Therefore A does not imply B, so the best we can say is $A \Leftarrow B$. For the third line if $xy = 0$ then $x = 0$ or $y = 0$ and so $x \geq 0$ or $y \geq 0$, giving $A \Rightarrow B$. The reverse implication is false, for if $x = 1$ and $y = 1$ then B is true and A false. In the last statement neither implies the other. (For if $x = y = 0$ A is true and B false, while if $x = y = 1$ then B is true and A false.) □

In these cases we have given more explanation than many people would find necessary. Notice the rôle played by the counterexamples. In (ii) we showed that $A \Leftarrow B$ and we thought that A did not imply B. Now it might be, particularly in a more complicated example, that the implication is true, but we cannot see why, so the construction of the example where A is true and B is false shows that it is possible for this combination to occur, so there cannot be some ingenious proof lurking in the background.

Example 2.4 Show that $x^2 = 3x \Leftrightarrow x = 3$ or $x = 0$.

Solution:
$$x^2 = 3x \iff x^2 - 3x = 0$$
$$\iff x(x-3) \quad = \quad 0$$
$$\iff x = 0 \text{ or } x - 3 = 0$$
$$\iff x = 0 \text{ or } x = 3 . \ \square$$

The only point of note in the above example is that $x = 0$ is a solution of the equation $x^2 = 3x$. If we start with $x^2 = 3x$ and divide by x, to obtain $x = 3$, we need to notice that the operation is legitimate only if $x \neq 0$; we cannot divide by 0.

Example 2.5 It is obvious that $x = y \Rightarrow x^2 = y^2$ but perhaps less obvious that the reverse implication is false. A systematic way of looking at this is to regard $x^2 = y^2$ as an equation to be solved and move everything to one side.

$$x^2 = y^2 \iff x^2 - y^2 = 0$$
$$\iff (x-y)(x+y) = 0$$
$$\iff x - y = 0 \text{ or } x + y = 0$$
$$\iff x = y \text{ or } x = -y .$$

That makes it clear what is true and we can more easily see how to construct an example to show that $x^2 = y^2$ does not imply that $x = y$. Again we are tacitly assuming here that the implication holds for all real numbers x and y. If this is not the case and we have additional information about x and y (that is, in addition to satisfying $x^2 = y^2$) then perhaps we could deduce $x = y$. For example, if x and y are both positive and $x^2 = y^2$ then we deduce that x and y are both positive and ($x = y$ or $x = -y$). Since $x = -y$ is impossible if x and y are both positive, we deduce that, in this case $x = y$. $\square$

Example 2.6 By writing $x^2 + xy + y^2 = (x + \frac{1}{2}y)^2$ plus a term not involving x, prove that for all real x and y, $x^2 + xy + y^2 = 0 \iff x = 0$ and $y = 0$.
Deduce that $x^3 = y^3 \iff x = y$.

Solution: For all real x and y $x^2 + xy + y^2 = (x + \frac{1}{2}y)^2 + \frac{3}{4}y^2$. Now $(x + \frac{1}{2}y)^2$ and $\frac{3}{4}y^2$ are non-negative so if their sum is zero they must *both* be zero, so

$$x^2 + xy + y^2 = 0 \Rightarrow (x + \frac{1}{2}y)^2 + \frac{3}{4}y^2 = 0$$
$$\Rightarrow (x + \frac{1}{2}y)^2 = 0 \text{ and } \frac{3}{4}y^2 = 0$$
$$\Rightarrow x + \frac{1}{2}y = 0 \text{ and } y = 0$$
$$\Rightarrow x = 0 \text{ and } y = 0 .$$

Obviously $x = 0$ and $y = 0 \Rightarrow x^2 + xy + y^2 = 0$, so we have shown that for all real x and y, $x^2 + xy + y^2 = 0 \iff x = 0$ and $y = 0$.

Now
$$x^3 = y^3 \Rightarrow x^3 - y^3 = 0$$
$$\Rightarrow (x - y)(x^2 + xy + y^2) = 0$$
$$\Rightarrow x - y = 0 \text{ or } x^2 + xy + y^2 = 0$$
$$\Rightarrow x = y \text{ or } x = y = 0$$

$$\Rightarrow x = y.$$

We have shown that $x^3 = y^3 \Rightarrow x = y$.

Clearly $x = y \Rightarrow x^3 = y^3$. $\square$

The process of writing $x^2 + xy + y^2$ in the form $(x + \frac{1}{2}y)^2$ plus a term not involving x is quite general, for quadratic expressions.   For example, $x^2 + 3xy + y^2 = (x + (3/2)y)^2 - (1/4)y^2$.   This is called **completing the square**. If a and b are independent of x we write $x^2 + ax + b$ as $(x + \frac{1}{2}a)^2 + (b - \frac{1}{4}a^2)$.

Problems

1. Solve the following equations.   You should ensure that you have found all solutions

    (i) $\sqrt{(1 + 2x)} = 1 - \sqrt{x}$.

    (ii) $\sqrt{(1 + x)} = 1 + \sqrt{(1 - x)}$. (You may find it helpful to notice that
              $\{(1 + \sqrt{3})/2\}^2 = 1 + (\sqrt{3})/2$, and $\{(\sqrt{3} - 1)/2\}^2 = 1 + (\sqrt{3})/2$.

    (iii) $\sin x = \cos x$. (Hint: Use the equation $\sin^2 x + \cos^2 x = 1$.)

2.   In each pair of statements below insert the most appropriate logical connection ($\Rightarrow$, $\Leftarrow$, $\Leftrightarrow$ or none) between the pair. The statements are supposed to hold for all real numbers x.

$x = 3$ .... $x^3 = 3x^2$; $x = 3$ .... $x^3 = 9x$;

$x^2 \neq 0$ .... $x^2 > 0$; $x^2 + 2x + 1 = 1$ .... $x = 0$ or $x = -2$;

$x^2 + 2x + 1 = 1$ .... $x = 0$ or $x = -1$ or $x = -2$;

$x^2 + 2x + 1 = 1$ .... $x = 0$.

3.   Prove that for all real x and y
$$x^4 = y^4 \Leftrightarrow x = y \text{ or } x = -y$$
(that is, we do not have more than the two possibilities on the right).

4.   Many people are unhappy with the logic of the statement
$$x = 1 \Rightarrow x = 1 \text{ or } x = -1, \qquad (*)$$
and others like it. This question provides another route to the result.
     Show that $x - 1 = 0 \Rightarrow (x - 1)(x + 1) = 0$ and use this to deduce (*).

5.   (i)   It is clear that if $x = y$ then $\frac{x}{1+x^2} = \frac{y}{1+y^2}$.   It is not so clear whether the two sides of the equation can be equal for certain values of $y \neq x$.
         Show that
$$\frac{x}{1+x^2} = \frac{y}{1+y^2} \Rightarrow x + xy^2 = y + x^2y \Rightarrow (x - y)(1 - xy) = 0.$$

Deduce that $\dfrac{x}{1+x^2} = \dfrac{y}{1+y^2} \Leftrightarrow x = y$ or $x = \dfrac{1}{y}$.

(ii) (Harder) Suppose that $a = \dfrac{x}{1+x^2}$ for some real number x. For most values of a part (i) provides another value of y, distinct from x, for which $a = \dfrac{y}{1+y^2}$. For which values of a is there no such second value y for which $a = \dfrac{y}{1+y^2}$ and $y \neq x$?

6. Show that $\dfrac{x}{1+2x^2} = \dfrac{y}{1+2y^2} \Leftrightarrow x = y$ or $x = \dfrac{1}{2y}$.

3

Mathematical Induction

"Great fleas have little fleas, upon their backs to bite 'em, and little fleas have lesser fleas, and so ad infinitum."   Augustus de Morgan

3.1 Introduction

We began by asking the reader to believe that, for every natural number n, $1 + 2 + 3 + ... + n = \frac{1}{2}n(n + 1)$,   and shortly afterwards we produced a demonstration that this is true.   While this proof had the advantage of being simple, it depended on noticing that we could rearrange the order of the sum so that the sum of corresponding terms in the two versions was constant.   In this section we shall develop a more general method, capable of dealing with a wide range of problems. We shall use the logical ideas of Chapter 2 and we shall begin by accepting the normal properties of the number system as given.   For example, we shall assume without further proof that if x, y and z are three real numbers, $x + y = y + x$, $xy = yx$, $x + (y + z) = (x + y) + z$ and so on.   At some stage we shall have to be more precise and write out exactly what these properties are which we are assuming, but that can wait until we need to do it.

3.2 Arithmetic Progressions

An **arithmetic progression** is a sequence which starts with a number a and then increases with a fixed step d: $a,\ a + d,\ a + 2d, ...$.   Let s_n denote the sum of the first n terms, so that

$$s_n = a + (a + d) + (a + 2d) + ... + (a + (n - 1)d) \ .$$

We wish to find a simple expression for s_n.

We can do this by writing down s_n in its natural order and then with the terms in the reverse order.

$$s_n = a + (a + d) + (a + 2d) + ... + (a + (n - 1)d) \ ,$$
$$s_n = (a + (n - 1)d) + (a + (n - 2)d) + (a + (n - 3)d) + ... + a \ .$$

Then, adding corresponding terms,

$$2s_n = \underbrace{(2a + (n-1)d) + (2a + (n-1)d) + ... + (2a + (n-1)d)}_{n \text{ terms}}$$
$$= n(2a + (n-1)d) \ .$$
$$\therefore \qquad s_n = na + \frac{1}{2}n(n-1)d \ .$$

We have proved the following little result.   (The polite name for a little result, usually a step towards something more interesting, is a Lemma.)

Lemma 3.1 The sum of the first n terms of the arithmetic progression a, $a+d$, $a+2d$, ... is $na + \frac{1}{2}n(n-1)d$. $\square$

The result just proved and others like it are often much clearer if we use the Σ notation for summation. The general term of the arithmetic progression above is $a + (k-1)d$, this being the kth term, so that $a + (a+d) + ... + (a+(n-1)d)$ is the sum of the terms when $k = 1$, $k = 2$, ..., and $k = n$ all added together, and we write it as

$$\sum_{k=1}^{n} \left(a + (k-1)d \right) .$$

More generally if f is some formula involving k, $\sum_{k=1}^{n} f(k)$ denotes $f(1) + f(2) + ... + f(k)$, the sum of the values obtained when k is successively substituted by $1, 2, ..., n$, or whatever values are indicated by the Σ sign. Thus $\sum_{k=3}^{5} f(k) = f(3) + f(4) + f(5)$. Notice that the k here is a "dummy" variable and that $\sum_{k=1}^{n} f(k)$ does not depend on k; we could replace k throughout by any other symbol which suits us except one which already has another meaning in this expression. $\sum_{n=1}^{n} n$ will not do, since the symbol n has two meanings.

The other common simple type of sequence is the **geometric progression**, which is a sequence of the form $a, ar, ar^2, ar^3, ...$ r being called the **common ratio**. Again there is a formula for the sum of the first n terms. Let $s_n = a + ar + ... + ar^{n-1}$, the sum of the first n terms. Then

$$
\begin{aligned}
s_n &= a + ar + ... + ar^{n-1} \\
rs_n &= \quad\quad ar + ... + ar^{n-1} + ar^{n}
\end{aligned}
$$

so
$$(1-r)s_n = a - ar^{n} = a(1 - r^{n}) .$$

Therefore, provided $r \neq 1$ (to avoid dividing by zero),

$$s_n = a \frac{1-r^{n}}{1-r} .$$

(Notice the form of this: the first term of the progression multiplies the whole expression and the power of r in the numerator is the number of terms. If $r > 1$ it is usually neater to write $s_n = a \frac{r^{n}-1}{r-1}$ so that the denominator is positive. In the case where $r = 1$ the formula makes no sense, but in this case all the terms summed are equal so the answer is easy.) We have proved:

Lemma 3.2 If $r \neq 1$ and n is a natural number

$$a + ar + ... + ar^{n-1} = \sum_{k=1}^{n} ar^{k-1} = a \frac{1-r^{n}}{1-r} . \quad \square$$

This result has an extremely useful consequence (which we call a corollary). Suppose we wish to factorise $x^{n} - y^{n}$. Then (assuming $x \neq 0$) $x^{n} - y^{n} = x^{n}(1 - \frac{y^{n}}{x^{n}}) = x^{n}(1 - (\frac{y}{x})^{n})$. Here we need to notice that we can use the formula in Lemma 3.2 with $r = y/x$ and $a = 1$. Then

$$1+\frac{y}{x}+\ldots+\frac{y^{n-1}}{x^{n-1}}=\frac{1-\left(\frac{y}{x}\right)^{n}}{1-\frac{y}{x}} \qquad \left(\text{provided } \frac{y}{x}\neq 1\right).$$

So $1-(\frac{y}{x})^{n}=(1-\frac{y}{x})(1+\frac{y}{x}+\ldots+\frac{y^{n-1}}{x^{n-1}})$ and multiplying through by x^{n} gives

$$x^{n}-y^{n}=(x-y)(x^{n-1}+x^{n-2}y+\ldots+xy^{n-2}+y^{n-1}) \ . \qquad (*)$$

We proved this for all real numbers satisfying the two conditions $x\neq 0$ and $x\neq y$. In these two cases the result is trivial to check, so we notice independently that they are true and the restriction can be removed. The main virtue of (*) is that $x^{n}-y^{n}$ has a factor of $x-y$, although the expression for the second factor is also useful.

These two examples, the arithmetic and geometric progressions, are useful but they give a slightly misleading impression in that it is easy to prove the formulae directly once you spot the trick. A more widely applicable way of proving that a formula is true for all natural numbers is to use mathematical induction. Notice that we wish to prove that a result is true for *all* natural numbers, so we could not do this by testing each value in turn.

3.3 The Principle of Mathematical Induction

Suppose that $P(n)$ is some statement about the natural number n. Then if

(i) $P(1)$ is true, and

(ii)   for every natural number k, $P(k) \Rightarrow P(k+1)$,

we conclude that for every natural number n, $P(n)$ is true.

There are several things to notice here. We need to show that statement (ii) is true, that is, to show that for every k, $P(k) \Rightarrow P(k+1)$. In other words, if the statement $P(k)$ is true for a natural number k it follows that it is true for the next natural number. To show this we assume that k is some unspecified natural number and assume that $P(k)$ is true; then we deduce that $P(k+1)$ must also be true. The fact that k is unspecified but a natural number means that the implication $P(k) \Rightarrow P(k+1)$ *must* be true for all such k.

The next observation is that statement (ii) above says that *if* $P(k)$ is true *then* so is $P(k+1)$. In other words, we have to show that $P(k+1)$ is true on the assumption that $P(k)$ is (and that this deduction holds for general k). The conclusion of the induction, that for all natural numbers n $P(n)$ is true, is a different statement – saying that $P(n)$ is true without the "if" condition that is in (ii). Notice, in particular, that (ii) is not the statement (for every k, $P(k)$ is true) $\Rightarrow P(k+1)$, for that is obvious whatever the statement P is.

Finally, notice that $P(n)$ must be a statement about n. That is, it must involve n rather than have n as a "dummy" variable. The statement "$n \geq 1$" is about n, while "For all n, $n \geq 1$" is not, since we could equally well write the second one as "For all m, $m \geq 1$". The words "for all" introduce a "dummy" variable where we can use any symbol we like as long as we use it consistently and do not employ the same symbol elsewhere in the argument to mean something else.

Example 3.1 Prove by induction that $\sum_{r=1}^{n} r^2 = \frac{1}{6}n(n+1)(2n+1)$.

Solution: The first thing is to invent a suitable statement about n, as distinct from the formula, or expression, $\frac{1}{6}n(n+1)(2n+1)$. Let $P(n)$ be the statement "$\sum_{r=1}^{n} r^2 = \frac{1}{6}n(n+1)(2n+1)$".

Then $P(1)$ is $\sum_{r=1}^{1} r^2 = \frac{1}{6}1(2)(3)$, which is true, since both sides equal 1.

Now suppose that k is some natural number and that $P(k)$ is true. Then $\sum_{r=1}^{k} r^2 = \frac{1}{6}k(k+1)(2k+1)$. Therefore

$$\sum_{r=1}^{k+1} r^2 = \sum_{r=1}^{k} r^2 + (k+1)^2$$
$$= \frac{1}{6}k(k+1)(2k+1) + (k+1)^2 \quad \text{(because } P(k) \text{ is true)}$$
$$= \frac{1}{6}(k+1)(k(2k+1) + 6(k+1))$$
$$= \frac{1}{6}(k+1)(2k^2 + 7k + 6) = \frac{1}{6}(k+1)(k+2)(2k+3)$$
$$= \frac{1}{6}(k+1)((k+1)+1)(2(k+1)+1) ,$$

from which it follows that $P(k+1)$ is true.

The last paragraph shows that if k is a natural number, $P(k) \Rightarrow P(k+1)$.

Therefore, by induction, for all natural numbers n, $P(n)$ is true, i.e. $\sum_{r=1}^{n} r^2 = \frac{1}{6}n(n+1)(2n+1)$. $\square$

The proof just given is fairly typical, in that proving $P(1)$ is true is easy and the substance of the proof is in the $P(k) \Rightarrow P(k+1)$ step. Notice that the summing up "Therefore, by induction, ..." is needed because induction has been used, and it provides the reason why we can show the general conclusion.

Example 3.2 Show that if n is a natural number then $n^3 + 2n$ is a multiple of 3.

Proof: We have to show that for all natural numbers n, $n^3 + 2n$ is a multiple of 3. We shall do this by induction. Let $P(n)$ be "$n^3 + 2n$ is a multiple of 3".

$1^3 + 2.1 = 3$, so that $P(1)$ is true.

Suppose now that k is some natural number and that $P(k)$ is true. Then $k^3 + 2k = 3m$ for some integer m. Therefore

$$(k+1)^3 + 2(k+1) = k^3 + 3k^2 + 3k + 1 + 2k + 2$$
$$= (k^3 + 2k) + 3k^2 + 3k + 3$$
$$= 3m + 3k^2 + 3k + 3$$
$$= 3(m + k^2 + k + 1) ,$$

which shows that $(k+1)^3 + 2(k+1)$ is a multiple of 3, that is, $P(k+1)$ is true.

Therefore if k is a natural number, $P(k) \Rightarrow P(k+1)$.

By induction, for all natural numbers n, $P(n)$ is true, that is, $n^3 + 2n$ is a multiple of 3. $\square$

3.4 Why All the Fuss About Induction?

Suppose we have shown that $P(1)$ is true, and that for every k, $P(k) \Rightarrow P(k+1)$. Then $P(1) \Rightarrow P(2)$ and $P(1)$ is true, so $P(2)$ is true. Also, putting $k = 2$ we see (since for all k, $P(k) \Rightarrow P(k+1)$) that $P(2) \Rightarrow P(3)$. But $P(2)$ is known to be true, so this tells us $P(3)$ is true. We could go on this way as long as we like. Is it not obvious that for all natural numbers n, $P(n)$ is true?

The answer is "yes and no". The procedure in the last paragraph could be used to prove $P(17)$ or, after spending a lot of time and paper $P(1\,234\,567\,890)$, but no matter how many steps we take we will not have dealt with every natural number. But, you say, it obviously *can* deal with any particular natural number, can't it? A philosopher might turn this around and ask exactly what you mean by a natural number, which would lead us along an interesting path but not one we wish to follow here. Essentially the idea is that if you start with 1 and repeatedly add 1, giving 1, $1+1$, $1+1+1$, ... , then the principle of mathematical induction is tantamount to saying that this sequence includes all the natural numbers. You can decide that is "obvious", or you can decide to list it as an assumption. The main thing is that from now on we agree to accept the idea. Appealing to the principle of induction is essentially reminding the reader of this agreement.

3.5 Examples of Induction

Example 3.3 Prove that for all natural numbers n, $6^n - 1$ is a multiple of 5.

Solution: We let $P(n)$ be "$6^n - 1$ is a multiple of 5".

$P(1)$ is "$6 - 1$ is a multiple of 5" which is obviously true.

Suppose $6^k - 1$ is a multiple of 5, so that $6^k - 1 = 5m$ for some integer m. (Note that it matters that m is an integer!) Then

$$6^{k+1} - 1 = 6.6^k - 1 = 6(5m+1) - 1 = 30m + 5 = 5(6m+1)$$

which shows that $6^{k+1} - 1$ is a multiple of 5. We have proved $P(k) \Rightarrow P(k+1)$ (and that this holds for all natural numbers k).

Therefore by induction, for all natural numbers n, $6^n - 1$ is a multiple of 5. □

In the two examples we have given we have established the result for all natural numbers. It may be that a result is true for all natural numbers n with $n \geq N$, in which case we prove the result is true for N and that, for each $k \geq N$, if the result is true for k then it is true for $k+1$. (In symbols, if $P(n)$ is a statement about n we prove that (i) $P(N)$ is true and (ii) for all natural numbers $k \geq N$, $P(k) \Rightarrow P(k+1)$, then we conclude that for all natural numbers $n \geq N$ $P(n)$ is true.)

Problem We can adopt the modified form of induction as another assumption, or we can deduce that the form of induction just stated is a consequence of the normal form. (Hint: Define the statement $Q(n)$ to be $P(n+N-1)$.) □

There are, however, forms of induction which are more substantially different from the original. It is not difficult to see the need for this, in fact, for there are plenty of cases where proving $P(k+1)$ directly from $P(k)$ would be awkward - for example if $P(k)$ is some statement about the factors of k, for there is no obvious relation between the factors of $k + 1$ and those of k. To avoid this we prove:

Theorem 3.3 Complete induction. Suppose that $P(n)$ is some statement about n and that

(i) $P(1)$ is true, and

(ii)   For every natural number k, if $P(1)$, $P(2)$, ... and $P(k)$ are all true then $P(k + 1)$ is true.

Then for all natural numbers n, $P(n)$ is true.

Remark: We could write (ii) more formally as, "For every natural number k, $(P(1)$ and $P(2)$ and ... and $P(k)) \Rightarrow P(k + 1)$."

Proof: Let $Q(n)$ be the statement "$P(1)$ and $P(2)$ and ... and $P(n)$".

$Q(1)$ is true by (i) (since $Q(1)$ is the same as $P(1)$).

Let k be a natural number and suppose that $Q(k)$ is true. Then, since $Q(k)$ means "$P(1)$ and $P(2)$ and ... and $P(k)$", $P(1)$, $P(2)$, ... and $P(k)$ are all true. Therefore by (ii), $P(k + 1)$ is true. Therefore "$P(1)$ and $P(2)$ and ... and $P(k + 1)$" is true, so that $Q(k + 1)$ is true. We have shown that $Q(k) \Rightarrow Q(k + 1)$.

By ordinary induction, for all natural numbers n, $Q(n)$ is true. But $Q(n)$ means $P(1)$, $P(2)$, ... $P(n)$ are all true, so in particular $P(n)$ is true, hence for all natural numbers n, $P(n)$ is true, as required. $\square$

Example 3.4 Every natural number greater than 1 can be expressed as the product of prime numbers; we allow a prime number to be considered as the "product" of one prime number.

[A natural number is said to be a **prime number** if it is larger than 1 and cannot be expressed as the product of two smaller natural numbers. For technical reasons we do not consider 1 to be a prime number.]

Solution: The first problem is to find a suitable statement to use in the induction. Rather than modify the form of induction to apply to $n \geq 2$ we shall let $P(n)$ be "$n = 1$ or n is a product of prime numbers."

Obviously $P(1)$ is true.

Now let k be a natural number and suppose that $P(1)$, $P(2)$, ..., $P(k)$ are all true. $k + 1$ is either a prime number (and thus the "product" of one prime) or it is not prime. If it is not prime then $k + 1 = a b$ where a and b are natural numbers smaller than $k + 1$ (definition of a prime number). Since a and b are both smaller than $k + 1$ and $a b = k + 1$ neither a nor b equals 1. It follows that a and b are each a product of primes, because $P(a)$ and $P(b)$ are true. Therefore $k + 1 = a b$ is a product of primes, and so $P(k + 1)$ is true. (We have proved that for all natural numbers k, $(P(1)$ and $P(2)$ and ... $P(k)) \Rightarrow P(k + 1)$.)

Therefore by induction, for all natural numbers n, $P(n)$ is true; in particular every natural number greater than 1 is a product of prime numbers. $\square$

Example 3.5 The **Fibonacci numbers** $a_1, a_2, a_3, ...$ are defined by setting $a_1 = a_2 = 1$ and for $n \geq 2$ $a_{n+1} = a_n + a_{n-1}$. Prove that for all natural numbers n,

$$a_n = \tfrac{1}{\sqrt{5}}\left\{\left(\tfrac{1+\sqrt{5}}{2}\right)^n - \left(\tfrac{1-\sqrt{5}}{2}\right)^n\right\}.$$

Solution: Let $P(n)$ be "$a_n = \tfrac{1}{\sqrt{5}}\left\{\left(\tfrac{1+\sqrt{5}}{2}\right)^n - \left(\tfrac{1-\sqrt{5}}{2}\right)^n\right\}$ ". Then it is simple to check that $P(1)$ and $P(2)$ are true. Therefore $P(1) \Rightarrow P(2)$ (because $P(2)$ is true; think about it – or see the comment below.)

Now suppose $k \geq 2$ and that $P(1)$, $P(2)$, ..., $P(k)$ are all true. Then

$$a_{k+1} = a_k + a_{k-1} \qquad \text{(because } k \geq 2)$$

$$= \tfrac{1}{\sqrt{5}}\left\{\left(\tfrac{1+\sqrt{5}}{2}\right)^k + \left(\tfrac{1+\sqrt{5}}{2}\right)^{k-1} - \left(\tfrac{1-\sqrt{5}}{2}\right)^k - \left(\tfrac{1-\sqrt{5}}{2}\right)^{k-1}\right\}$$

$$\text{(because } P(k) \text{ and } P(k-1) \text{ are true)}$$

$$= \tfrac{1}{\sqrt{5}}\left\{\left(\tfrac{1+\sqrt{5}}{2}\right)^{k-1}\left(\tfrac{1+\sqrt{5}}{2}+1\right) - \left(\tfrac{1-\sqrt{5}}{2}\right)^{k-1}\left(\tfrac{1-\sqrt{5}}{2}+1\right)\right\}$$

$$= \left\{\left(\tfrac{1+\sqrt{5}}{2}\right)^{k-1}\left(\tfrac{1+\sqrt{5}}{2}\right)^2 - \left(\tfrac{1-\sqrt{5}}{2}\right)^{k-1}\left(\tfrac{1-\sqrt{5}}{2}\right)^2\right\}$$

$$\left[\text{because } \left(\tfrac{1+\sqrt{5}}{2}\right)^2 = \tfrac{1+2\sqrt{5}+5}{4} = \tfrac{3+\sqrt{5}}{2} = \tfrac{1+\sqrt{5}}{2}+1\right.$$

$$\left. \text{and } \left(\tfrac{1-\sqrt{5}}{2}\right)^2 = \tfrac{1-2\sqrt{5}+5}{4} = \tfrac{1-\sqrt{5}}{2}+1\right]$$

$$= \tfrac{1}{\sqrt{5}}\left\{\left(\tfrac{1+\sqrt{5}}{2}\right)^{k+1} - \left(\tfrac{1-\sqrt{5}}{2}\right)^{k+1}\right\}.$$

Therefore $P(k+1)$ is true.

By induction, then, for all natural numbers n, $P(n)$ is true. (This uses complete induction, noting that $P(1)$ and $P(2)$ and ... and $P(k) \Rightarrow P(k+1)$ was proved in the last paragraph for all $k \geq 2$, while $k = 1$ was done separately.) $\square$

Comment: We said above that because $P(2)$ is true, it follows that $P(1) \Rightarrow P(2)$. This is because "$P(1) \Rightarrow P(2)$" means "if $P(1)$ is true, then $P(2)$ is true" so that if we already know $P(2)$ to be true, then it is still true if $P(1)$ is also true. Of course, we are giving away information here, for $P(2)$ is also true in this case if $P(1)$ is false. We know more than we need here.

To illustrate this, we know $1 = 1$. Then it is certainly true that *if* $x = 0$ then $1 = 1$.

The reader may be wondering where on earth the numbers $(1 \pm \sqrt{5})/2$ came from in the Fibonacci sequence, for it seems that these have been plucked from thin air like a rabbit from a hat. This is a more appropriate expression than it might seem, for Fibonacci introduced his numbers as a way of modelling the breeding of rabbits. The model is oversimplified, of course, for a_n increases indefinitely with n, and the world is not yet overrun by rabbits.

Although we shall not pursue the rabbit-breeding connection, sequences defined by a rule such as

$$a_{n+1} = a\,a_n + b\,a_{n-1} \;,$$

where a and b are constants, are quite interesting. They are also useful from time to time in various connections, so we shall pay them a little more attention (and see where $\frac{1+\sqrt5}{2}$ comes from!).

Let us begin by noticing that if we specify a_1 and a_2 and demand that

$$a_{n+1} = aa_n + ba_{n-1} \quad \text{(for all } n \geq 2) \;, \tag{1}$$

then we have determined the value of a_n for all natural numbers n (even though we may not yet have a satisfactory formula for a_n). For this we can use complete induction. a_1 and a_2 are given, so they are determined. If $k \geq 2$ and we know $a_1, a_2, ..., a_k$ then, in particular, we know a_{k-1} and a_k. Equation (1) then allows us to determine a_{k+1}. That is, if we know $a_1, ..., a_k$ then we know a_{k+1} $(k \geq 2)$. Complete induction now tells us that we have determined a_n for all natural numbers n. (Notice that the step $k = 1$ to $k = 2$ is valid because we are given a_2.) To put this in simpler terms, knowing that a_n satisfies equation (1), together with the two extra pieces of information provided by knowing a_1 and a_2, fixes all the values of a_n. This tells us that we are asking a sensible question – but it has not answered it!

It would be easier if the equation (1) were in the form $b_{n+1} = x\,b_n$ for some constant x, for then each term is obtained from the previous one by multiplying by x, and we can see the answer straight away: $b_n = b_1 x^{n-1}$ or $b_n = Ax^n$ where A is a constant (b_1/x in this case). Now we might try to arrange this by choosing a constant y and setting $b_n = a_n - y\,a_{n-1}$. Then if $n \geq 2$

$$\begin{aligned}
b_{n+1} &= a_{n+1} - y\,a_n = (a-y)a_n + b\,a_{n-1} \qquad \text{using (1)}\\
&= (a-y)(b_n + y\,a_{n-1}) + b\,a_{n-1}\\
&= (a-y)b_n + (b + ay - y^2)a_{n-1} \;.
\end{aligned}$$

Therefore to arrange that, for all $n \geq 2$,

$$b_{n+1} = (a - y)b_n \tag{2}$$

we need to choose y so that $b + ay - y^2 = 0$. If we choose such a value of y then we see from (2) that for each natural number $k \geq 3$, $b_k = (a - y)b_{k-1}$. We then use (2) again to express b_{k-1} as $(a - y)b_{k-2}$ which is valid as long as $k - 2 \geq 2$, i.e. $k \geq 4$. We carry on with this as long as possible, that is, until the suffix on b is 2, for then (2) is no longer helpful. This gives

$$b_n = (a - y)b_{n-1} = (a - y)^2 b_{n-2} = \; ... \; = (a - y)^{n-2} b_2 \;,$$

for all $n \geq 2$. (Notice that in the case $n = 2$ we need to interpret $(a - y)^0$ as 1.)

This process of repeated use of equation (2) (or repeated substitution) may appear mysterious. The informal way of thinking of it is to notice that each time we reduce the suffix on b by 1 we increase the power of $(a - y)$ by 1, so the total of the suffix and the power stays constant at n. Also the process stops once we reduce the suffix on b to 2. (The formal way to do this is to prove that for all $n \geq 2$ $b_n = (a - y)^{n-2} b_2$ by induction. If you find the informal argument unsatisfactory, the formal proof may clear matters up.)

For our purposes it is enough to notice that there is a constant A such that for $n \geq 2$ $b_n = A(a - y)^n$ (where $A = b_2/(a - y)^2$ in the notation above), provided $a - y \neq 0$ as otherwise we have divided by zero in forming A. In fact we could notice that even if $a - y = 0$ $b_n = A(a - y)^n$ for all $n \geq 3$ (for in this case $b_3 = (a - y)b_2 = 0$ and $b_n = 0$ for all $n \geq 3$).

Example 3.6 Suppose that $a_1 = a_2 = 1$ and $a_{n+1} = a_n + a_{n-1}$, the Fibonacci sequence. Then in this case $a = b = 1$ so we solve $y^2 = y + 1$ (equivalent to $1 + y - y^2 = 0$) to obtain $y = \frac{1}{2}(1 \pm \sqrt{5})$. Let

$$b_n = a_n - \tfrac{1}{2}(1 + \sqrt{5})a_{n-1}$$

so that for all $n \geq 2$

$$b_{n+1} = a_{n+1} - \tfrac{1}{2}(1 + \sqrt{5})a_n$$

$$= \tfrac{1}{2}(1 - \sqrt{5})a_n + a_{n-1} \quad (\text{since } a_{n+1} = a_n + a_{n-1})$$

$$= \tfrac{1}{2}(1 - \sqrt{5})\left(a_n + \tfrac{2}{1-\sqrt{5}}a_{n-1}\right)$$

$$= \tfrac{1}{2}(1 - \sqrt{5})\left(a_n - \tfrac{1}{2}(1 + \sqrt{5})a_{n-1}\right)$$

$$= \tfrac{1}{2}(1 - \sqrt{5})b_n \quad .$$

Therefore for all $n \geq 2$ $b_n = \left(\frac{1-\sqrt{5}}{2}\right)^{n-2} b_2$ and since $b_2 = a_2 - \tfrac{1}{2}(1 + \sqrt{5})a_1$ $= \tfrac{1}{2}(1 - \sqrt{5})$, we see that $b_n = \left(\frac{1-\sqrt{5}}{2}\right)^{n-1}$ for $n \geq 2$.

Now if we choose the other solution of $y^2 = ay + b$, that is, $y = (1 - \sqrt{5})/2$, we may go through similar calculations. For $n \geq 2$ let $c_n = a_n - \tfrac{1}{2}(1 - \sqrt{5})a_{n-1}$, so that $c_{n+1} = a_{n+1} - \tfrac{1}{2}(1 - \sqrt{5})a_n = \tfrac{1}{2}(1 + \sqrt{5})c_n$ and for $n \geq 2$

$$c_n = \left(\frac{1+\sqrt{5}}{2}\right)^{n-1} \quad .$$

This more or less solves the problem for we now know that for all $n \geq 2$,

$$a_n - \tfrac{1}{2}(1 + \sqrt{5})a_{n-1} = b_n = \left(\frac{1-\sqrt{5}}{2}\right)^{n-1}$$

$$a_n - \tfrac{1}{2}(1 - \sqrt{5})a_{n-1} = c_n = \left(\frac{1+\sqrt{5}}{2}\right)^{n-1} \quad .$$

Therefore

$$-\sqrt{5}a_n = \tfrac{1}{2}(1 - \sqrt{5})a_n - \tfrac{1}{2}(1 + \sqrt{5})a_n = \left(\frac{1-\sqrt{5}}{2}\right)b_n - \left(\frac{1+\sqrt{5}}{2}\right)c_n = \left(\frac{1-\sqrt{5}}{2}\right)^n - \left(\frac{1+\sqrt{5}}{2}\right)^n$$

giving the formula $a_n = \frac{1}{\sqrt{5}}\left\{\left(\frac{1+\sqrt{5}}{2}\right)^n - \left(\frac{1-\sqrt{5}}{2}\right)^n\right\}$ which we proved by induction. $\square$

This is the general method of solving these equations where we can find two distinct real solutions of the auxiliary equation $y^2 - ay - b$. Suppose that y_1 and y_2 are distinct real solutions of this equation, so that $y_1^2 = ay_1 + b$ and

$y_2^2 = ay_2 + b$. Then $y_1^2 - y_2^2 = a(y_1 - y_2)$ (subtracting the second equation from the first) so, because $y_1^2 - y_2^2 = (y_1 + y_2)(y_1 - y_2)$ and since $y_1 - y_2 \neq 0$ we may divide by $y_1 - y_2$ to obtain $y_1 + y_2 = a$. Also

$$a^2 = (y_1 + y_2)^2 = y_1^2 + 2y_1y_2 + y_2^2 = ay_1 + b + 2y_1y_2 + ay_2 + b$$
$$= a(y_1 + y_2) + 2y_1y_2 + 2b = a^2 + 2y_1y_2 + 2b$$

giving $y_1y_2 = -b$. These equations will be used in the calculation in Example 3.7 below.

Example 3.7 Suppose that a, b, α and β are given real numbers and that we define a_n for all natural numbers n by

$$a_1 = \alpha, \quad a_2 = \beta \quad a_{n+1} = aa_n + ba_{n-1} \quad (n \geq 2) .$$

Then if $b \neq 0$ and the auxiliary equation

$$y^2 = ay + b$$

has two *distinct* real solutions y_1 and y_2 we have for all natural numbers n,

$$a_n = A y_1^n + B y_2^n$$

where the constants A and B are chosen so that

$$A y_1 + B y_2 = \alpha ,$$
$$A y_1^2 + B y_2^2 = \beta . \quad \square$$

Remark: To see why this turns out the way it does, notice that if we set $b_n = a_n - y_1 a_{n-1}$ and $c_n = a_n - y_2 a_{n-1}$ then for $n \geq 2$

$$b_{n+1} = a_{n+1} - y_1 a_n = (a - y_1)a_n + ba_{n-1}$$
$$= y_2 a_n \, y_1 y_2 a_{n-1}$$
$$= y_2 (a_n - y_1 a_{n-1})$$
$$= y_2 b_n ,$$

so that $b_n = Cy_2^n$ for some C. Similarly $c_{n+1} = y_1 c_n$ $(n \geq 2)$ and $c_n = Dy_1^n$ and we solve the simultaneous equations

$$a_n - y_1 a_{n-1} = b_n = Cy_2^n$$
$$a_n - y_2 a_{n-1} = c_n = Dy_1^n$$

to find a_n. Notice that the condition $b \neq 0$ guarantees that the formula for a_{n+1} does actually depend on a_{n-1}.

Solution: We can choose A and B so that $Ay_1 + By_2 = \beta$. Straightforward arithmetic, multiplying the first equation by y_2 and subtracting, gives

$$A = \frac{-\alpha y_2 + \beta}{y_1(y_1 - y_2)} \quad B = \frac{\alpha y_1 - \beta}{y_2(y_1 - y_2)}$$

noticing that the denominators are non-zero because $y_1 \neq y_2$ and $y_1y_2 = -b \neq 0$. It follows that with this choice of A and B $a_n = Ay_1^n + By_2^n$ is true for $n = 1$ and $n = 2$.

Then for $k \geq 2$, assume the result is true for $n = 1, 2, ..., k$.

$$a_{k+1} = aa_k + ba_{k-1}$$
$$= a(Ay_1^k + By_2^k) + b(Ay_1^{k-1} + By_2^{k-1})$$
$$= A(ay_1 + b)y_1^{k-1} + B(ay_2 + b)y_2^{k-1}$$
$$= Ay_1^{k+1} + By_2^{k+1}$$

(using $y_1^2 = ay_1 + b$ and $y_2^2 = ay_2 + b$).

The result follows for all natural numbers n by (complete) induction. □

Example 3.8 Suppose that $a_1 = 1$, $a_2 = 3$ and that for all $n \geq 2$, $a_{n+1} = 3a_n$ $2a_{n-1}$. Find a simple formula for a_n.

Solution: The auxiliary equation is $x^2 = 3x - 2$ (or $x^2 - 3x + 2 = 0$) which has solutions $x = 1$ and $x = 2$. From the example above we see that a_n has the form $A1^n + B2^n$, so to ensure that a_1 and a_2 are correct we need to solve $A + 2B = 1$, $A + 4B = 3$, giving $B = 1$, $A = -1$. Therefore $a_n = 2^n - 1$ for all $n \geq 1$. □

We can pursue this further. The case where the auxiliary equation has no real solutions presents no new difficulty to those who are familiar with complex numbers, for it has two distinct complex solutions. This is what motivates the solution quoted in Problem 13, which, however, avoids complex numbers.

The interesting case, however, occurs when $a^2 + 4b = 0$, for then the auxiliary equation has only one solution (or, more colloquially, its two solutions are equal). If $x^2 = ax + b$ then it is easy to check that for whatever constant A we choose $a_n = A x^n$ does have the property that $a_{n+1} = aa_n + ba_{n-1}$. However, with only one parameter A at our disposal this may not allow us to choose A so that both a_1 and a_2 have the prescribed values. There seems to be a solution "missing".

Suppose that $b = -\frac{1}{4}a^2$, so that we wish to solve

$$a_{n+1} = aa_n - \frac{1}{4}a^2 a_{n-1} \qquad (n \geq 2) \qquad\qquad (*)$$

where a_1 and a_2 are specified (and $a \neq 0$, for then the result is obvious.) We consider the auxiliary equation $x^2 = ax - \frac{1}{4}a^2$, which has the unique solution $x = \frac{1}{2}a$ (for $x^2 - ax + \frac{1}{4}a^2 = (x - \frac{1}{2}a)^2 = 0$). Since we know one solution is of the form $A(\frac{1}{2}a)^n$, we set $a_n = (\frac{1}{2}a)^n b_n$. Then equation (*) shows us that for $n \geq 2$, $(\frac{1}{2}a)^{n+1} b_{n+1} = a(\frac{1}{2}a)^n b_n - (\frac{1}{4}a^2)(\frac{1}{2}a)^{n-1} b_{n-1}$ so, dividing by $(\frac{1}{2}a)^{n+1}$, $b_{n+1} = 2b_n - b_{n-1}$ and $b_{n+1} - b_n = b_n - b_{n-1}$.

Therefore $b_3 - b_2 = b_2 - b_1$ and, by induction, $b_n - b_{n-1} = b_2 - b_1$, , a constant, B say. Therefore for all $n \geq 2$, $b_n = b_{n-1} + B$ and we see that $b_n = b_1 + (n-1)B = A + nB$ where $A = b_1 - B$, another constant. Therefore b_n is of the form $A + nB$ so a_n is of the form $a_n = (A + nB)(\frac{1}{2}a)^n$. We now choose A and B so that a_1 and a_2 have the correct values, that is, so that A and B satisfy

$$(A + B)(\tfrac{1}{2}a) = a_1$$

$$(A + 2B)(\tfrac{1}{2}a)^2 = a_2 \ .$$

It is simple to check that A and B can be chosen (because $a \neq 0$). Then we prove by induction that with this choice of A and B $a_n = (A + nB)(\tfrac{1}{2}a)^n$. (We have just chosen A and B to make this true for $n = 1$ and 2. If we assume it is true for $n = 1, 2, \ldots k$, where $k \geq 2$, then

$$\begin{aligned}
a_{k+1} &= a a_k - \tfrac{1}{4}a^2 a_{k-1} \\
&= a(A + kB)(\tfrac{1}{2}a)^k - \tfrac{1}{4}a^2(A + (k-1)B)(\tfrac{1}{2}a)^{k-1} \\
&= (A + (k+1)B)(\tfrac{1}{2}a)^{k+1} \ ,
\end{aligned}$$

that is, the result holds for $n = k + 1$.

It follows by induction that the formula holds for all natural numbers n. We have just shown that the following result is true.

Example 3.9 Suppose that a, b, α and β are all given real numbers and that we define a_n for all natural numbers by

$$a_1 = \alpha \ , \qquad a_2 = \beta \ , \qquad a_{n+1} = a a_n + b a_{n-1} \qquad (n \geq 2) \ .$$

Then if $b \neq 0$ and the auxiliary equation

$$y^2 = a y + b$$

has only one real solution y_0 (or, if you prefer, its "two" solutions are equal) a_n is given by $a_n = (A + nB)y_0^n$ where the constants A and B are chosen so that

$$\begin{aligned}
a_1 &= (A + B)y_0 \ , \\
a_2 &= (A + 2B)y_0^2 \ .
\end{aligned}$$

(It is a simple exercise in induction to check that this statement is true - once you notice that if $y^2 = ay + b$ has only one solution y_0 then $y_0 = \tfrac{1}{2}a$ and $b = -y_0^2$. However, checking that the result is true is not a very satisfying way of seeing where it comes from.) □

3.6 The Binomial Theorem
In some of the work above we have used expressions like $(x + y)^2$ and their equivalent form $x^2 + 2xy + y^2$. There is a general form of this, for $(x + y)^n$. Now it is not difficult to multiply out and check that

$$(x + y)^3 = x^3 + 3x^2 y + 3xy^2 + y^3$$

$$(x + y)^4 = x^4 + 4x^3 y + 6x^2 y^2 + 4xy^3 + y^4$$

and so on. The general case will have

$$(x + y)^n = c_0 x^n + c_1 x^{n-1} y^1 + c_2 x^{n-2} y^2 + \ldots + c_{n-1} x y^{n-1} + c_n y^n$$

for some coefficients $c_0, \ldots, c_n$. (This is clearer if we write $(x + y)^n$ as $(x + y)(x + y) \ldots (x + y)$; multiplying out, we will have many terms, each with some ys (say k) and some xs $(n-k)$ multiplied together. This actually shows us that each of $c_0, \ldots, c_n$ is a natural number and $c_0 = c_n = 1$. In fact if we choose the y term in k of the brackets and x in the other $n-k$ we will obtain a contribution of $x^{n-k}y^k$ to the right hand side, and each way of selecting k different brackets out of the n will give another such term, so c_k is actually the number of ways of choosing k objects out of n.)

Notation Let n and k be integers with $0 \le k \le n$. We define $n!$ (pronounced "n factorial") to be $1.2.3 \ldots n$ if $n \ge 1$ and set $0! = 1$ so that $n! = n(n-1)!$ if $n \ge 1$. Then we define $\binom{n}{k}$ (pronounced "n choose k") by

$$\binom{n}{k} = \frac{n!}{k!(n-k)!}$$

Theorem 3.4 The Binomial Theorem. Let n be a natural number. Then, adopting the notation $x^0 = y^0 = 1$, for all real numbers x and y,

$$(x+y)^n = \sum_{k=0}^{n} \binom{n}{k} x^k y^{n-k} \quad.$$

Proof: The result is obvious for $n = 1$, for $\binom{1}{0} = \binom{1}{1} = 1$.

Suppose the result is true for some n and let x and y be real numbers.

Then $\quad (x+y)^{n+1} = (x+y)(x+y)^n = (x+y)\sum_{k=0}^{n}\binom{n}{k}x^k y^{n-k}$

$$= \sum_{k=0}^{n}\binom{n}{k}(x^{k+1}y^{n-k} + x^k y^{n+1-k})$$

$$= \sum_{l=1}^{n+1}\binom{n}{l-1}x^l y^{n+1-l} + \sum_{k=0}^{n}\binom{n}{k}x^k y^{n+1-k}$$

$$= \binom{n}{0}y^{n+1} + \sum_{k=1}^{n}\left\{\binom{n}{k-1}+\binom{n}{k}\right\}x^k y^{n+1-k} + \binom{n}{n}x^{n+1}$$

Now for $1 \le k \le n$, $k-1$ is a natural number or zero and

$$\binom{n}{k-1}+\binom{n}{k} = \frac{n!}{(k-1)!(n-k+1)!} + \frac{n!}{k!(n-k)!}$$

$$= \frac{n!}{k!(n-k+1)!}(k+(n-k+1)) \quad \text{(put over common denominator)}$$

$$= \frac{(n+1)n!}{k!(n-k+1)!} = \binom{n+1}{k} \quad.$$

Also $\binom{n}{0} = 1 = \binom{n+1}{0}$ and $\binom{n}{n} = 1 = \binom{n+1}{n+1}$. Therefore

$$(x+y)^{n+1} = \sum_{k=0}^{n+1}\binom{n+1}{k}x^k y^{n+1-k} \quad.$$

Since x and y were typical real numbers, this holds for all real x and y and so we have shown that if the result holds for n then it holds for $n + 1$.

Therefore, by induction, the result holds for all natural numbers n. ☐

Note: The way we have done the proof above we have let $P(n)$ be "For all real x and y, $(x+y)^n = \sum_{k=0}^{n}\binom{n}{k}x^k y^{n-k}$" (and abbreviated still further by calling this "the result"). In this sense we proved $P(1)$ and that $P(n) \Rightarrow P(n + 1)$, the usual form of induction.

We have shown above that $\sum_{k=1}^{n} k = \frac{1}{2}n(n+1)$ and that $\sum_{k=1}^{n} k^2 = \frac{1}{6}n(n+1)(2n+1)$. We might guess from this that the expression for $\sum_{k=1}^{n} k^3$ is a polynomial of degree 4 in n, that is, an expression of the form $An^4 + Bn^3 + Cn^2 + Dn + E$ for some constants A, B, C, D and E. It is not hard to evaluate $\sum_{k=1}^{n} k^3$ for $n = 1, 2, 3, 4, 5$ so that, at the expense of some tedious calculation, we could work out the values of A, B, C, D and E. The polynomial turns out to be $\frac{1}{4}n^2(n+1)^2$, and we can prove that this is correct by induction. In fact, we can do rather better by seeking less information. Notice that the coefficient of n^2 in $\sum_{k=1}^{n} k$ is $\frac{1}{2}$, the coefficient of n^3 in $\sum_{k=1}^{n} k^2$ is $\frac{1}{3}$ and that n^4 in $\sum_{k=1}^{n} k^3$ is $\frac{1}{4}$. We might guess that the coefficient of n^{r+1} in $\sum_{k=1}^{n} k^r$ is $\frac{1}{r+1}$.

Example 3.10 Show that if r is a natural number then there are constants $a_0, \dots, a_r$ such that, for all natural numbers n,

$$\sum_{k=1}^{n} k^r = \frac{1}{r+1}n^{r+1} + a_r n^r + \dots + a_0 .$$

Notation: We shall call an expression of the form
$$a_r x^r + a_{r-1} x^{r-1} + \dots + a_1 x + a_0 ,$$
where $a_0, \dots, a_r$ are constants, a **polynomial**. If $a_r \neq 0$ then we say this polynomial has **degree** r.

Solution: We prove this result by induction on r. For this we need to be a little careful, for we need to formulate correctly the statement to be proved.

For each natural number r, let $P(r)$ be the statement "There are constants $a_0, \dots, a_r$ such that for all natural numbers n

$$\sum_{k=1}^{n} k^r = \frac{1}{r+1}n^{r+1} + a_r n^r + \dots + a_0 ."$$

Then $P(1)$ is true because we have already proved that for all natural numbers n, $\sum_{k=1}^{n} k = \frac{1}{2}n^2 + \frac{1}{2}n \left(= \frac{1}{2}n(n+1) \right)$.

Now suppose that r is a natural number and that for $s = 1, 2, \dots, r$ $P(s)$ is true. For all natural numbers k, by the Binomial Theorem,

$$(k+1)^{r+2} = k^{r+2} + (r+2)k^{r+1} + \binom{r+2}{2}k^r + \ldots + 1$$

$$= k^{r+2} + (r+2)k^{r+1} + \text{a polynomial of degree } r \text{ in } k \ .$$

Therefore

$$(k+1)^{r+2} - k^{r+2} = (r+2)k^{r+1} + \text{a polynomial of degree } r \text{ in } k \ . \ \ (*)$$

Now we sum from $k = 1$ to n on both sides, where n is a natural number.

$$\sum_{k=1}^{n}\{(k+1)^{r+2} - k^{r+2}\} = \sum_{k=2}^{n+1} k^{r+2} - \sum_{k=1}^{n} k^{r+2}$$

$$= (n+1)^{r+2} - 1^{r+2} \ .$$

(This is the clever step; all the intermediate terms have cancelled.) Now on the right hand side of (*) suppose the polynomial of degree r in k is $a_0 + a_1 k + \ldots + a_r k^r$. Then

$$\sum_{k=1}^{n}(a_0 + a_1 k + \ldots + a_r k^r) = a_0 \sum_{k=1}^{n} 1 + a_1 \sum_{k=1}^{n} k^1 + \ldots + a_r \sum_{k=1}^{n} k^r$$

$$= \text{a polynomial of degree } r+1 \text{ in } n \ .$$

(This uses the fact that we assumed that $P(1)$, ..., $P(r)$ are true.) Therefore

$$(n+1)^{r+2} - 1 = (r+2)\sum_{k=1}^{n} k^{r+1} + \text{a polynomial of degree } r+1 \text{ in } n \ .$$

But $(n+1)^{r+2} - 1 = n^{r+2} + (\text{another})$ polynomial of degree $r+1$ in n, so

$$(r+2)\sum_{k=1}^{n} k^{r+1} = n^{r+2} + \text{a polynomial of degree } r+1 \text{ in } n \ ,$$

and $$\sum_{k=1}^{n} k^{r+1} = \tfrac{1}{r+2} n^{r+2} + \text{a polynomial of degree } r+1 \text{ in } n \ .$$

As this holds for all natural numbers n, $P(r+1)$ is true.

Therefore $((P(1) \text{ and } P(2) \text{ and } \ldots \text{ and } P(r)) \Rightarrow P(r+1))$.

By induction, then, $P(r)$ is true for all natural numbers r. $\square$

Caution: Notice the induction in the last proof is on the value of r, not n.

Problems

1. Show that for all natural numbers n,
$$1^3 + 2^3 + \cdots + n^3 = \tfrac{1}{4}n^2(n+1)^2 \ ; \ 1.2 + 2.3 + \cdots + n(n+1) = \tfrac{1}{3}n(n+1)(n+2);$$

$$\sum_{k=1}^{n} \frac{1}{k(k+1)} = 1 - \frac{1}{n+1} \quad \text{and} \quad \sum_{k=1}^{n} \frac{1}{k(k+1)(k+2)} = \frac{1}{4} - \frac{1}{2(n+1)(n+2)}$$

2. Show that for all natural numbers n,
$$1^2 + 3^2 + \ldots + (2n-1)^2 = \tfrac{1}{3}n(4n^2 - 1)$$
$$1^2 - 2^2 + 3^2 - \ldots + (2n-1)^2 - (2n)^2 = -n(2n+1)$$
$$1^3 - 2^3 + 3^3 - \ldots + (2n-1)^3 - (2n)^3 = -n^2(4n+3) \ .$$

3.   Prove that the following equations hold for all natural numbers n:

$$\left(1-\tfrac{1}{2}\right)\left(1-\tfrac{1}{3}\right)\cdots\left(1-\tfrac{1}{n+1}\right) = \tfrac{1}{n+1} \;,\; \left(1+\tfrac{1}{2}\right)\left(1+\tfrac{1}{3}\right)\cdots\left(1+\tfrac{1}{n+1}\right) = \tfrac{n+2}{2} \;,$$

$$\left(1-\tfrac{1}{4}\right)\left(1-\tfrac{1}{9}\right)\cdots\left(1-\tfrac{1}{(n+1)^2}\right) = \tfrac{1}{2}\left(1+\tfrac{1}{n+1}\right) \;.$$

4.   Show that, for all natural numbers n:

         (i) 10^n leaves a remainder of 1 on division by 9 ,

         (ii) 4^{2n} leaves a remainder of 6 on division by 10 ,

         (iii) 4^{2n+1} leaves a remainder of 4 on division by 10 .

5.   Show that, for all natural numbers n, $10^n + (-1)^{n-1}$ is a multiple of 11.

6.   Let $a_1 = 1$, $a_2 = 2$ and $a_{n+1} = 4a_n - 3a_{n-1}$ for all $n \geq 2$. Find a formula for a_n. (Use the examples in the chapter.)

7.   Let $a_1 = a_2 = 1$ and $a_{n+1} = 5a_n - 6a_{n-1}$ for all $n \geq 2$. Find a general formula for a_n.

8.   Let $a_1 = a_2 = 1$ and $a_{n+1} = a_n - \tfrac{1}{4}a_{n-1}$ for all $n \geq 2$. Find a formula for a_n.

9.   The binomial coefficients $\binom{n}{k}$ for $0 \leq k \leq n$ are all positive integers. From

the formula $\binom{n}{k} = \dfrac{n!}{k!(n-k)!} = \dfrac{n(n-1)\ldots(n-k+1)}{k!}$ this is not at all

obvious, for it is not clear that each factor in the denominator cancels with something in the numerator.

         Use the equation $\binom{n+1}{k} = \binom{n}{k} + \binom{n}{k-1}$ for $(1 \leq k \leq n)$ to show that

if $\binom{n}{k}$ is an integer for $0 \leq k \leq n$, then $\binom{n+1}{k}$ is an integer for

$1 \leq k \leq n$. By noticing that $\binom{n+1}{0} = \binom{n+1}{n+1} = 1$ deduce that $\binom{n+1}{k}$ is an

integer for $0 \leq k \leq n+1$.

         Finish off the proof by induction that for all natural numbers n if k is

an integer and $0 \leq k \leq n$ then $\binom{n}{k}$ is an integer.

10.   In Problem 4 you showed that 10^n leaves a remainder of 1 on division by 9. Show that if $a_0, \ldots, a_m$ are integers between 0 and 9 then the difference between $a_m 10^m + a_{m-1} 10^{m-1} + \ldots + a_1 10 + a_0$ and $a_m + a_{m-1} + \ldots + a_0$ is a multiple of 9. Hence show that an integer, written in the usual decimal notation, is a multiple of 9 if and only if the sum of its digits in its decimal is a multiple of 9.

11.   Use Problem 5 to show that the difference between
$$a_m 10^m + a_{m-1} 10^{m-1} + \dots + a_0 \quad \text{and} \quad a_0 - a_1 + a_2 - a_3 + \dots + (-1)^m a_m$$
is a multiple of 11.   Deduce that $a_m 10^m + \dots + a_0$ is divisible by 11 if and only if the related number $a_0 - a_1 + a_2 - \dots + (-1)^m a_m$ is.

(So 27 819 is divisible by 11 because $9 - 1 + 8 - 7 + 2 \ (= 11)$ is a multiple of 11.)

12.   Show that if n is a natural number and $x \neq 1$
$$1 + 2x + 3x^2 + \dots + nx^{n-1} \quad = \quad \frac{1 - (n+1)x^n + nx^{n+1}}{(1-x)^2} \quad .$$

13*. Suppose that a, b, α and β are given real numbers and that we define a_n for all natural numbers by
$$a_1 = \alpha, \qquad a_2 = \beta, \qquad a_{n+1} = a\, a_n + b\, a_{n-1} \qquad (n \geq 2) \ .$$
Suppose also that $a^2 + 4b < 0$, so that $y^2 = ay + b$ has no real solutions. Show that $a_n = (\surd(-b))^n (A\cos(n\theta) + B\sin(n\theta))$ where θ is chosen so that
$$\surd(-b)\cos\theta = \tfrac{1}{2}a$$
$$\surd(-b)\sin\theta = \surd(-\tfrac{1}{4}a^2 - b) \ .$$
(You should check that there is such a value of θ.) A and B are chosen to satisfy $A\cos\theta + B\sin\theta = \alpha / \surd(-b)$
$$A\cos 2\theta + B\sin 2\theta = \beta / (-b) \ ,$$
which is possible because $\sin\theta \neq 0$.

(You will need the formulae $\cos(x+y) = \cos x \cos y - \sin x \sin y$ and $\sin(x+y) = \sin x \cos y + \cos x \sin y$.)

14.* Some sequences can be misleading.   Let $\binom{n}{k} = \dfrac{n(n-1)\cdots(n-k+1)}{k!}$ when n and k are non-negative integers.   For $0 \leq k \leq n$ this gives the binomial coefficients already defined, while if $k > n$ we have $\binom{n}{k} = 0$.

Now let $a_n = 1 + \binom{n}{2} + \binom{n}{4} + \binom{n}{6}$.   By direct calculation we find that the first seven terms are 1, 2, 4, 8, 16, 32, 64.   Who would not expect 128 for the next?   But the next term is 127.

Use the binomial theorem, and the fact that $(1 + (-1))^n = 0$, to show that if n is a natural number,
$$\binom{n}{0} + \binom{n}{2} + \dots + \binom{n}{n} \ = 2^{n-1} \qquad \text{if } n \text{ is even, and}$$
$$\binom{n}{0} + \binom{n}{2} + \dots + \binom{n}{n-1} = 2^{n-1} \qquad \text{if } n \text{ is odd} \ .$$

Hence show that if k is a natural number and b_n is given by

$$b_n = \binom{n}{0} + \binom{n}{2} + \binom{n}{4} + \cdots + \binom{n}{2k}$$

then $b_n = 2^{n-1}$ for $n = 1, 2, \ldots, 2k+1$, but that $b_{2k+2} = 2^{2k+1} - 1$.

You can make this sequence mimic the sequence of powers of 2 for any fixed number of steps, but eventually the two sequences will differ.

4

Sets and Numbers

"Mathematicians are a species of Frenchman: if you say something to them, they translate it into their own language and presto! it is something entirely different."

Goethe.

4.1 Sets

We have seen how induction is used as a technique for proving results about natural numbers. In this chapter we shall develop a few more techniques to add to our repertoire and apply these to number systems other than the natural numbers. One useful idea is that of a set. At first this seems to be little more than a neat notation for various mathematical objects, but it will acquire a more substantial value quickly. Even if sets were only a good notation we should not dismiss them lightly, for a good notation may be an immense help in thinking about a problem. Anyone who doubts this should try expressing two numbers in Roman numerals, XLIX and CLV say, and multiply them without using any notation other than the Roman!

A **set** is a specified collection of objects. The objects which belong to a given set are called its **elements** or **members**. The elements of a set can be of any sort whatever provided they are clearly specified in some way. For example, we could consider "the set of all odd positive integers," or "the set of all rabbits alive today". Naturally, some sets are of more interest in mathematics than others and we shall not pay much attention to sets of rabbits. The sorts of set that is likely to be of interest to us are sets of numbers, or of functions or of other mathematical objects.

We can specify a set in several ways. The most obvious is to list the members, in which case we write the members between curly brackets and separate them with commas, e.g. $\{1, 2, 3\}$ denotes the set whose elements are the numbers 1, 2 and 3, and no others. We can specify a set in words (e.g. "the set of all integers") or by using some property.

$$\{x: \ x \text{ is an integer and } x^2 + x < 100\}$$

means "the set of all x such that x is an integer and $x^2 + x < 100$". (The colon is replaced by a bar by some authors as $\{x| \ x \text{ is an integer and } x^2 + x < 100\}$.) The only criterion necessary for defining a set is that given an object it is in principle possible to determine whether or not it is a member. "In principle" here means that we could say that x is a member if the equation $y^3 + y = x$ has three real solutions, even if we are, perhaps temporarily, unable to solve that equation or otherwise determine how many solutions it has. A little common sense is needed with all this: the notation $\{1, 2, 3\}$ would be problematical if we wished to construct a set one of whose members was a comma or a bracket!

We use the symbol $\in$ to denote when an **object belongs to** or **is an element of** a set, for example $1 \in A$ which we pronounce "one belongs to A" or "1 is an element of A". Thus $1 \in \{1, -1\}$. The symbol $\notin$ means "is not an element of" so $2 \notin \{1, -1\}$.

4.2 Standard Sets
We have some standard notations for sets which crop up very frequently.

$\mathbb{N}$ the set of **natural numbers** (so $\mathbb{N} = \{1, 2, 3, ...\}$)

$\mathbb{Z}$ the set of **integers** (so $\mathbb{Z} = \{0, 1, -1, 2, -2, ...\}$). The $\mathbb{Z}$ stands for the German, Zahl.

$\mathbb{Q}$ the set of **rational numbers**. (Q for quotient.)

$\mathbb{R}$ the set of **real numbers**.

In writing sets like $\mathbb{N}$ and $\mathbb{Z}$ in the form with dots, we have to be sure that it is obvious what the rule for membership is and not leave it as a puzzle for the reader. $\{1, 7, 9, 12, 14\frac{1}{2}, ...\}$ is not acceptable.

We have already presumed something here in using the symbol $=$. We need to define what it means to say that two sets are equal.

Definition Two sets are said to be equal if they have the same elements.

An object is either in a set or not in it, so the sets $\{1, 2, 3\}$, $\{3, 2, 1\}$ and $\{1, 1, 2, 3\}$ are all equal. The order in which we write the elements and the fact that we have written some more than once are immaterial.

We sometimes need to refer to the set with no elements at all. (This is a set in the same way that an empty box is still a box, even though it contains nothing.) We may refer to the set with no elements without knowing it, for we might consider $\{x : x$ is an integer and $x^2 = 2\}$, which has no elements because no x satisfies the conditions. In more complicated cases it may not be obvious whether the conditions for belonging to a set can be satisfied. The set with no elements is called the **empty set**, and denoted by the Danish letter $\varnothing$.

In dealing with sets, we sometimes find that one set contains all of the elements of another. For example, every natural number is an integer, so the set of all integers contains all the members of the set of all natural numbers. There is a notation for this.

Definition A **subset** of a set A is a set all of whose elements belong to A. If B is a subset of A we write $B \subset A$.

Notice that whatever the set A is, we always have $A \subset A$, and $\varnothing \subset A$. The last statement is rather trivial: $\varnothing \subset A$ is because all of the elements of $\varnothing$ (and there are none) belong to A. Putting it another way, if $x \in \varnothing$ then $x \in A$, which is trivially satisfied because no x satisfies the condition $x \in \varnothing$. Notice that if $A \subset B$ and $B \subset A$ then $A = B$; this is actually the commonest way of showing that two sets are equal. Also notice that $A \subset B$ is the same thing as saying that $x \in A \Rightarrow x \in B$.

Notation We define subsets of the standard sets in an abbreviated form.   For example $\{\, x \in \mathbb{R} : x^2 = 2 \,\}$ means the set of all x which belong to $\mathbb{R}$ and satisfy the equation $x^2 = 2$.

Example 4.1 Suppose that A is a subset of the set of natural numbers (in symbols $A \subset \mathbb{N}$), and that A has the properties
      (i) $1 \in A$
and (ii) $n \in A \Rightarrow n+1 \in A$.
      Show that $A = \mathbb{N}$.

Solution: We are given that $A \subset \mathbb{N}$ so we need only prove that $\mathbb{N} \subset A$.
    Let $P(n)$ be the statement "$n \in A$". By (i) $P(1)$ is true, and by (ii) for all $n \in \mathbb{N}$, $P(n) \Rightarrow P(n+1)$.   By induction, then, for all $n \in \mathbb{N}$, $n \in A$, that is $\mathbb{N} \subset A$.
    We now know that $A \subset \mathbb{N}$ and $\mathbb{N} \subset A$ so $\mathbb{N} = A$. □

    We could develop the theory of sets more, but we shall postpone that until we need it.   As we shall see in due course, the virtue of the idea of a set is that it allows us a way of discussing and proving things about otherwise awkward quantities.   We can easily define the set $\{x : x \in \mathbb{Q} \text{ and } x^2 < 2\}$, and we may notice that 0 and 1 belong to this set but 2 does not.   Now if we can deduce properties of this set then some of these will be related to the condition $x^2 < 2$, and the largest members of the set will presumably have x^2 close to 2.   This might provide a way of discussing numbers which satisfy $x^2 = 2$, which, as we shall see, appear not to be fractions (i.e. rational numbers).   Experience shows that this is an approach which is useful: the elements of a set which are the largest, or smallest, or are distinguished in some other way, may allow us to discuss a problem which is otherwise awkward.   This will emerge as we proceed.

4.3 Proof by Contradiction
At first sight this seems an extremely perverse way of proving anything.   In practice it turns out to be useful, mainly when the sort of property we are dealing with is such that knowing that it is false gives us more specific information than knowing that it is true, or where a contradiction argument allows us to move from simple information to more complicated information, rather than extract a simple conclusion from a complicated starting point.

Example 4.2 Let $n \in \mathbb{N}$ and n^2 be divisible by 2.   Prove that n is divisible by 2.

Comment: The problem here is that we are given information about n^2 and have to extract information about n.   It would be easier to move from information about n to information about n^2.
    Notice that an integer x is **divisible by** y if there is an integer z with $x = yz$. We say that y is a **divisor** or **factor** of x.

Solution: Suppose that $n \in \mathbb{N}$, n^2 is divisible by 2 and the result is false, that is, suppose that n is **not** divisible by 2.

Since every integer can be written in the form $2q + r$ where q is an integer and $r = 0$ or 1, there must be an integer q with $n = 2q + 1$, for n is not divisible by 2. Then

$$n^2 = (2q + 1)^2 = 4q^2 + 4q + 1 = 2(2q^2 + 2q) + 1$$

so n^2 is not divisible by 2. But this is nonsense, for we began with the knowledge that n^2 is divisible by 2. Therefore our unjustified assumption that n is not divisible by 2 must be wrong (for it has led to this nonsense). Therefore n is divisible by 2. □

Proving Results Case by Case

There is a point in the example above that is worth further comment. We noticed that every integer leaves a remainder of 0 or 1 on division by 2, and dealt with the two cases separately. In this example we wanted to show that one of the possibilities did not arise. Had we been interested in the remainders on division by 3 then we would have three cases, $r = 0$, 1 or 2, to consider. This is a generally applicable technique to split up the situation into cases and deal with them in turn. We need to be careful that we have considered every possible case, of course. If the number of cases is large the technique may be unappealing and we might try to avoid plodding through many cases. Nevertheless, even if there are regrettably many cases to be checked, it is a possible way of getting the results we wish if no better method seems to be available.

We have assumed here that if d is an integer greater than 1, then if n is an integer we can write n in the form $n = dq + r$ where q and r are integers and $r \in \{1, 2, \ldots d - 1\}$. This is sometimes called division with remainder. In due course we will prove this but accept it for now for the purposes of illustration.

Example 4.3 There is no rational number whose square is 2.

Proof: Suppose the result is false, that is, suppose there is a rational number x with $x^2 = 2$.

Since x is rational there are integers p and q, with q positive, such that $x = p/q$. If p and q have a common factor, r say, which is greater than 1, then there are integers p' and q' for which $p = rp'$ and $q = rq'$. Then $x = rp'/(rq') = p'/q'$ and $q' < q$. Now if p' and q' have a common factor greater than 1 we can remove it in the same way to obtain $x = p''/q''$. At each such step we reduce the denominator by at least 1 (for $q'' < q' < q$ and they are integers) so the process cannot go on indefinitely; it must stop after at most $q - 1$ steps. Assume that we have carried this out as often as we can, so we obtain $x = a/b$ where a and b are integers with no number greater than 1 which divides them both.

Then $a^2/b^2 = 2$ so $a^2 = 2b^2$. Therefore a^2 is an integer which is divisible by 2 so, by the last example, a must be divisible by 2 also. Then $a = 2a'$, for some integer a' so that $4(a')^2 = 2b^2$ and $b^2 = 2(a')^2$. This shows us that b^2 is a multiple of 2 and so, by the example above again, b is a multiple of 2. We have shown that a and b are both multiples of 2. This is a contradiction, for we are given above that there is no number greater than 1 which divides both a and b.

It follows from the contradiction that there can be no rational number whose square is 2. (In other words $\sqrt{2}$ is irrational.) ☐

This result, that $\sqrt{2}$ is not a rational number, was discovered by the ancient Greeks and is attributed to Hippasus. It is said that Hippasus discovered this when on board a ship and the result so disturbed his fellow travellers that they threw him overboard!

Irrational numbers turn out to have useful and interesting properties. Some particularly useful numbers, such as π or Euler's constant e, are irrational, although we are not yet in a position to show this.

Suppose we have an angle θ, and we increase this by steps of θ, giving $2\theta, 3\theta$ and so on. Once n is large enough $n\theta$ will be greater than a complete revolution and it is more convenient to measure it by subtracting $360°$ or 2π radians. For larger values of n we may need to subtract a multiple of $360°$ (or 2π radians). So the angle we would wish to measure is $n\theta - k360°$ where k is the integer chosen so that $n\theta - k360°$ is at least $0°$ and less than $360°$. In this case it is simpler to divide by $360°$ and consider θ as a proportion of a complete revolution, as $x = \theta/360°$. Then $n\theta - k360° = (nx - k)360°$ where k is an integer chosen so that $nx - k$ is at least zero and less than 1. In this sense we measure the "fractional part" of nx, by subtracting off the largest integer k so that $nx - k$ is at least 0.

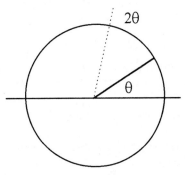

An interesting point in all this is to notice when the pattern of angles $x, 2x, 3x, \dots$ repeats itself and when it does not. If x is rational, say m/n, then $nx = m$ and so nx is an integer, whence its fractional part is 0. Then $(n + 1)x = m + x$, so its fractional part is x. That is, after $n + 1$ steps we arrive back at x and the pattern then repeats itself every n steps. (It might be that the pattern repeats more often than that, for we have not troubled to check whether nx is the first occasion on which the fractional part is zero.) Moreover, if x is not rational the pattern does not repeat. For if mx and nx represent the same angle in the original problem, mx and nx have the same fractional part, where we are supposing $m \neq n$ here (or else there is no repetition). Therefore $mx - nx = k$ for some integer k and $x = k/(m - n)$, a rational number. So the pattern of angles $\theta, 2\theta, 3\theta, \dots$ repeats if, and only if, θ is a rational multiple of a complete revolution.

Example 4.4 Show that the sum of two rational numbers is rational, and that the sum of one rational and one irrational number is irrational. What can you say about the sum of two irrational numbers?

Solution: Let x and y be rational. Then for some integers a, b, c and d, with b and d non-zero, $x = a/b$ and $y = c/d$. Therefore

$$x + y \; = \; \tfrac{a}{b} + \tfrac{c}{d} \; = \; \tfrac{ad+bc}{bd} \; ,$$

which is the ratio of two integers, so $x + y$ is rational.

Now let x be rational and y irrational. We know the form of x but nothing very specific about y. If the conclusion were false, that is, $x + y$ were rational, we would have more specific information about $x + y$ than we have about y, so we proceed this way, by contradiction.

Let x be rational and y irrational. Suppose that $x + y$ is rational. Then $y = (x + y) - x = (x + y) + (-x)$, the sum of two rational numbers. Therefore y is rational, which is a contradiction. The contradiction shows that our assumption that $x + y$ is rational is wrong, so $x + y$ is irrational.

What can we say about the sum of two irrational numbers? $\tfrac{1}{2}\sqrt{2} + \tfrac{1}{2}\sqrt{2} = \sqrt{2}$ is the sum of two irrational numbers, and $\tfrac{1}{2}\sqrt{2}$ is irrational (for $\tfrac{1}{2}\sqrt{2} = \tfrac{1}{\sqrt{2}}$ and if $\tfrac{1}{\sqrt{2}}$ were rational then $\sqrt{2} = 1/(1/\sqrt{2})$) would also be rational, a contradiction). So the sum of two irrational numbers *can be* irrational. The sum of two irrational numbers can also be rational (e.g. $\sqrt{2} + \sqrt{(-2)} = 0$). $\square$

Example 4.5 There are infinitely many prime numbers.

Solution: Suppose the result were false, that is, suppose that there were only finitely many prime numbers, N say. Call them $p_1, p_2, p_3, ..., p_N$. Then let $n = p_1 p_2 p_3 ... p_N. + 1$. Now n is a natural number so, by Example 3.4, n is the product of prime numbers (including the possibility of being a "product" of one prime number,) so it has at least one prime factor. Call this prime factor p. Now n is not a multiple of p_1, for there is a remainder of 1 when we divide by p_1 so $p \neq p_1$. Similarly $p \neq p_2$, $p \neq p_3$, ... and $p \neq p_N$. Therefore p is a prime number not in the list $p_1, ..., p_N$. . This is a contradiction since $p_1, ..., p_N$ were supposed to be all the prime numbers. $\square$

This result is fairly simple once you are used to proof by contradiction. Most other results about prime numbers are far from simple to prove!

Lemma 4.1 Every non-empty set of natural numbers has a smallest element.

Comment: The result may look odd. Notice that a non-empty set of integers (e.g. $\mathbb{Z}$ itself) need not have a smallest element, nor has $\{x : x$ is rational and $0 < x\}$. This property of automatically having a smallest element is useful.

Proof: Let $A \subset \mathbb{N}$ and $A \neq \varnothing$.

Suppose that the result is false, that is, suppose that A does not have a smallest element. Then $1 \notin A$ (for if $1 \in A$ it would be the smallest element). Also if $1, 2, 3, ..., n$ all do not belong to A then $n + 1 \notin A$, for otherwise $n + 1$ would be the smallest element. By induction, then, with the statement $P(n)$ being $n \notin A$, we see that for all $n \in \mathbb{N}$ $n \notin A$. Therefore no natural numbers belong to A and so $A = \varnothing$, a contradiction.

It follows that A must have a smallest element. $\square$

This result looks rather esoteric, but it actually allows us to prove some useful things. As an example, let us prove the result about division with remainder, which we have so far accepted without proof.

Lemma 4.2 Division with Remainder. Let d be a natural number greater than 1. Every integer n can be expressed in the form $n = dq + r$ where q is an integer and $r \in \{0, 1, ..., d-1\}$. The integers d and r with these properties are unique.

Proof: Let n be an integer and define the set A by

$$A = \{x \in \mathbb{N} : \text{for some integer } y, \ x = n - dy\} \ .$$

A is a subset of $\mathbb{N}$ and $A \neq \varnothing$ (For if $n > 0$, $n = x - d0 \in A$. If $n = 0$ then $x = n - d(-1) = d \in A$, while if $n < 0$ then $x = n - dn = (-n)(d - 1) \in A$ because $-n$ and $d - 1$ are natural numbers.) Therefore, by Lemma 4.1, A has a smallest element; call it r_0. Since $r_0 \in A$, $r_0 > 0$ and there is an integer y_0 for which $r_0 = n - dy_0$. Now we cannot have $r_0 > d$, for if that were true $r_0 - d > 0$ and $r_0 - d = n - d(y_0 + 1)$ so that $r_0 - d \in A$. But r_0 was the smallest element of A whence this is a contradiction. Therefore $r_0 \leq d$.

If $r_0 < d$, then because it is a natural number $r_0 \in \{1, 2, ... d - 1\}$ and we set $r = r_0, q = y_0$ to obtain $n = dq + r$. If $r_0 = d$ then set $r = 0$ and $q = y_0 + 1$ so that $n = d(y_0 + 1) = dq + r$.

The above work shows that r and q exist. Their uniqueness is easier. Suppose that $n = dq_1 + r_1 = dq_2 + r_2$ where q_1, q_2, r_1 and r_2 are integers and r_1 and r_2 both belong to $\{0, 1, ..., d - 1\}$. Then $d(q_1 - q_2) = r_2 - r_1$. Now if $q_1 > q_2$, $q_1 - q_2$ is a positive integer, hence $q_1 - q_2 \geq 1$ and $d(q_1 - q_2) \geq d$. But $r_2 - r_1 \leq r_2 < d$ so we cannot have and $d(q_1 - q_2) = r_2 - r_1$. Similarly if $q_1 < q_2$ we cannot satisfy $d(q_1 - q_2) = r_2 - r_1$ so we are left with $q_1 = q_2$, showing that $r_2 - r_1 = d(q_1 - q_2) = 0$ and $r_1 = r_2$ also. □

What we have proved above should come as no surprise, of course. In the spirit of the Lemma we could, putting $d = 12$, notice that $75 = 12 \times 6 + 3$ and the like. The choice of remainders in this case from the numbers $\{0, 1, ..., 11\}$ is the "standard" one but for particular purposes (which will not concern us here) it may be useful to vary this; we could have opted to use $\{1, 2, ..., 12\}$ as our set of remainders, for example. Meantime, however, we shall prove two other properties of the integers. The value of Theorem 4.3 below is in the least obvious part, the expression for the greatest common divisor. First, however, we must say what the words mean.

Definition Let a and b be two integers. A **common divisor** of a and b is an integer which is a divisor (or factor) of both a and b.

Example 4.6 The numbers 6 and 27 have a common divisor of 3, and so do 6 and -27. Notice that every number is a divisor of 0 (for $0 = 0.x$) so we see that 7 is a common divisor of 0 and 7. □

Theorem 4.3 Let a and b be two natural numbers. Then among all the common divisors of a and b there is a greatest common divisor. If d is the greatest common divisor of a and b then there are integers x and y for which $d = ax + by$.

Proof: Let A be the set

$$A = \{z: \text{there are integers } x \text{ and } y \text{ for which } z = ax + by \text{ and } z > 0\} \ .$$

A is a set of natural numbers. It is also non-empty, for $a = a1 + b0$ so $a \in A$. By Lemma 4.1, then, A has a smallest element; call it c. Then by definition c is a natural number and there are integers x_0 and y_0 with $c = ax_0 + by_0$.

Let d be a common divisor of a and b, so that $a = da'$ and $b = db'$ for integers a' and b'. Then $c = d(a'x_0 + b'y_0)$ and we see that c is a multiple of d. If d is positive we have $c = dc'$ for some positive integer c' (since $c > 0$) so that $c' \geq 1$ and $c \geq d$. If d is not positive, then $d \leq 0$ and $0 < c$ so $d \leq c$ in this case too. Therefore all common divisors d of a and b satisfy $d \leq c$.

Next we notice that c is a common divisor of a and b. Suppose that c were not a divisor of a. Then $a = cq + r$ for some integers q and r with $r \in \{1, 2, ..., c - 1\}$. Therefore $a = (ax_0 + by_0)q + r$, and $r = a(1 - x_0q) + b(-y_0q)$ showing that $r \in A$, for by choice $r > 0$. But $r < c$ and c was the smallest element of A, a contradiction. Therefore c must be a divisor of a (giving $a = cq + r$ where $r = 0$: this r does not belong to A). Similarly c is a divisor of b, which proves that c is a common divisor of a and b.

We have shown that c is a common divisor and that every common divisor is less than or equal to c. Therefore c is the greatest common divisor of a and b. $\square$

By trial and error, we may check that the greatest common divisor of 24 and 42 is 6. Then $6 = 24a + 42b$ for some integers a and b; $a = 2$ $b = -1$ is one example of this.

A useful consequence of this, particularly in arguments showing that a particular number is or is not rational is

Theorem 4.4 Let a and b be natural numbers and p be a prime number. If p divides ab then either p divides a or p divides b.

Caution: The corresponding result is false if we divide by a number which is not prime. For example, 6 divides 4×9 but 6 does not divide either 4 or 9.

Proof: Let p divide ab. We need to prove that either p divides a or p divides b. Suppose that p does not divide a. Then the greatest common divisor of p and a is 1, for if d is the greatest common divisor then d divides p, so $d = 1$ or $d = p$; we cannot have $d = p$ because p does not divide a. By Theorem 4.3 there are integers x and y for which $px + ay = 1$. Then $bpx + aby = b$. Since ab is a multiple of p and bpx obviously is, the left hand side is a multiple of p, hence so is b.

We have proved that if p does not divide a then it divides b, that is, either it divides a or it divides b. $\square$

Example 4.7 We showed earlier that $\sqrt{2}$ is irrational and a similar argument allows us to prove $\sqrt{3}$ is also irrational. These two rely on the preliminary result that if n is an integer and if n^2 is divisible by 2 (or 3) then n is divisible by 2 (or 3, respectively). The corresponding result fails for 4, that is, n^2 can be divisible by 4 even though n is not. (2^2 is divisible by 4 but 2 is not.) This has the consequence that our previous method of proof will not show that $\sqrt{4}$ is irrational, which is reassuring! □

4.4 Sets Again

We have so far made only slight use of sets. This will change as we proceed and more situations arise where they will prove useful. The idea of a set is useful in dealing with the infinite, which can be tricky to cope with. For example, each natural number is finite when considered on its own, but the set $\mathbb{N}$ of all of them is not finite. Given any individual natural number it is easy to produce another natural number greater than it (add 1), but we cannot produce one natural number which is greater than all of the members of the set $\mathbb{N}$. The set has some properties distinct from the properties of the members. A **finite set** is one which is either empty or where the elements can be paired with the numbers 1, 2, ..., n for some n. (Less formally, we count the number of elements and if the process stops the set is finite.) Clearly $\mathbb{N}$ is not finite, that is, it is an infinite set, even though the objects in it, considered as individuals, are finite. Infinite sets have some odd properties which are not shared by finite sets, which is why we need some care in dealing with them.

Suppose that A and B are two finite sets, with $A \subset B$. Then if the two sets have the same number of elements they are equal. (For B contains A, so the number of elements in B is the number in A plus the number of elements of B which are not in A. If A and B have the same finite number of elements, there must be no elements of B which are not in A.) This property is false for infinite sets, and it is rather spectacularly false. Let $B = \mathbb{N}$ and A be the set of all even natural numbers. Then we can match up all the elements of A with elements of B, by matching $n \in B$ with $2n \in A$. We have matched the elements of the two sets so we might think that they have the same "size". But $A \subset B$ and here $A \neq B$. We need to be cautious not to assume that infinite sets have the same properties as finite sets.

4.5 Where We Have Got To – and The Way Ahead

We have now proved a number of results, including the last three about integers and divisibility. Some readers will regard some of these as obvious, others may not. In a sense this is inevitable for we have not written out a list of things we are prepared to assume, or take for granted. We might accept "the ordinary properties of arithmetic" as given, and many people would accept the result of Lemma 4.2 as one of these "ordinary properties". What we have shown above is that we do not need to assume the result about division with remainder: we can prove it, and thus reduce the list of what we are assuming. There is an aesthetic satisfaction in reducing the number of assumptions, of course, but this is taking us into the realms of the philosophy of mathematics, which we do not wish to consider in depth here.

A more immediate point is to notice the content of the proofs of Lemmas 4.1 and 4.2 and Theorems 4.3 and 4.4. They make use of proof by contradiction, sometimes overall, sometimes in small components of the proof. The later results also use the result that a non-empty set of natural numbers has a smallest element. These techniques occur again and again!

The type of result we have discussed, about the integers and their properties, is really a part of number theory. This has the virtue that it is easy to appreciate many of the results and the problems, but, as those who go on to study number theory will discover, it may not be at all easy to answer the questions involved. We shall digress briefly into one number-theoretic problem to illustrate some of the points we shall need about proof. The question we shall consider is when an integer is the sum of two squares. The results here will not be needed later.

4.6 A Digression

We can easily check, by trial and error, that $2 = 1^2 + 1^2$, $4 = 2^2 + 0^2$, $5 = 2^2 + 1^2$ and that 3, 6 and 7 are not the sum of two squares. It is also easy to prove (but much harder to think in the first place that it might be worth proving) that if x and y are both the sum of two squares then so is xy. To see this notice that

$$\begin{aligned}(a^2 + b^2)(c^2 + d^2) &= a^2c^2 + b^2d^2 + b^2c^2 + a^2d^2 \\ &= (ac + bd)^2 + (ad - bc)^2 .\end{aligned}$$

From this we see that 4, 10 and 25 are all the sum of two squares and so is any product of these with 2 or 5. In fact, all numbers which are the product of a power of 2 and a power of 5 must be the sum of two squares. (We mean powers here which are non-negative integers.)

Now let us look at powers of 3. The number 3 itself is not the sum of two squares but $9 (= 3^2 + 0^2)$ is, while 27 is not and $81 (= 9^2 + 0^2)$ is. Trial and error suggests that 3^n is the sum of two squares if n is even and not if n is odd (n here being a positive integer). For even n, $n = 2m$ and $3^{2m} = (3^m)^2 + 0^2$. For odd m, the remaining case, we need to pull another rabbit out of the hat! Notice that if x is an integer then x^2 leaves a remainder of 0 or 1 on division by 4 (check that!) and hence the sum of two squares can leave a remainder of $0 + 0$, $0 + 1$ or $1 + 1$ on division by 4, *but not a remainder of 3*. It is now a simple matter to check by induction that if n is odd 3^n leaves a remainder of 3 on division by 4, so it is not the sum of two squares.

We have now shown (or, more accurately, could have shown if we had filled in the details) that a number whose factors are purely 2s, 3s and 5s, that is, one of the form $2^{\alpha_1} 3^{\alpha_2} 5^{\alpha_3}$; where α_1, α_2, and α_3 are non-negative, is the sum of two squares provided α_2 is even. It is important to notice that although we might guess, correctly, that these are the only numbers of the form $2^{\alpha_1} 3^{\alpha_2} 5^{\alpha_3}$ which are the sum of two squares, we have not yet given any reason for this. All we know about numbers of the form $2^{\alpha_1} 3^{\alpha_2} 5^{\alpha_3}$ for α_2 odd is that we seem unable to show that they are the sum of two squares. That is not the same thing as knowing that they are not.

The clue in this case is again simple in hindsight, but not simple to think of in the first place: we look at remainders on division by 3. If x is an integer then x^2

leaves a remainder of 0 or 1 on division by 3 so the possible remainders of $x^2 + y^2$ on division by 3 are $0 + 0$, $0 + 1$ or $1 + 1$, that is, all possible remainders 0, 1 or 2. However, the only way in which $x^2 + y^2$ can leave a remainder of 0 on division by 3 is if x^2 and y^2 separately leave a remainder of 0 on division by 3, which means that x and y must be divisible by 3. (That needs some checking.) Therefore if $x^2 + y^2$ is divisible by 3 there are integers u and v with $x = 3u$ and $y = 3v$ giving $x^2 + y^2 = 9(u^2 + v^2)$. This is the outline of a proof that if a number which is the sum of two squares is divisible by 3 then it is divisible by 9 and, moreover, if we divide the number by 9 the result is also the sum of two squares. (In the notation above, $(x^2 + y^2)/9 = u^2 + v^2$.) So if $2^{\alpha_1} 3^{\alpha_2} 5^{\alpha_3}$ is the sum of two squares and is divisible by 3 (that is, $\alpha_2 \geq 1$) then it is divisible by 9 (hence $\alpha_2 \geq 2$) and $2^{\alpha_1} 3^{\alpha_2 - 2} 5^{\alpha_3}$ is the sum of two squares. Pursuing this allows us to show that α_2 is even if $2^{\alpha_1} 3^{\alpha_2} 5^{\alpha_3}$ is the sum of two squares.

   To take this further requires some general results which would take us too far from the techniques we shall need in the rest of the book. We can decompose a typical positive integer x into its prime factors and write the number as

$$x = 2^{\alpha_0} p_1^{\alpha_1} p_2^{\alpha_2} \dots p_r^{\alpha_r} q_1^{\beta_1} q_2^{\beta_2} \dots q_s^{\beta_s}$$

where $p_1, \dots, p_r, q_1, \dots, q_s$ are prime numbers, the p_j leaving a remainder of 1 on division by 4 and the q_j leaving a remainder of 3 on division by 4. The result is that x is the sum of two squares if, and only if, $\beta_1, \beta_2, \dots, \beta_s$ are all even. To establish the "if" part we need to show that every prime number which leaves a remainder of 1 on division by 4 is the sum of two squares, which is done very cleverly in a paper by D. Zagier ("A One-Sentence Proof That Every Prime $p \equiv 1$ (mod 4) Is a Sum of Two Squares", *American Mathematical Monthly*, **97** (1990), 144.) For the "only if" part, to show that the only numbers which are sums of two squares have the β_j even, we need a development of the idea we used to deal with powers of 3.

   Notice the nature of the argument here: we show by one method that the set of natural numbers which are the sum of two squares includes all those of a certain form, that is, the set is at least as large as the set of those of the stated form. Then, separately and in this case by quite different arguments, we show that the set of all sums of squares is no larger than the one we think. If we call the two sets A and B then we prove $A \subset B$ and $B \subset A$, the two proofs using different ideas. In a complicated case like this we are reduced to working out fragments of the solution, and our logic is essential to the process of assembling these fragments into a complete argument. Finally, notice that these ideas of proof have allowed us to show the connection between two quite different things: sums of squares and prime factors. We have done (or would have done if we had filled in the large amount of missing detail) one of the things a mathematician looks to do, classified the integers into those which have some property and those which do not, in this case into those which are sums of squares and those which are not.

Problems

1.   Prove that if n is an integer and n^2 is a multiple of 3 then n is a multiple of 3. Deduce that $\sqrt{3}$ is irrational.

2.   Prove that $\sqrt[3]{2}$ (that is the number x satisfying $x^3 = 2$) is irrational. (You will need to show that if n is an integer and n^3 is a multiple of 2 then n is a multiple of 2.)

3.   Show that $\sqrt{6}$ is irrational.

4.   Show that the product of two rational numbers is rational. Deduce that if a is rational *and non-zero* and b is irrational then ab is irrational.

5.   Show that if x^2 is irrational then x is irrational. Use some of the earlier problems in this section and the examples in the chapter to show that $(\sqrt{2} + \sqrt{3})^2$ is irrational, and hence that $\sqrt{2} + \sqrt{3}$ is irrational.

6*.   Show that if a and b are two rational numbers, both non-zero, then $a\sqrt{2} + b\sqrt{3}$ is irrational. (This shows in a sense that $\sqrt{2}$ and $\sqrt{3}$ are irrational in "different ways" in that the irrationality of one cannot cancel the irrationality of the other. $\sqrt{2}$ and $\sqrt{8}$ are not independent in this way in that $2\sqrt{2} + (-1)\sqrt{8} = 0$.)

7.   Show that if x and y are integers then x^2 must leave a remainder on division by 7 of 0, 1, 2 or 4, and deduce that if $x^2 + y^2$ is divisible by 7 then both x and y are divisible by 7. Hence show that if the number z is the sum of two squares and is divisible by 7 then z is divisible by 49 and $z/49$ is the sum of two squares. (This can be taken further to show that if a positive integer is the sum of two squares then the power of 7 in its prime factorisation is even.)

8*.   Suppose that p and q are integers satisfying $p^3 + pq^2 + q^3 = 0$. Show that if one of p and q is divisible by 2 so is the other. Show also that if p and q are both odd numbers then they cannot satisfy the given equation. Deduce that there is no rational number satisfying $x^3 + x + 1 = 0$.

9*.   Let a, b and c be three natural numbers. Let
$$A = \{ax + by + cz : x, y \text{ and } z \text{ belong to } \mathbb{Z}, \text{ and } ax + by + cz > 0\} .$$
Show that $A \neq \varnothing$ and let d be the smallest element of A. (Notice that we need to use Lemma 4.1 to be sure that such a number d does exist.) Show that a is a multiple of d (and by symmetry, so are b and c). Deduce that $A = \{nd : n \in \mathbb{N}\}$. We have shown that d is a common divisor of a, b and c; use the fact that $d \in A$ to show that it is the greatest common divisor.

10.   Use Theorem 4.4 to show that if p is a prime number then $\sqrt{p}$ is irrational.

5

Order and Inequalities

"Good order is the foundation of all things." Edmund Burke.

5.1 Basic Properties
As well as the properties arising from the arithmetic operations of addition and multiplication, the real numbers have an order, in that we think of some numbers as being larger than others. This is reinforced by our picture of the real numbers describing the points on a line, for the line corresponds to an ordering. Results about the order, for example, where we show that one particular number is less than another, are often called **inequalities**. This is not a particularly good name, but it emphasises that we are not dealing with equations where the two sides are equal and therefore interchangeable. With inequalities the two sides are definitely not interchangeable.

The order properties are not quite as simple to deal with as the arithmetical properties. This is partly caused by the fact that some of the properties involve "one-way" implications, so we need to pay attention to the logic to avoid errors.

Definitions We write $x < y$ to mean that "x is less than y". This has exactly the same meaning as "y is greater than x", which we write $y > x$. The symbol $\leq$ means "is less than or equal to" that is, $x < y$ or $x = y$ (with the corresponding interpretation of $x \geq y$).

The basic properties of the order are:

For all real numbers x, y and z:
1. Exactly one of the three statements $x < y$, $x = y$, $x > y$ holds;
2. If $x < y$ and $y < z$ then $x < z$;
3. If $x < y$ then $x + z < y + z$;
4. If $x < y$ and $z > 0$ then $xz < yz$.

Notice that the third property is "reversible", in that we can apply it by adding the number $-z$ to $x + z$ and $y + z$, so by this rule
$$x + z < y + z \quad \Rightarrow \quad (x + z) + (-z) < (y + z) + (-z)$$
$$\Rightarrow \quad x < y \ .$$
Putting this less formally, we are not "giving away" any information when we use the third property, for we can recover the original information from the result. This is quite different from the second property for if we know that $x < y$ and $y < z$, then we deduce that $x < z$. This conclusion tells us nothing at all about y, so we cannot recover the information we had in the first place from the statement $x < z$ on

its own. We have "given away" information. (In formal terms $y < z \Rightarrow x < z$ but $x < z \not\Rightarrow x < y$ and $y < z$.)

The last property needs care. We use this to deduce information by multiplying both sides of the inequality $x < y$ by the same number z, but *notice that the deduction is allowable if $z > 0$.* This condition, that z be greater than 0, is awkward in practice.

Using the basic properties

The symbols x, y and z appearing in the basic properties are "dummy" variables and we may substitute other numbers or symbols for them, provided we do this consistently. For example, property 2 says that

For all real numbers x, y and z, if $x < y$ and $y < z$ then $x < z$.

If we choose three real numbers, we may substitute them consistently for x, y and z, that is we replace every occurrence of x with one of the numbers, every occurrence of y with another of the numbers and every occurrence of z with the third. So, for example, if $2 < 3$ and $3 < 4$ then $2 < 4$. We may also replace the symbols x, y and z with other symbols, provided every occurrence of x is replaced by the same symbol, every occurrence of y is replaced by the same symbol and so on. So we can legitimately deduce that

For all real numbers a, b and c, if $a < b$ and $b < c$ then $a < c$.

There is no reason why we need the new symbols all to be distinct; it would be legitimate to replace x by z for example, provided all occurrences of x are replaced. This would give

For all real numbers z, y and z, if $z < y$ and $y < z$ then $z < z$.

(This could be tidied up, of course, but in fact it is not a useful thing to know.) Notice that we have lost information this way, for, while we can replace z in this statement, we must replace all zs with the same letter or symbol, so we cannot recover the original statement from it.

5.2 Consequences of the Basic Properties

These properties above are the basic assumptions. We can deduce others from them, but there are relatively few general properties. Readers should be wary of properties which we do not prove: they may well be false even if they look plausible! This is perhaps the first area in which proof becomes directly important (other than as a matter of checking the author's correctness), for it will happen that the reader needs to consider a slight modification of the results stated, and he or she needs to be able to decide if it is correct.

Example 5.1 $x < y$ and $z < t \Rightarrow x + z < y + t$.

Solution: Suppose that $x < y$ and $z < t$.

     Then $x + z < y + z$ (by property 3, since $x < y$).

     Also $y + z < y + t$ (by property 3, since $z < t$: add y

to both sides.)

Therefore $x + z < y + t$ 　　　　　　　　　　(for $x + z < y + z$ and $y + z < y + t$ so we use property 2.)

We have shown that $x < y$ and $z < t \Rightarrow x + z < y + t$. 　In fact, since we have assumed nothing about x, y, z or t except that they are real numbers, the deduction holds for all real x, y, z and t, so we could have stated the result more fully as

"For all real numbers x, y, z and t, $x < y$ and $z < t \Rightarrow x + z < y + t$." $\square$

In the deduction above notice that the property we have used, property 3, says that for all x, y and z, $x < y \Rightarrow x + z < y + z$. 　What is important here is the form of the process (add the same number to both sides of the inequality). 　The actual labels do not matter, so it is equally correct to say that $y < z \Rightarrow y + x < z + x$. 　(If that seems mysterious, notice that we could substitute neutral variables in 3, say a, b and c in place of x, y and z respectively. 　Then $a < b \Rightarrow a + c < b + c$. 　Then we can substitute y, z and x respectively for a, b and c.)

Example 5.2 For all real x and y, $x < y \Leftrightarrow -y < -x$.
(Putting this loosely, to make it memorable, if we multiply both sides of $x < y$ by -1 we need to reverse the inequality to obtain the correct conclusion $-x > -y$.)

Solution: Suppose that x and y are real numbers. 　Then

$$x < y \;\Rightarrow\; x + (-x - y) < y + (-x - y) \qquad (\text{property 3})$$
$$\Rightarrow\; -y < -x \qquad\qquad\qquad\qquad (\text{rearranging}) \; .$$

Also 　　　　$-y < -x \Rightarrow -y + (x + y) < -x + (x + y) \Rightarrow x < y$.

Therefore 　$x < y \Leftrightarrow -y < -x$. $\square$

In this case the result has the $\Leftrightarrow$ connection, so we have not lost any information. 　This is not true in Example 5.1, in the sense that it is possible to have $x < y$ and $x + z < y + t$ in some cases where $z < t$ is false. 　(Try $x = 1$, $y = 3$, $z = 1$, $t = 0$.)

The next example, and more particularly what is not true in the same spirit, shows why we need care with the order properties.

Example 5.3 If x, y, z and t are real numbers then
$$x < y \text{ and } z < t \Rightarrow x - t < y - z \; .$$

Solution: This is easy, given Examples 5.1 and 5.2. 　For if $x < y$ and $z < t$ then $-t < -z$ (by Example 5.2), so by Example 5.1 $x + (-t) < y + (-z)$, as required. $\square$

Notice what we have *not* shown. 　The result obtained by subtracting the corresponding side of one inequality from the other would be $x - z < y - t$ but this is false in general. 　(By that we mean that it is not always true, although there may be particular cases for which it is true.) 　Let $x = 1$, $y = 2$, $z = 3$ and $t = 5$. 　Then

$x < y$ and $z < t$ but $x - z = -2$ and $y - t = -3$, so it is false that $x - z < y - t$.

At this point we need to take stock of how we deduce things in practice. In solving equations we have used the technique of taking two equations and adding the corresponding sides to produce a new equation. This is informally expressed as "adding the two equations" or the like and is justified by noting that if $x = z$ and $y = t$ then $x + y = z + t$ (because the corresponding numbers on each side are equal). Also if $y = t$ then $-y = -t$, so $x + (-y) = z + (-t)$ or $x - y = z - t$. With the order properties the numbers we are discussing are not usually equal and there is not the same symmetry. Most notably if $x < y$ then $-y < -x$, that is, the negative sign reverses the sense of the inequality. Example 5.1 shows that, informally speaking, we can "add two inequalities" (adding the terms on the corresponding side of the $=$ sign). *However, nothing in our rules (assumed or derived) lets us "subtract inequalities".*

So far, we have used only the additive properties of order (regarding subtraction as undoing addition). The corresponding results about multiplying can be done by imitating what we did for addition, with the added complication that property 4 allows us to multiply both sides of an inequality by the same positive number. (A number is called **positive** if it is greater than 0, and negative if it is less than 0.)

Notation If $x < y$ and $y < z$ we abbreviate this and write $x < y < z$. Similarly $a > b > c$ means $a > b$ and $b > c$.

Example 5.4 For all real numbers x, y, z and t

$$0 < x < y \text{ and } 0 < z < t \Rightarrow 0 < xz < yt$$

Solution: Suppose that $0 < x < y$ and $0 < z < t$.

Then $xz < yz$ (since $x < y$ and $z > 0$, using property 4) .

Also $y > 0$ (property 2, since $0 < x$ and $x < y$)

and therefore $yz < yt$ (property 4 again, since $y > 0$ and $z < t$) .

Therefore $xz < yt$ (for $xz < yz$ and $yz < yt$, using property 2) .

It remains to show that $xz > 0$. Since $x > 0$ and $z > 0$ we may use property 4 to deduce that $xz > 0z$ and since $0z = 0$ we see that $xz > 0$ (i.e. $0 < xz$). $\square$

Comment: Notice that the first half of the proof is analogous to Example 5.1 except that we multiply instead of adding, but we need a digression to show that $y > 0$, which we need in order to use property 4 to show that $yz < yt$.

Example 5.5 For all real x, $x^2 \geq 0$.

Solution: We need some comment here before starting. All of the earlier examples have begun with information that one number is less than another, which we have used to deduce that certain other inequalities hold. We do not have that sort of information here, and we cannot use properties 2 to 4 without it. We are forced, therefore, to use property 1, even if, at first glance, it does not look hopeful.

Let $x \in \mathbb{R}$. Then by property 1 $x > 0$ or $x = 0$ or $x < 0$.

In the first case, $x > 0$, Example 5.4 shows us that $x^2 > 0$　(for $0 < x$ and $0 < x$ so $0 < x.x$). Therefore $x^2 \geq 0$.

In the second case $x = 0$ so $x^2 = 0$, and it follows that $x^2 \geq 0$.

In the last case $x < 0$ so that (by Example 5.2) $-0 < -x$, that is, $-x > 0$. Example 5.4 now shows that $(-x)^2 > 0$ and since $(-x)^2 = x^2$, we have $x^2 > 0$. Therefore $x^2 \geq 0$.

In all three cases, one of which must occur, $x^2 \geq 0$. □

There are some consequences of this result which may come as a surprise. Since $1^2 = 1$ we notice that $1 \geq 0$ (because $1^2 \geq 0$) and therefore $1 > 0$ because $1 \neq 0$. We might reasonably have accepted that $1 > 0$ as part of our basic assumptions. This result shows that we do not need to make such an assumption.

Example 5.4 shows us that if $x \in \mathbb{R}$ then $x^2 > 0$ or $x^2 = 0$ and therefore $1 + x^2 > 1$ or $1 + x^2 = 1$, and in either case $1 + x^2 > 0$. Therefore $1 + x^2$ cannot be 0 if x is a real number. Putting this another way, the combination of the arithmetic properties and the order properties prevent there being a solution of the equation $1 + x^2 = 0$. There are number systems, in particular the complex numbers, in which this equation has a solution but in order to gain that we must sacrifice some other properties. To avoid having to change our basic assumed properties of the number system we shall restrict attention in this book to real numbers.

Let us finish the string of results about order with one last general example.

Example 5.6 For all real numbers x and y

$$x > 0 \Rightarrow \frac{1}{x} > 0 \text{ and } 0 < x < y \Rightarrow \frac{1}{y} < \frac{1}{x}.$$

Solution: Let $x > 0$. Then $x \neq 0$ so that $\frac{1}{x}$ exists. By property 1 $\frac{1}{x} > 0$ or $\frac{1}{x} = 0$ or $\frac{1}{x} < 0$. The second case cannot arise (for if $\frac{1}{x} = 0$ then $1 = x \cdot \frac{1}{x} = x \cdot 0 = 0$ which is a contradiction). Neither can the third case be true for if $\frac{1}{x} < 0$ then by property 4, $\frac{1}{x} \cdot x < 0 \cdot x$, that is, $1 < 0$. But $1 > 0$ and property 1 tells us that exactly one of $1 > 0$, $1 = 0$, $1 < 0$ is true whence $1 < 0$ is false. This contradiction shows that $\frac{1}{x} < 0$ is impossible.

We are left only with the possibility $\frac{1}{x} > 0$ and since we know one of the three possibilities is true, $\frac{1}{x} > 0$.

Now suppose $0 < x < y$. Then (property 2) $y > 0$, so by what we have just shown $\frac{1}{x} > 0$ and $\frac{1}{y} > 0$. Then $x\frac{1}{x} < y\frac{1}{x}$ (property 4) so we deduce that $1 < \frac{y}{x}$. Also by property 4 $1\frac{1}{y} < \frac{y}{x} \cdot \frac{1}{y}$, that is, $\frac{1}{y} < \frac{1}{x}$, as desired. □

Notice that in Example 5.6 the order of $\frac{1}{x}$ and $\frac{1}{y}$ is the opposite of that of x and y (and that what we did was restricted to positive x and y).

　　　We have now established enough rules to do all that we need, but we shall have to be cautious with the order properties.　The main awkwardness lies in the multiplication properties usually being related to having to know that certain numbers are positive.　This may cause us to consider several cases if we do not know whether or not the numbers we have to consider are positive.

　　　There are many plausible results concerning the order properties which are either false, or true only under some restrictions.　For this reason we need to treat these properties with some care and we often need to work out results to cover the situation in hand.　The basic properties at the beginning of the chapter together with the results of the examples will often be useful but of even greater use are the techniques used in the examples.　If you like to regard the examples as little theorems then the calculations done there (the "proofs" if you like) are good illustrations of the sort of technique we can deploy in dealing with the order properties.

Example 5.7 Let $x < -2$.　Show that $\frac{1}{x} > -\frac{1}{2}$.

　　　This does not fit neatly into the examples we have done since $x < 0$ (because $-2 < 0$).　One way of proceeding is to convert to positive numbers:

$$x < -2 \Rightarrow -x > -(-2) = 2 \qquad \Rightarrow 0 < 2 < -x$$

$$\Rightarrow \frac{1}{-x} < \frac{1}{2}$$

$$\Rightarrow \frac{1}{x} > -\frac{1}{2} . \qquad \square$$

　　　Relating x^2 to y^2 when we know, for example, that $x < y$ is less easy than we might think if we allow negative numbers to be considered.　It is easy to use Example 5.4 with $z = x$ and $t = y$ to show that, if $0 < x < y$, then $x^2 < y^2$.　(Alternatively, knowing that x and y are positive, we could deduce from $x < y$ that $x^2 < xy$ and $xy < y^2$ and hence $x^2 < y^2$.　There are other routes, but the main thing is to be able to produce a reliable result without undue use of memory.)　If x and y are both negative then $x < y < 0 \Rightarrow 0 < -y < -x \Rightarrow (-y)^2 < (-x)^2 \Rightarrow y^2 < x^2$: in this case the relative order of x^2 and y^2 is different from that of x and y.　The trickiest situation is, of course, where we do not know whether the numbers involved are greater or less than 0.

Example 5.8 Let $a < x < b$.　Then $x^2 < \max(a^2, b^2)$, where by $\max(u, v)$ we mean the maximum of u and v (that is, $\max(u, v) = u$ if $u \geq v$ and $\max(u, v) = v$ if $u < v$).

Solution: All would be simple if we knew whether $x > 0$, $x = 0$ or $x < 0$, but we do not.

　　　Let $a < x < b$.　Then if $x > 0$, $0 < x < b$ so $x^2 < b^2$ (using the working above).　Since $b^2 < \max(a^2, b^2)$ or $b^2 = \max(a^2, b^2)$ we deduce that $x^2 < \max(a^2, b^2)$.

　　　If $x < 0$ then $a < x < 0$ so $0 < -x < -a$ and it follows that $(-x)^2 < (-a)^2$, that is, $x^2 < a^2$.　Since $a^2 \leq \max(a^2, b^2)$, $x^2 < \max(a^2, b^2)$ again.

　　　If $x = 0$ then $b > 0$ so $b^2 > 0$, that is, $x^2 < b^2$.　Once again

$b^2 \le \max(a^2, b^2)$ and $x^2 < \max(a^2, b^2)$.

Whichever case occurs for x, we deduce that $x^2 < \max(a^2, b^2)$. $\square$

This example is easier to visualise if we draw a graph. There are three cases, depending on whether a and b are both positive, both negative or both zero. Notice that in the case where a and b straddle 0 we need to look to see which branch of the graph comes highest for $a < x < b$.

There are more explicit ways in which negative numbers cause problems with inequalities. Occasionally these prove to be a real difficulty, but in most of the situations we shall encounter we can sidestep the problems. First, however, we need to see what the difficulties can be.

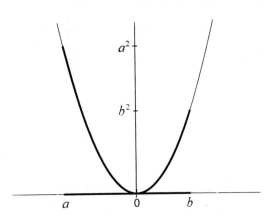

Fig 5.1 $a < x < b \implies x^2 < \max(a^2, b^2)$

Example 5.9 Show that $0 < x < 3 \implies -1 \le x^2 - 2x < 3$.

Solution: The simplest way to proceed is to notice that

$$0 < x < 3 \quad \implies \quad 0 < x^2 < 9$$

$$\text{and} \quad 0 < x < 3 \quad \implies \quad -3 < -x < 0 \implies -6 < -2x < 0$$

(using Examples 5.8 and 5.2). Then, using Example 5.1, $0 - 6 < x^2 - 2x < 9 + 0$, giving the result $0 < x < 3 \implies -6 < x^2 - 2x < 9$.

This is perfectly correct, and is routine, and for many purposes may give enough information to do whatever we wish to do. In this case, however, we are asked for more specific information than we have found. The difficulty is not hard to find: the values of x which make x^2 close to its upper limit of 9 (those x close to 3) are not the same as the values of x which make $-2x$ close to its upper limit. We have "given away" information. In this case we resort to the trick of completing the square to treat the x^2 and the $-2x$ term together, by noticing in this case that $x^2 - 2x = (x-1)^2 - 1$.

Then $0 < x < 3 \implies -1 < x - 1 < 2$

$$\implies (x-1)^2 < \max((-1)^2, 2^2) = 4 \qquad \text{(Example 5.8)}$$

$$\implies 0 \le (x-1)^2 < 4 \qquad \text{(Example 5.5)}$$

$$\implies 0 - 1 \le x^2 - 2x < 4 - 1 \qquad \text{(Property 3)}$$

$$\implies -1 \le x^2 - 2x < 3 . \qquad \square$$

The technique of completing the square is often useful in handling inequalities involving quadratic expressions. We shall need to use these ideas occasionally but for the moment let us concentrate on a rather different notion, the relative size of various quantities.

Example 5.10 Show that for all $n \in \mathbb{N}$, $2^n > n$.

Solution: The result $2^n > n$ is true for $n = 1$.
Suppose that $n \in \mathbb{N}$ and $2^n > n$. Then

$$2^{n+1} = 2.2^n > 2n \qquad \text{(because } 2 > 0, \text{ using property 4)}$$

$$= n + n$$

$$\geq n + 1 \qquad \text{(because } n \geq 1 \text{ if } n \in \mathbb{N}) .$$

Therefore $2^n > n \Rightarrow 2^{n+1} > n + 1$.
By induction we have, for all $n \in \mathbb{N}$, $2^n > n$. $\square$

This is a simple comparison of a relatively complicated quantity, 2^n, with a much simpler one. By substituting a few values of n we can see that 2^n is very much greater than n if n is at all large but the value of a result like this is its simplicity, not that it provides a good estimate of 2^n. From the knowledge that $2^n > n$ we can deduce that $4^n > n^2$ or that $\frac{1}{n} > (\frac{1}{2})^n$, for example. The last statement is potentially useful, for we might have the feeling that $(\frac{1}{2})^n$ can be made as small as we wish by choosing n large enough. This inequality confirms this belief and even tells us, albeit crudely, how large is "enough". If we wish $(\frac{1}{2})^n$ to be less than some positive number a, then if $n > \frac{1}{a}$ so that $\frac{1}{n} < a$ then $(\frac{1}{2})^n < a$ also.

5.3 Bernoulli's Inequality
The inequality $2^n > n$ is actually a special case of a general result, attributed to a member of the Bernoulli family (of whom several were well-known mathematicians in the seventeenth and eighteenth centuries).

Theorem 5.1 Bernoulli's Inequality
If $x > -1$ and $n \in \mathbb{N}$ then $(1 + x)^n \geq 1 + nx$.
A more commonly useful form of this is:
If $y > 0$ and $n \in \mathbb{N}$ then $y^n \geq 1 + n(y - 1)$, and in particular

$$y^n > n(y - 1) .$$

Proof: Clearly $(1 + x)^n \geq 1 + nx$ is true for $n = 1$ since both sides are equal.
Suppose that $n \in \mathbb{N}$, $x > -1$, and $(1 + x)^n \geq 1 + nx$. Then

$$(1+x)^{n+1} = (1+x)(1+x)^n \geq (1+x)(1+nx) \qquad \text{(by property 4, for } 1 + x > 0)$$

$$= 1 + (n+1)x + nx^2 \geq 1 + (n+1)x$$

so if the property holds for n, then it holds for $n+1$. By induction, then, for all
$n \in \mathbb{N}$ and $x > -1$, $(1+x)^n \geq 1 + nx..$

Now let $y > 0$ and set $x = y - 1$. Then $x > -1$ so that by the first part

$$y^n = (1+x)^n \geq 1 + nx = 1 + n(y-1). \quad \square$$

Although there are odd occasions when the first form of Bernoulli's Inequality
is useful, its main value is the second form, and then usually when $y > 1$ so that
$1 + n(y - 1)$ increases with n. (If we put $y = \frac{1}{2}$ we obtain
$(\frac{1}{2})^4 > 1 + 4(-\frac{1}{2}) = -1$, which is obviously true, so in this case the inequality does
not enlarge our knowledge.)

We can nevertheless adapt Bernoulli's Inequality to deal with numbers which
are less than 1 and still obtain useful information.

Example 5.11 Suppose that $0 < x < 1$. Then for $n \in \mathbb{N}$

$$x^n < \frac{1}{n}\frac{x}{1-x} \ .$$

Solution: Let $0 < x < 1$. Then $\frac{1}{x} > 1$ and by Bernoulli's Inequality,

$$\frac{1}{x^n} = (\tfrac{1}{x})^n \geq 1 + n(\tfrac{1}{x} - 1) > n(\tfrac{1}{x} - 1) = n\frac{1-x}{x} > 0 \ .$$

Therefore $x^n < \frac{1}{n}\frac{x}{1-x} \ . \square$

We can use Bernoulli's Inequality (usually in the form $y^n > n(y - 1)$ for
$y > 1$) to obtain more interesting results. For example, since $\sqrt{2} > 1$, we deduce
from Bernoulli's Inequality that $(\sqrt{2})^n > 1 + n(\sqrt{2} - 1) > n(\sqrt{2} - 1)$. Then
(noticing that $\sqrt{2} - 1 > 0$) $(\sqrt{2})^n (\sqrt{2})^n > n^2(\sqrt{2} - 1)^2$ or, more simply, $2^n > Cn^2$
where C is a constant, $(\sqrt{2} - 1)^2$ in this case. So not only is it true that for all
$n \in \mathbb{N}$ $2^n > n$, but $2^n > Cn^2$ for some constant C. We can pursue this further,
for it gives us an idea of how fast differing quantities grow. We have compared
powers of 2 to n and to n^2. Clearly we could try comparing 2^n with n^k for
some fixed k. We might also consider d^n for some other $d > 1$, and compare
that with n^k.

Example 5.12 If $d > 1$ then there is a positive constant C with the property that
for all $n \in \mathbb{N}$ $d^n > Cn$.

Solution: This is just a case of using Bernoulli's Inequality, for it tells us that
$d^n \geq 1 + d(n - 1) > Cn$ where $C = d - 1 > 0$. $\square$

It would be neater if we could choose $C = 1$ in Example 5.11, but a little trial
and error shows that this is not always correct. For example, if we set $d = 1.1$
then $d^2 = 1.21 \not> 2$. However since $d^n > Cn$ we can obtain information about d^{2n}
in two ways: by noticing that $2n$ is a natural number, so that $d^{2n} > C(2n)$ or by

using our work on inequalities to deduce that (because $d^n > Cn > 0$) $d^{2n} = (d^n)^2 > C^2n^2$. Now $C^2n^2 = (\frac{1}{2}C^2n)2n$ and if n is chosen large enough, $\frac{1}{2}C^2n > 1$ so, for these n, $d^{2n} > 2n$. That is $d^m > m$ for large even values of m. We need to look more carefully to see if we can make a more general rule here.

If $d > 1$ then $\sqrt{d} > 1$. (We have not actually shown this, and we should be cautious until we convince ourselves. Suppose $\sqrt{d} \leq 1$ then since $\sqrt{d} > 0$ we have $0 < \sqrt{d} < 1$ so $0 < (\sqrt{d})^2 < 1^2$, that is, $d \leq 1$ which is a contradiction.) Now we can use Bernoulli's Inequality to show that for all $n \in \mathbb{N}$

$$\left(\sqrt{d}\right)^n > nC$$

where $C > 0$. (Actually $C = \sqrt{d} - 1$.) Therefore

$$((\sqrt{d})^n)^2 = d^n > C^2n^2 = (C^2n)n$$

so that for all $n \geq 1/C^2$, $C^2n \geq 1$ and $d^n > C^2n.n \geq n$. Putting this less precisely (but perhaps more clearly) if $d > 1$ then "eventually" $d^n > n$, or for all large enough n, $d^n > n$.

What we have shown is that for large enough n, d^n grows at least as fast as n (where d is a fixed number, greater than 1). How does d^n compare with n^2? If we return to our simplest case, 2^n, and compare 2^n with n^2 we see that $2^1 > 1^2$, $2^2 = 2^2$, $2^3 < 3^2$, $2^4 = 4^2$, $2^5 > 5^2$, $2^6 > 6^2$, and it is not hard to show by induction that $2^n > n^2$ for all $n \geq 5$. That is, 2^n is "eventually" greater than n^2. Also $2^n > n^3$ for all $n \geq 10$ (again, by induction). Is there a general rule here?

Example 5.13 If k is a natural number, then for all sufficiently large natural numbers n, $2^n > n^k$.

Solution: Choose $x = 2^{\frac{1}{k+1}}$, that is, the positive number x for which $x^{k+1} = 2$. (It is tempting to try $2^{\frac{1}{k}}$ but, as we shall see, that is not good enough.) Then $x > 1$, for if not we have $0 < x \leq 1$ so $0 < x^{k+1} < 1^{k+1} = 1$ which contradicts $x^{k+1} = 2$. By Bernoulli's Inequality then, for all $n \in \mathbb{N}$

$$x^n > n(x-1) = Cn$$

where $C = x - 1 > 0$. Then for all $n \in \mathbb{N}$,

$$(x^n)^{k+1} = (x^{k+1})^n > C^{k+1}n^{k+1} = (C^{k+1}n)n^k .$$

It follows from this that for all $n > 1/C^{k+1}$, $C^{k+1}n > 1$, so $\left(C^{k+1}n\right)n^k > n^k$ and

$$2^n = (x^{k+1})^n > n^k .$$

That is, for all "sufficiently large" n, $2^n > n^k$. $\square$

We could continue with this, expanding on the details, but we shall leave that to the problems. Notice the main thrust of this, however, that if $d > 1$ and k is a natural number, then d^n is "eventually" greater than n^k or, more precisely, for all $n \geq N$ (where N is some natural number) $d^n > n^k$. This gives us some "scales" on which numbers grow.   For natural numbers n, it is clear that $n^3 \geq n^2 \geq n$ so n^k grows faster as n increases the larger we choose k.   But if we choose $d > 1$ then d^n eventually exceeds each of the n^k.   So d^n grows "faster" than all the others.   That is, in spirit, what is being considered here, although the absolutely correct statements are a little more complicated by details such as "for all n greater than some number ...".

We have shown that if $d > 1$ then for all n greater than some N, $d^n > n$. This is false if $d < 1$.   If we restrict attention to positive d, to avoid complications arising from negative numbers, and consider $0 < d < 1$, then it is pretty obvious that d^n becomes smaller as n increases.   In fact we can go further by noticing that if $0 < d < 1$ then $1/d > 1$ so that our earlier work can be applied to $1/d$. Then, for some N, if $n \geq N$ $(1/d)^n > n$ hence $1/d^n > n$ and $d^n < 1/n$.   Here d^n becomes small for large values of n and this gives us some idea of how small. Similarly, because $1/d > 1$, if we choose k to be a natural number, then for all sufficiently large n, $(1/d)^n > n^k$ so that $d^n < 1/n^k$.   The values of d^n become small "rapidly" as n increases.

The behaviour of the powers of a fixed positive number d depends crucially on whether $d > 1$ or $d < 1$.   If $d > 1$ then d^n becomes larger as n increases while if $0 < d < 1$, d^n becomes smaller as n increases.   The paragraphs above spell this out in more detail.

5.4 The Modulus (or Absolute Value)

We have seen several situations where negative numbers introduce complications into the order properties.   For example if $0 < x < a$ and $0 < y < b$ then $xy < ab$, but if we do not know that $x > 0$ and $y > 0$ then the information that $x < a$ and $y < b$ is, on its own, not enough for us to draw any conclusions about the order of xy and ab.   For example, if we put $x = y = 2$ and $a = b = 3$ then $xy < ab$; but if we have $x = y = -4$ and $a = b = 3$ then $xy > ab$.   In some cases this is an inconvenient obstruction to progress.   This is particularly true where in some sense x and y are small and we wish to know how small xy has to be.   The modulus, or absolute value, of a number is useful here.

Definition Let x be a real number.   We define the **modulus** of x, denoted by $|x|$, by

$$|x| = \begin{cases} x & \text{if} \quad x \geq 0 \\ -x & \text{if} \quad x < 0 \ . \end{cases}$$

Notice that for all $x \in \mathbb{R}$, $|x| \geq 0$ and that $|x| = 0$ if and only if $x = 0$.   Also notice that $|x| = -x$ in the case where $x = 0$ so that we can conclude that $|x| = -x$ for all $x \leq 0$.   This may occasionally save a case in some deduction.

The modulus is best thought of as the "size" of a number, ignoring its positive or negative sign.   This is another idea which proves useful, for $|x - y|$ is then either $x - y$ or $y - x$, whichever is non-negative, and can be thought of as the distance between x and y.   In this sense, $|x|$ is the distance of x from 0.

Lemma 5.2 The modulus has the following properties:

(i)   For all real x, $-|x| \leq x \leq |x|$

(ii)   For all real x, $|x| \le a \Leftrightarrow -a \le x \le a$
(iii)   For all real x and y, $|x + y| \le |x| + |y|$
(iv)   For all real x and y, $|xy| = |x||y|$.

Proof: (i) Let $x \in \mathbb{R}$. Then either $x \ge 0$ or $x < 0$. If $x \ge 0$ then $0 \le x = |x|$ so $0 \le x \le |x|$ and $-|x| \le 0$ so that $-|x| \le x \le |x|$.

(In giving this argument we have modified the rules we have derived for deductions involving $x < y$ to those involving $x \le y$. For example $x \le y \Rightarrow -y \le -x$ and $x \le y$ and $y \le z \Rightarrow x \le z$. These can be checked by dealing with the cases $x < y$ and $x = y$ separately. For a full statement of these properties see Problem 10.)

(ii) Suppose $|x| \le a$. Then $-a \le -|x|$ so by part (i) $-a \le -|x| \le x \le |x| \le a$, and so $-a \le x \le a$ (since $-a \le -|x|$ and $-|x| \le x$). Therefore $|x| \le a \Rightarrow -a \le x \le a$.

Now suppose $-a \le x \le a$. Then $|x| = x$ if $x \ge 0$, in which case $|x| = x \le a$, so $|x| \le a$, while if $x < 0$, $|x| = -x$ therefore $-a \le -|x|$, giving $|x| \le a$ again. In either case $|x| \le a$.

(iii) Suppose that x and y are real numbers. Then by part (i) $-|x| \le x \le |x|$ and $-|y| \le y \le |y|$ giving

$$-(|x| + |y|) = -|x| - |y| \le x + y \le |x| + |y| \quad .$$

Then using (ii) with $a = |x| + |y|$ gives $|x + y| \le |x| + |y|$.

(iv) Since $|x| = +x$ or $-x$, we see that $|xy| = \pm xy$ and $|x||y| = \pm xy$. If x and y are both non-negative, or both negative, we choose the $+$ sign in both cases. If one is non-negative and the other negative we choose the $-$ sign in both cases. Either way $|xy| = |x||y|$. $\square$

The inequality $|x + y| \le |x| + |y|$ is called the **triangle inequality**, and we shall encounter it very frequently, usually in a context where we need to show that $|x + y|$ is "small" in some sense. The name "triangle inequality" is not particularly meaningful here, but arises from a similar inequality in connection with vectors. For better or worse, the name is universal.

Example 5.14 Let x and y be two real numbers with $|y| < |x|$. Then $x + y$ has the same sign as x. (To be specific here, we say that x and y have the same sign if they are either both positive or both negative.)

Comment: Let $x > 0$. Then clearly $x + y > 0$ if $y \ge 0$, so the tricky case occurs if $y < 0$. But if $|y| < x$ adding y moves us a shorter distance to the left than x is

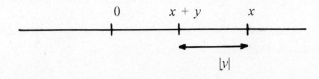

Fig 5.2

to the right of 0. That is "why" the result is true. The formal solution confirms this (of course) and gives us some techniques which may be useful in more complicated situations where a simple picture is not available.

Solution: Since $|x| > |y| \geq 0$, $|x| > 0$ so $x \neq 0$.
Let $x > 0$. Then $y \geq -|y|$ so $x + y \geq x - |y| > 0$ (since $x = |x| > |y|$).
Now let $x < 0$. Then $y \leq |y|$ so $x + y \leq x + |y|$ and since $|y| < |x| = -x$ $x + |y| < 0$, whence $x + y < 0$.
We have considered the two possible cases and whichever occurs $x + y$ has the same sign as x. $\square$

The triangle inequality is simple enough that we can use it in clever ways. For example, by substituting $-y$ for y we obtain $|x - y| = |x + (-y)| \leq |x| + |-y| = |x| + |y|$. A more substantial modification occurs in the following result.

Lemma 5.3 For all real x and y $\big||x| - |y|\big| \leq |x - y|$.

Proof: It is not easy to see how to start this. However, if $x, y \in \mathbb{R}$ then $\big||x| - |y|\big|$ equals $|x| - |y|$ or $|y| - |x|$.
Now $|x| = |(x - y) + y| \leq |x - y| + |y|$ (triangle inequality) so, adding $-|y|$ to both sides,
$$|x| - |y| \leq |x - y| .$$
Also, by interchanging the rôles of x and y
$$|y| - |x| \leq |y - x| = |x - y| .$$
Therefore whichever of $|x| - |y|$ or $|y| - |x|$ equals $\big||x| - |y|\big|$ we have
$$\big||x| - |y|\big| \leq |x - y| . \square$$

This result can be paraphrased as saying that two numbers are at least as far apart as their moduli.
We now come to an important type of problem. The underlying idea, which will emerge in later work, is that we wish to find out by how much we can allow x to vary from some given value without the values of some function of x varying by more than a prescribed amount. In the first case, we notice that $x^2 + 3x$ has the value 0 at $x = 0$ and we wish to find by how much we may let x vary from 0 and yet keep $x^2 + 3x$ within 1 of its value at 0.

Example 5.15 Find a positive δ such that $|x| < \delta \Rightarrow |x^2 + 3x| < 1$.

Solution: Before starting, notice that we are only asked to find a positive value of δ with the required property, not the greatest possible one. We may therefore impose any convenient additional assumptions.
Suppose that $|x| < \delta$. Then
$$|x^2 + 3x| = |x||x + 3| \leq \delta|x + 3| \qquad \text{(since } |x| < \delta \text{)}$$
$$\leq \delta(|x| + |3|) \qquad \text{(triangle inequality, noting } \delta > 0\text{)}$$
$$< \delta(\delta + 3)$$

$$\le \delta(1+3) = 4\delta \qquad \textit{IF } \delta \le 1$$
$$\le 1 \qquad\qquad \textit{IF } \delta \le 1/4$$

Therefore, had we chosen δ in the first place to be positive and satisfy the two "*IF*" conditions, we would obtain

$$|x| < 1/4 \Rightarrow |x^2 + 3x| < 1 .$$

The arbitrary assumption that $\delta \le 1$ is made purely for convenience, since it avoids a more complicated calculation in finding a δ such that $\delta^2 + 3\delta \le 1$. Had we supposed $\delta \le 2$ here, instead, we would have obtained a different, but equally correct, set of constraints on δ. The key step, though, is that in which we obtain $|x^2 + 3x| \le \delta |x + 3|$. From this point onwards it is enough to find a constant k, independent of x and δ, such that $|x| < \delta \Rightarrow |x + 3| < k$, whence $|x^2 + 3x| < k\delta$, and if we impose the additional constraint of $\delta \le 1/k$, we shall have solved the problem.

If we do the same problem again, differently, we might obtain

$$|x|<\delta \Rightarrow |x^2 + 3x| \le \delta|x + 3| \le \delta(|x|+3) < \delta(\delta + 3)$$
$$\Rightarrow |x^2 + 3x| < (7/2)\delta \qquad \textit{IF } \delta \le 1/2$$
$$\le 1 \qquad\qquad \textit{IF } \delta \le 2/7 .$$

Thus, setting $\delta = 2/7$, we obtain $|x| < 2/7 \Rightarrow |x^2 + 3x| < 1 . \ \square$

Problem The reader is cordially invited to show that the largest value of δ for which $|x| < \delta \Rightarrow |x^2 + 3x| < 1$ is $(\sqrt{13} - 3)/2$. The volume of calculation will be a demonstration of the value of the method above.

We may make the problem more useful by substituting a more general number on the right and considering x close to values other than 0.

Example 5.16 Given $\varepsilon > 0$ find a $\delta > 0$ for which
$$|x - 1| < \delta \Rightarrow |x^3 - 3x^2 - 2x + 4| < \varepsilon .$$

Solution: Let $|x - 1| < \delta$. Then
$$|x^3 - 3x^2 - 2x + 4| = |x - 1||x^2 - 2x - 4| \le \delta|x^2 - 2x - 4|$$
$$\le \delta(|x|^2 + |2x| + |-4|) \quad \text{(triangle inequality)}$$

Also, since $|x - 1| < \delta$, $|x| = |(x - 1) + 1| \le |x - 1| + 1 < \delta + 1$, so
$$|x^3 - 3x^2 - 2x + 4| \qquad < \delta((\delta + 1)^2 + 2(\delta + 1) + 4)$$
$$\le \delta(2^2 + 2.2 + 4) \qquad \textit{IF } \delta \le 1$$
$$= 12\delta \le \varepsilon \qquad\qquad \textit{IF } \delta \le \varepsilon/12$$

Therefore $|x - 1| < \min(1, \varepsilon/12) \Rightarrow |x^3 - 3x^2 - 2x + 4| < \varepsilon$.

In this case, of course, we have not enough information to decide which of 1 or $\varepsilon/12$ is the smaller. Notice, again, the crucial step of showing that the required expression is less than a multiple of δ, and then notice that this multiplying factor is less than or equal to some constant. $\square$

Example 5.17 Let $\varepsilon > 0$. Find a positive δ such that $|x - 1| < \delta$ and $|y - 1| < \delta$ together imply that $|xy - 1| < \varepsilon$.

Solution: Let $|x - 1| < \delta$ and $|y - 1| < \delta$. Then
$$|xy - 1| = |x(y - 1) + (x - 1)| \le |x||y - 1| + |x - 1| < \delta(|x| + 1)$$

$$\le \delta(|x - 1| + 2) < \delta(\delta + 2)$$

$$\le 3\delta \qquad \text{IF } \delta \le 1$$

$$\le \varepsilon \qquad \text{IF } \delta \le \varepsilon/3$$

Therefore $|x - 1| < \min(1, \varepsilon/3)$ and $|y - 1| < \min(1, \varepsilon/3) \Rightarrow |xy - 1| < \varepsilon$. □

Again, notice in the last example that we express the desired quantity, $|xy - 1|$, in terms of those which are 'small' in the sense that we can force such quantities to be as small as we need by a suitable choice of δ. In all these examples, one of the chain of inequalities is strict. If this were not so we would have to impose it at the end, e.g. $3\delta < \varepsilon$ (i.e. $\delta < \varepsilon/3$), in which case we would have to be careful to name a particular δ which satisfies all the conditions - noticing that $\delta = \varepsilon/3$ does not satisfy $3\delta < \varepsilon$. (At this level it is such trivia that may lead eventually to serious mistakes.)

In our nextl example, we need to cope with a denominator.

Example 5.18 Let $\varepsilon > 0$. Find a positive δ such that
$$|x| < \delta \Rightarrow \left| \frac{1}{1+x} - 1 \right| < \varepsilon .$$

Discussion: The expression to be shown less than ε is $|-x/(1 + x)|$ so, assuming $|x| < \delta$, we see that this is less than $\delta/|1 + x|$. The usual technique is to prove that the factor multiplying δ is less than some constant (subject to suitable conditions). To show $1/|1 + x| \le K$ it is enough to prove $|1 + x| \ge 1/K$ (assuming K positive). So we need to find a condition of the form $|x| < \delta' \Rightarrow |1 + x| \ge K'$ (putting K' for $1/K$). This is best spotted from a diagram (Fig. 5.3), since $|x| < \delta' \Rightarrow -\delta' < x < \delta' \Rightarrow 1 - \delta' < 1 + x < 1 + \delta'$

-δ'   0   δ'   0   1 - δ'   1   1 + δ'

Fig. 5.3

We notice here that for $x = 0$ (which will certainly satisfy $|x| < \delta$) $1 + x$ is positive (hence equal to its own modulus). We therefore choose δ' to keep $1 + x$ "away from zero". Choosing δ' so that $1 - \delta' = 1/2$ has the desired effect. (Any other positive number for $1 - \delta'$ would also be acceptable.)

Solution: $|x| < \delta \Rightarrow -\delta < x < \delta$ $\Rightarrow 1 - \delta < 1 + x$

$$\Rightarrow 1 + x > \tfrac{1}{2} \quad \text{IF } \delta \le \tfrac{1}{2}$$
$$\Rightarrow |1 + x| > \tfrac{1}{2}$$
$$\Rightarrow 1/|1 + x| < 2 .$$

Let $|x| < \delta$. Then

$$\left| \frac{1}{1+x} - 1 \right| = \left| \frac{-x}{1+x} \right| \le \frac{\delta}{|1+x|} < 2\delta \qquad \text{IF } \delta \le \tfrac{1}{2}$$

$$\le \varepsilon \qquad\qquad\qquad \text{IF } \delta \le \varepsilon / 2 .$$

Therefore $|x| < \min\left(\tfrac{1}{2}, \tfrac{1}{2}\varepsilon\right) \Rightarrow \left| \dfrac{1}{1+x} - 1 \right| < \varepsilon .$ □

Problems

1.   Show that if $x < y$ and $z < 0$ then $xz > yz$.   (Notice the reversal of the inequality sign!)

2.   Show that if $0 < x < y$ and $0 < z < t$ then $0 < \dfrac{x}{t} < \dfrac{y}{z}$.

3.   Show that if $0 < x < y < 1$ then $0 < \dfrac{x}{1-x} < \dfrac{y}{1-y}$.

4.   To show that $0 < x < y \Rightarrow \dfrac{x}{1+x} < \dfrac{y}{1+y}$ is more tricky than problem 3, since the effects of increasing x in the numerator and denominator oppose each other.   There are two reasonable methods.

    (i)   By noticing that $\dfrac{x}{1+x} = 1 - \dfrac{1}{1+x}$,   in which the variable appears only in the denominator prove that

$$0 < x < y \Rightarrow \tfrac{1}{1+x} > \tfrac{1}{1+y} \Rightarrow \tfrac{x}{1+x} < \tfrac{y}{1+y} .$$

    (ii)   Show that if x and y are positive

$$\tfrac{x}{1+x} < \tfrac{y}{1+y} \quad \Leftrightarrow \quad x(1+y) < y(1+x) \Leftrightarrow x < y$$

and hence show that $0 < x < y \Rightarrow \dfrac{x}{1+x} < \dfrac{y}{1+y}$.

5.   Show that $0 < x < y \Rightarrow 0 < x^n < y^n$ where n is a natural number, and deduce that $0 \le x \le y \Rightarrow 0 \le x^n \le y^n$.   Use this to show that if $x \ge 0$ and $y \ge 0$ then $x^n > y^n \Rightarrow x > y$.   (Caution!   Notice that prior assumption that $x, y \ge 0$. $(-3)^2 > 2^2$ but $-3 \not> 2$.)

6.   Show that if x and y are real numbers $x^2 + xy + y^2 \ge 0$,   and that $x^2 + xy + y^2 = 0 \Rightarrow x = y \, (= 0)$.   By factorising $y^3 - x^3$ or otherwise show that $x \le y \Leftrightarrow x^3 \le y^3$.

7.   Show that $0 < x < 1 \Rightarrow 1 - x < \dfrac{1}{1+x} < 1 - x + x^2$.

8.   Show that $(x \geq 0$ and $y \geq 0)$ or $(x \leq 0$ and $y \leq 0) \Rightarrow xy \geq 0$ and that $(x > 0$ and $y < 0)$ or $(x < 0$ and $y > 0) \Rightarrow xy < 0$. Deduce that $xy < 0 \Leftrightarrow (x > 0$ and $y < 0)$ or $(x < 0$ and $y > 0)$.

This can be extended to show that $x_1 x_2 ... x_r < 0$ if and only if an odd number of the factors x_i are negative, with the rest positive.

9.   By considering $(1 - x)(1 - y)$, or otherwise, show that if $x < 1$ and $y < 1$ then $x + y < 1 + xy$.

10.   Check as many of the following properties of the $\leq$ connection as you consider require it. (If you think they are all obvious do the last.) You should deduce these properties from the basic assumptions at the beginning of the chapter together with any other properties deduced from them.

For all real numbers x, y and z:

If $x \leq y$ and $y \leq z$ then $x \leq z$;

If $x \leq y$ then $x + z \leq y + z$;

If $x \leq y$ and $z \geq 0$ then $xz \leq yz$;

If $0 < x \leq y$ then $\frac{1}{y} \leq \frac{1}{x}$.

11.   Show that for all real x, $x^2 + 1 \geq 2x$ and deduce that if $x > 0$, $x + \frac{1}{x} \geq 2$.

Can you deduce anything about $x + \frac{1}{x}$ for $x < 0$?

12.   Show that $1 \leq x \leq 4 \Rightarrow -\frac{9}{4} \leq x^2 - 3x \leq 4$, and that $0 \leq x \leq 2 \Rightarrow -1 \leq x^2 - 2x \leq 0$. (Hint: Examples 5.5 and 5.8.)

13.   Show that $2 \leq x \leq 3 \Rightarrow -\frac{1}{4} \leq x^2 - 5x + 6 \leq 0$ and $-3 \leq x \leq 3 \Rightarrow -1 \leq x^2 + 2x \leq 15$.

14.   By factorising and using Problem 8 show that

$x^2 - 5x + 6 \leq 0 \Rightarrow 2 \leq x \leq 3$ and $x^2 - 5x + 6 > 0 \Rightarrow x > 3$ or $x < 2$.

(Caution! You must be careful to give reasons for disallowing conclusions you think are false.)

15.   Show that $1 \leq x \leq 3 \Rightarrow 0 \leq x - \frac{1}{x} \leq \frac{8}{3}$ and

$2 \leq x \leq 4 \Rightarrow 0 \leq x - \frac{4}{x} \leq 3$.

16.   We wish to show that $1 \leq x \leq 4 \Rightarrow 4 \leq x + \frac{4}{x} \leq 5$, which turns out to be slightly awkward. Show that $1 \leq x \leq 4 \Rightarrow x^2 - 4x + 4 \geq 0 \Rightarrow x + \frac{4}{x} \geq 4$

and that $1 \leq x \leq 4 \Rightarrow x^2 - 5x + 4 \leq 0 \Rightarrow x + \frac{4}{x} \leq 5$.

Deduce that $1 \leq x \leq 4 \Rightarrow 4 \leq x + \frac{4}{x} \leq 5$.

17. Use similar ideas to those of Problem 13 to show that $1 \le x \le 3 \Rightarrow$
$2 \le x + \frac{1}{x} \le \frac{10}{3}$ and $1 \le x \le 3 \Rightarrow 2\sqrt{2} \le x + \frac{2}{x} \le \frac{11}{3}$

18. We have proved that $0 \le x \le y \Rightarrow x^n \le y^n$, where n is a natural number.
Deduce that $0 \le u < v \Rightarrow u^{\frac{1}{n}} < v^{\frac{1}{n}}$.
Use this and Bernoulli's Inequality to show that if $x \ge 1$ and $n \in \mathbb{N}$

then $\qquad\qquad\qquad\qquad 1 \le x^{\frac{1}{n}} \le 1 + \frac{1}{n}x$.

(Notice that this tells us that $x^{\frac{1}{n}}$ is close to 1 for large enough values of n
and gives an estimate of how close. Use the fact that $x = (x^{\frac{1}{n}})^n$.)

19. Use problem 18 to show that if $n \in \mathbb{N}$, $1 \le n^{\frac{1}{n}} \le 2$. By choosing a suitable
value for x, show further that if $n \in \mathbb{N}$ $1 \le \sqrt{(n^{\frac{1}{n}})} \le 1 + \frac{1}{\sqrt{n}}$ and hence that

$1 \le n^{\frac{1}{n}} \le 1 + \frac{2}{\sqrt{n}} + \frac{1}{n} \le 1 + \frac{3}{\sqrt{n}}$.   (This is another interesting result, which

may be surprising: $n^{\frac{1}{n}}$ is close to 1 for large values of n.)

20. Show that for all $n \in \mathbb{N}$ $3^n > n^2$. (Do the induction for $n \ge 2$ and deal with
$n = 1$ separately.)

21. Show that for all $n \in \mathbb{N}$ $3^n \ge n^3$. (Prove it by induction for all $n \ge N$, where
you need to decide on N, and then deal with the remaining cases.)

22. Let $a^n = 2^n n^{-3}$. Show that $a_{n+1} > a_n \Leftrightarrow 2 > (1 + \frac{1}{n})^3 \Leftrightarrow n > 1/(2^{\frac{1}{3}} - 1)$.
Deduce that the smallest value of a_n (for $n \in \mathbb{N}$) is a_4 and hence that for all
$n \in \mathbb{N}$ $2^n \ge n^3/4$

23. Let x and y be real numbers with $|x| \le a$ and $|y| \le b$. Show that $|xy| \le ab$.
(Be careful if any of the numbers you use could be negative!)

24. We define $\max(x, y)$ to be the maximum of x and y. Let $\min(x, y)$ be the
minimum. Therefore

$$\max(x, y) \;=\; \begin{cases} x \text{ if } x \ge y \\ y \text{ if } y > x \end{cases} \qquad\qquad \min(x, y) \;=\; \begin{cases} y \text{ if } x \ge y \\ x \text{ if } y > x \end{cases} .$$

Show that, for all real x and y: (i) $|x| = \max(x, -x)$

(ii) $\max(x + z, y + z) = z + \max(x, y)$ (iii) $\max(x, y) = \frac{1}{2}(x + y + |x - y|)$

(iv) $\min(x, y) = \frac{1}{2}(x + y - |x - y|)$ (v) $\max(x, y) + \min(x, y) = x + y$.

25. Show that $a \le x \le b \Rightarrow |x| \le \max(|a|, |b|)$. (It may help to notice that if y is real $\min(x, y) \le y \le \max(x, y)$.)

26. Show that if $x < y$ and $z \in \mathbb{R}$ then $x|z| \le y|z|$. Give an example of three numbers x, y and z with $x < y$ and $x|z| = y|z|$ (to show that we cannot replace $\le$ by $<$ in general).

27. Show by induction that if $n \in \mathbb{N}$ and $x_1, ..., x_n \in \mathbb{R}$
$$|x_1 + x_2 + ... + x_n| \le |x_1| + |x_2| + ... + |x_n| \ .$$

28. In each case below, find a positive δ for which the implication holds:
    (i) $|x| < \delta \Rightarrow |2x^2 + x| < 1$,
    (ii) $|x| < \delta \Rightarrow |x^3 + 5x^2 - 3x| < 1$,
    (iii) $|x| < \delta \Rightarrow |x^3 - 2x| < 1$.

29. Let $\varepsilon > 0$. Find, in each case, a positive δ with the required property:
    (i) $|x| < \delta \Rightarrow |2x^2 + x| < \varepsilon$,   (ii) $|x| < \delta \Rightarrow |x^3 + 5x^2 - 3x| < \varepsilon$,
    (iii) $|x| < \delta \Rightarrow |x^3 - 2x^2| < \varepsilon$,   (iv) $|x - 1| < \delta \Rightarrow |x^3 - 3x^2 + 2| < \varepsilon$,
    (v) $|x + 2| < \delta \Rightarrow |x^3 - 2x + 4| < \varepsilon$.

30. Find a positive δ_1 such that $|x| < \delta_1 \Rightarrow |2 + x| \ge 1$. (You can either do this directly or use Lemma 5.3, noting that $2 + x = 2 - (-x)$.) Hence find a positive δ_2 for which $|x| < \delta_2 \Rightarrow \left| \dfrac{x+1}{x+2} - \dfrac{1}{2} \right| < \varepsilon$.

31. In each case below, find a positive δ such that the implication is true:
    (i) $|x - 1| < \delta \Rightarrow \left| \dfrac{1}{x} - 1 \right| < \varepsilon$,   (ii) $|x - 1| < \delta \Rightarrow \left| \dfrac{x+1}{x^2+1} - 1 \right| < \varepsilon$,

    (iii) $|x + 1| < \delta \Rightarrow \left| \dfrac{x+1}{x^2+2x} \right| < \varepsilon$.

32. Show that $|x + y| \ge 1 \Rightarrow |x| \ge \frac{1}{2}$ or $|y| \ge \frac{1}{2}$. More generally, prove that $|x_1 + x_2 + ... + x_n| \ge 1 \Rightarrow$ for some i with $1 \le i \le n$, $|x_i| \ge 1/n$.

33. There are many seemingly plausible results about moduli which are wrong. Give examples to show that, for arbitrary x, y and a,
$x \le y \not\Rightarrow |x| \le |y|$, $\quad x \le y \not\Rightarrow |x + a| \le |y + a|$, $\quad |x + y| \le 1 \not\Rightarrow |x| \le 1$ or $|y| \le 1$.

34. (i) Show that if x and y are non-negative, $x - y = (\sqrt{x} - \sqrt{y})(\sqrt{x} + \sqrt{y})$ and use this to show that if $n \in \mathbb{N}$ $\sqrt{(n+1)} - \sqrt{n} = \dfrac{1}{\sqrt{(n+1)} + \sqrt{n}}$. Deduce that
$$\dfrac{1}{2\sqrt{n}} \ge \sqrt{(n+1)} - \sqrt{n} \ge \dfrac{1}{2\sqrt{(n+1)}} \ .$$

(ii)   Let $s_n = 1 + \frac{1}{\sqrt{2}} + \frac{1}{\sqrt{3}} + \ldots + \frac{1}{\sqrt{n}}$.   Prove   that   for   all $n \in \mathbb{N}$ $2(\sqrt{(n+1)} - 1) \le s_n \le 2\sqrt{n}$.   Show that if $n > 120$ then $s_n > 20$ while if $s_n > 20$ then $n > 100$.

35. Let $a_n = \frac{1.3.5 \ldots (2n-1)}{2.4.6 \ldots (2n)}$.   We wish an estimate of the size of a_n for large

values of n .
(i) By grouping the terms in pairs (one each from numerator and denominator)
show that for all $n \in \mathbb{N}$ $a_n \le 1$.   Group the terms in pairs a different way to

show that $a_n \ge \frac{1}{2n}$.

(ii) Show that if $m \in \mathbb{N}$ then $\frac{(m-1)(m+1)}{m \quad m} < 1$.   Use this to show that

$$a_n^2 = \frac{1.1.3.3.5.5 \ldots}{2.2.4.4.6.6} \frac{(2n-1)(2n-1)}{(2n)(2n)} < \frac{1(2n-1)}{2n \quad 2n}$$

and deduce that $a_n < \frac{1}{\sqrt{(2n)}}$.   By grouping the terms differently, use the fact

that $\frac{m \quad m}{(m-1)(m+1)} > 1$ for $m \ge 2$ to show that $a_n^2 \ge \frac{1.1}{2.2n}$ and hence that

$a_n \ge \frac{1}{\sqrt{(4n)}}$.   We have shown that $\frac{1}{\sqrt{(4n)}} \le a_n < \frac{1}{\sqrt{(2n)}}$.

36. Show that if $1 \le k \le n$ and k and n are natural numbers then

$$\frac{1}{n^k} \binom{n}{k} = \frac{\left(1 - \frac{1}{n}\right)\left(1 - \frac{2}{n}\right) \ldots \left(1 - \frac{k-1}{n}\right)}{k!} \le \frac{1}{k!}$$

where $\binom{n}{k}$ denotes the binomial coefficient as in the Binomial Theorem.

Deduce that if $n \in \mathbb{N}$ then $\left(1 + \frac{1}{n}\right)^n \le 1 + \frac{1}{1!} + \frac{1}{2!} + \ldots + \frac{1}{n!}$.   By showing

that $n! \ge 2^{n-1}$ (for $n \in \mathbb{N}$)   deduce that for all $n \in \mathbb{N}$

$$1 \le \left(1 + \frac{1}{n}\right)^n \le 3 .$$

6

Decimals

6.1 Decimal Notation

We easily become so used to the standard decimal way of expressing a number that we tend to take this for granted and forget that the idea has some content. By the number

$$2345$$

we mean

$$2 \times 1000 + 3 \times 100 + 4 \times 10 + 5,$$

and, more generally, we denote a number of the form

$$\sum_{k=0}^{n} a_k 10^k \tag{6.1}$$

by $a_n a_{n-1} \ldots a_0$ where each of the numbers a_k is one of the "digits" $0, 1, \ldots, 9$. It is obvious that any number of the form (6.1) is a non-negative integer. It is not quite so obvious that every positive integer can be expressed this way. An important feature is that there is *exactly one* standard way of expressing a number in the form (6.1), provided we insist that the coefficient of the highest power of 10 is non-zero. (For if we do not make this restriction we can add on extra terms in which the a_k are all zero.)

There are some crucial arithmetical properties of the expression (6.1). Firstly, notice that each term a_k satisfies $0 \le a_k \le 9$ so that

$$0 \le \sum_{k=0}^{n} a_k 10^k \ \le \ \sum_{k=0}^{n} 9.10^k \ = \ 9.\frac{10^{n+1}-1}{10-1}$$

$$= 10^{n+1} - 1 < 10^{n+1}$$

and if we also demand that $a_n \ne 0$ so that $a_n \ge 1$ then

$$10^n \ \le \ \sum_{k=0}^{n} a_k 10^k \ < \ 10^{n+1} .$$

This is what determines the value of n, for if x is a natural number there is exactly one value of n with $10^n \le x < 10^{n+1}$ (but notice that n may be 0). From this we notice that $1 \le x/10^n < 10$ and we choose a_n to be the largest integer which does not exceed $x/10^n$, that is, a_n is an integer and $a_n \le x/10^n < a_n + 1$. Then $1 \le a_n < 10$ so that $a_n \in \{0, 1, \ldots, 9\}$. Also

$$0 \le x / 10^n - a_n < 1$$

giving

$$0 \le x - a_n 10^n < 10^n .$$

In effect we have subtracted from x the largest multiple of 10^n which gives a non-negative result. Now $x - a_n 10^n$ is smaller than x so we can apply the same process, but subtracting a multiple of 10^{n-1}. It will be easier if we first define some notation.

Definition Let x be a real number. Then $[x]$, the **integer part** of x, is the unique integer $[x]$ which satisfies

$$[x] \le x < [x] + 1 \ .$$

With this notation we see that a_n above is $[x/10^n]$, and this gives us the largest integer multiple of 10^n which does not exceed x. Now $0 \le x - a_n 10^n < 10^n$ and we want to subtract a multiple of 10^{n-1}, so we first divide through by 10^{n-1}, giving

$$0 \le (x - a_n 10^n) / 10^{n-1} < 10 \ .$$

Let $a_{n-1} = [(x - a_n 10^n) / 10^{n-1}]$, so that

$$a_{n-1} \le (x - a_n 10^n) / 10^{n-1} < a_{n-1} + 1$$

and we have, in turn

$$a_{n-1} 10^{n-1} \le x - a_n 10^n < a_{n-1} 10^{n-1} + 10^{n-1} \ .$$

Proceeding in this way gives us the result we would expect:

Theorem 6.1 Let x be a non-negative integer. Then there are an integer $n \ge 0$, and "digits" $a_0, a_1, ..., a_n$, all chosen from the set $\{0, 1, 2, ..., 9\}$ such that

$$x = \sum_{k=0}^{n} a_k 10^k \ .$$

If $x > 0$ and we demand that $a_n \ne 0$ then $n, a_1, a_2, ..., a_n$ above are unique. □

Essentially we have proved this, in that the process of constructing $a_n, a_{n-1}, ...$ above reduces the difference between x and the sum of the terms at each step. A full proof is a little long-winded, partly because of the need to deal with the uniqueness clearly. (See Appendix.)

The use of the integer 10 in what we have done is not essential. If d is a natural number and $d > 1$ then we know from Bernoulli's Inequality that $d^m > m(d - 1)$ so that we can choose m for which $d^m > x$. ($m = x$ would do.) It is then possible to repeat all of this in expansions in powers of d so that we express x in the form $\sum_{k=0}^{n} a_k d^k$ where each of the numbers a_k belongs to the set $\{0, 1, ..., d-1\}$. The most important case of this when $d \ne 10$ is when $d = 2$ (giving binary expansions of numbers). The case $d = 16$ is used in computing. There are one or two specific places in mathematics where $d = 3$ is used. The results for general d are identical in spirit to those for $d = 10$ so we shall not confuse matters by varying d; we shall stick to $d = 10$.

6.2 Decimals of Real Numbers

The idea that we can express every positive integer exactly in the form $\sum_{k=0}^{n} a_k 10^k$ for suitable values of $n, a_0, a_1, ..., a_n$, can be pursued further. In essence what we have done is to notice that $0 \le x < 10^{n+1}$ and divide this interval into 10 equal pieces (of length 10^n) and choose the coefficient a_n in this way, then we move on to $x - a_n 10^n$ which is between 0 and 10^n. We divide that interval into 10, and so on.   We can use the same idea for numbers x which are not integers so that $x - \Sigma_{k=0}^{n} a_k 10^k$ will not be exactly zero but will lie between 0 and 1, an interval which we can divide into 10.   We can deal with the integer part , $[x]$, as above. What is left is the "fractional part" $x - [x]$.   Now $0 \le x - [x] < 1$, so we shall consider such numbers.

Let $0 \le y < 1$.   Then we wish to find an integer b_1 so that

$$\frac{b_1}{10} \le y < \frac{b_1 + 1}{10} ,$$

the largest multiple of $\frac{1}{10}$ not exceeding y.   This is true if and only if

$$b_1 \le 10y < b_1 + 1 . \tag{6.2}$$

From (6.2) we see that $b_1 = [10y]$, and $0 \le 10y - b_1 < 1$.   Then we wish the largest multiple of $\frac{1}{100}$ to add to $b_1 10^{-1}$ so that the result does not exceed y , that is, we wish b_2 so that

$$b_1 10^{-1} + b_2 10^{-2} \le y < b_1 10^{-1} + (b_2 + 1)10^{-2} .$$

This is true if and only if

$$b_2 \le \left(y - b_1 10^{-1}\right)10^2 < b_2 + 1 ,$$

showing that

$$b_2 = [(y - b_1 10^{-1})10^2] .$$

We proceed in this way.

Lemma 6.2 Suppose that $0 \le y < 1$.   Then there is a sequence $b_1, b_2, b_3, ...$ of integers, all chosen from the set $\{0, 1, ..., 9\}$ with the property that, for all natural numbers n,

$$0 \le y - \Sigma_{k=1}^{n} b_k 10^{-k} < 10^{-n} . \tag{6.3}$$

The numbers b_n are defined by:

$$b_1 = [10y]$$

$$b_{n+1} = \left[10^{n+1}(y - \Sigma_{k=1}^{n} b_k 10^{-k})\right] .$$

Proof: We define b_1 by $[10y]$.   Then since $0 \le y < 1$, $0 \le 10y < 10$, whence $[10y]$ is an integer and it satisfies $0 \le [10y] \le 10y < 10$, so $b_1 = [10y] \in \{0, 1, ..., 9\}$.   By choice $b_1 \le 10y < b_1 + 1$ so that

$$0 \le y - b_1 10^{-1} < 10^{-1} ,$$

showing that we can define b_1 and (6.3) holds for $n = 1$.

Suppose that we have defined $b_1, b_2, ..., b_n$, and that (6.3) holds.   Then $0 \le 10^{n+1}(y - \Sigma_{k=1}^{n} b_k 10^{-k}) < 10$ so that if we set

$$b_{n+1} = \left[10^{n+1} \left(y - \sum_{k=1}^{n} b_k 10^{-k} \right) \right]$$

then $b_{n+1} \in \{0, 1, \ldots, 9\}$. Moreover

$$b_{n+1} \le 10^{n+1} \left(y - \sum_{k=1}^{n} b_k 10^{-k} \right) < b_{n+1} + 1$$

whence $b_{n+1} 10^{-(n+1)} \le y - \sum_{k=1}^{n} b_k 10^{-k} < b_{n+1} 10^{-(n+1)} + 10^{-(n+1)}$

and $0 \le y - \sum_{k=1}^{n+1} b_k 10^{-k} < 10^{-(n+1)}$.

In other words b_{n+1} also has the required properties. It follows by induction that b_n is defined for all $n \in \mathbb{N}$ and satisfies (6.3). □

Corollary The integers $b_1, b_2, \ldots, b_n$ are the only sequence of numbers all chosen from $\{0, 1, \ldots, 9\}$ which satisfy (6.3).

Proof: Suppose that $n \in \mathbb{N}$, $c_1, \ldots, c_n$ all belong to $\{0, 1, \ldots, 9\}$ and that

$$0 \le y - \sum_{k=1}^{n} c_k 10^{-k} < 10^{-n} .$$

Then for each $m = 1, 2, \ldots, n$,

$$\sum_{k=m+1}^{n} c_k 10^{-k} \le y - \sum_{k=1}^{m} c_k 10^{-k} < \sum_{k=m+1}^{n} c_k 10^{-k} + 10^{-n} \qquad (6.4)$$

and $0 \le \sum_{k=m+1}^{n} c_k 10^{-k} \le \sum_{k=m+1}^{n} 9.10^{-k} = 10^{-m} - 10^{-n}$.

Then, from (6.4) we have

$$0 \le y - \sum_{k=1}^{m} c_k 10^{-k} < 10^{-m} \qquad (6.5)$$

for $m = 1, 2, \ldots, n$. Putting $m = 1$ gives $0 \le y - c_1 10^{-1} < 10^{-1}$ which shows that $c_1 \le 10y < c_1 + 1$. Therefore $c_1 = [10y] = b_1$.

Now suppose that $r \in \{1, 2, \ldots, n - 1\}$ and that $c_k = b_k$ for $k = 1, 2, \ldots, r$. Then from (6.5) with $r + 1$ in place of m

$$0 \le y - \sum_{k=1}^{r+1} c_n 10^{-k} < 10^{-(r+1)}$$

and since $c_k = b_k$, for $k \le r$, we see that

$$c_{r+1} 10^{-(r+1)} \le y - \sum_{k=1}^{r} b_k 10^{-k} < c_{r+1} 10^{-(r+1)} + 10^{-(r+1)}$$

which shows that

$$c_{r+1} \le 10^{r+1} \left(y - \sum_{k=1}^{r} b_k 10^{-k} \right) < c_{r+1} + 1 .$$

Since c_{r+1} is an integer (by assumption) we have

$$c_{r+1} = \left[10^{r+1} \left(y - \sum_{k=1}^{r} b_k 10^{-k} \right) \right] = b_{r+1} .$$

It follows that $c_k = b_k$ for $k = 1, 2, \ldots, n$. □

Example 6.1 Let $y = \frac{1}{3}$. Then $[10y] = [\frac{10}{3}] = 3$. In this case $b_1 = 3$ (in the notation above) and $0 < \frac{1}{3} - 3\frac{1}{10} < \frac{1}{10}$ or, more expressively, $\frac{3}{10} < \frac{1}{3} < \frac{4}{10}$.
Then $y - b_1 10^{-1} = \frac{1}{30}$ and $[10^2(y_1 - b_1 10^{-1})] = [\frac{10}{3}] = 3$, giving $b_2 = 3$. We can check that $b_3 = b_4 = b_5 = ... = 3$ so that the sequence here is $3, 3, 3, ...$. Notice that there is no natural number n for which $\Sigma_{k=1}^n b_k 10^{-k}$ equals $1/3$, although we can make it as close as we wish to $1/3$ by choosing n sufficiently large. □

Notice what we have shown. If $0 \le y < 1$ then there is a sequence of integers $b_1, b_2, ...$ so that whatever $n \in \mathbb{N}$ we choose $0 \le y - \Sigma_{k=1}^n c_k 10^{-k} < 10^{-n}$. So the larger we choose n , the closer the sum becomes to y. Putting this another way, we can ensure that $\Sigma_{k=1}^n b_k 10^{-k}$ is as close as we like to y if we choose n large enough. We might then consider that if we added all the terms the two would "become" equal. Of course, there is a practical difficulty in adding infinitely many numbers, but in fact we do not have to. What we notice is that we can ensure that the sum is within any specified distance of its "limit" if we take enough terms.

Definition Suppose that for each $n \in \mathbb{N}$, s_n is a real number. Then we say that "s_n **tends to** s **as** n **tends to infinity**" and write $s_n \to s$ as $n \to \infty$ if s_n can be made as close as we wish to s if n is chosen sufficiently large. That is, no matter how close we wish s_n to be to s , we can ensure this by choosing n large enough.

Example 6.2 $\frac{1}{n} \to 0$ as $n \to \infty$.

Solution: We have to show that $\frac{1}{n}$ can be made as close to 0 as we wish if n is chosen sufficiently large.

Suppose we wish $\frac{1}{n}$ to be within some prescribed distance of 0. We shall call the prescribed distance ε , as this is traditional. So whatever ε is (and as it is a distance it is positive) we need the distance of $\frac{1}{n}$ from 0 , that is, $\frac{1}{n}$ to be less than ε. Now $\frac{1}{n} < \varepsilon$ if and only if $n > \frac{1}{\varepsilon}$. Therefore in this case if ε is our distance then $\frac{1}{n}$ is less than ε away from zero if we choose $n > \frac{1}{\varepsilon}$.

We have shown that $\frac{1}{n} \to 0$ as $n \to \infty$. □

Example 6.2 shows that the definition of $s_n \to s$ as $n \to \infty$ is more complicated than we might have thought. It is also a little vague. If we continue to use ε as our measure of distance, then we mean that no matter how small we choose the distance ε , s_n is less than ε away from s if n is sufficiently large. More precisely, we mean that

for every positive number ε , there is a natural number N
such that for every $n > N$ $|s_n - s| < \varepsilon$.

This is in the range 0 to 10^{-n} permitted by Lemma 6.2. Therefore $\frac{1}{9} = \sum_{k=1}^{\infty} 10^{-k}$.

More interestingly, notice that if we put the first term 0 and the rest 9, to give $\sum_{k=2}^{\infty} 9.10^{-k}$, then

$$\sum_{k=2}^{n+1} 9.10^{-k} = \frac{9}{100} \cdot \frac{1 - 10^{-n}}{1 - \frac{1}{10}}$$

$$= \frac{1}{10} \cdot \left(1 - 10^{-n}\right) \to \frac{1}{10} \text{ as } n \to \infty .$$

That is, in the usual decimal notation , $\cdot0999\ldots$ corresponds to a series whose sum is $\frac{1}{10}$. (Here we are letting $\cdot b_1\, b_2\, b_3\, \ldots$ represent $\sum_{k=1}^{\infty} b_k 10^{-k}$.) However, this is not the expression the process we used earlier gives for $\frac{1}{10}$, for if we wish to choose $b_1 \in \{0, 1, \ldots, 9\}$ so that $0 \le \frac{1}{10} - b_1 10^{-1} < 10^{-1}$ then $b_1 = 1$ and then $b_k = 0$ for $k > 1$. We have found a number with two *different* expressions in the form $\sum_{k=1}^{\infty} c_k 10^{-k}$ where $c_k \in \{0, 1, \ldots, 9\}$. It turns out that this is the only way in which we can obtain two different versions of the decimal expansion of a number.

Let $0 \le y < 1$. Then by the **standard decimal expansion** of y we shall mean the expression of y as in Lemma 6.2,

$$y = \sum_{k=1}^{\infty} b_k 10^{-k}$$

where

$$b_1 = [10y], \quad b_{n+1} = \left[10^{n+1}\left(y - \sum_{k=1}^{n} b_k 10^{-k}\right)\right] .$$

Theorem 6.3 Suppose that $0 \le y < 1$ and that $y = \sum_{k=1}^{\infty} c_k 10^{-k}$ where for each k , $c_k \in \{0, 1, \ldots, 9\}$. Then *either* this is the standard decimal expansion of y, that is, for all k, $c_k = b_k$ *or* y is an integer multiple of 10^{-N} for some N. In the second case there is an integer M for which

$$\begin{aligned}
c_k &= b_k & (k = 1, 2, \ldots, M - 1), \\
c_M &= b_M - 1 , \\
c_k &= 9 & \text{for } k > M , \\
b_k &= 0 & \text{for } k > M .
\end{aligned}$$

Illustration: $\frac{1}{10} = \cdot0999\ldots = \cdot100\ldots.$ (In this case $M = 1$).

Proof: Let $y = \sum_{k=1}^{\infty} c_k 10^{-k}$, where each $c_k \in \{0, 1, \ldots, 9\}$. Then if $n \in \mathbb{N}$

$$y - \sum_{k=1}^{n} c_k 10^{-k} = \sum_{k=n+1}^{\infty} c_k 10^{-k} . \tag{6.6}$$

Because each c_k satisfies $0 \le c_k \le 9$, we have, using Example 6.4

Decimals of Real Numbers

$$0 \leq \sum_{k=n+1}^{\infty} c_k 10^{-k} \leq \sum_{k=n+1}^{\infty} 9.10^{-k} = 10^{-n} . \qquad (6.7)$$

Therefore, combining (6.6) and (6.7), we have

$$0 \leq y - \sum_{k=1}^{n} c_k 10^{-k} \leq 10^{-n} . \qquad (6.8)$$

Now we know, from the Corollary to Lemma 6.2, that if the right hand inequality in (6.8) is strict for some value of n, that is, if

$$0 \leq y - \sum_{k=1}^{n} c_k 10^{-k} < 10^{-n}$$

then $c_k = b_k$ for $k = 1, 2, ..., n$. Now, looking at (6.8) we see that either there are some values of n for which

$$y - \sum_{k=1}^{n} c_k 10^{-k} = 10^{-n} \qquad (6.9)$$

or, for all $n \in \mathbb{N}$,

$$0 \leq y - \sum_{k=1}^{n} c_k 10^{-k} < 10^{-n} .$$

In the second case we see that, for all $n \in \mathbb{N}$, $c_k = b_k$ for $k = 1, 2, ..., n$ so that in particular, for all $n \in \mathbb{N}$ $c_n = b_n$, and we have the standard expansion.

Now suppose that the other case holds, that is, (6.9) is true for some values of n. Let M be the smallest such value of n (which will exist since (6.9) is true for n belonging to some non-empty subset of $\mathbb{N}$). It is clear from (6.9) that y is an integer multiple of 10^{-M}, since

$$y = \sum_{k=1}^{M} c_k 10^{-k} + 10^{-M} .$$

Moreover, because M is the smallest value of n for which (6.9) holds, either $M = 1$ (so $M - 1 \notin \mathbb{N}$ or $M - 1 \in \mathbb{N}$ but (6.9) does not hold for $M - 1$, whence

$$0 \leq y - \sum_{k=1}^{M-1} c_k 10^{-k} < 10^{-(M-1)} . \qquad (6.10)$$

In the second case we see that $c_k = b_k$ for $k = 1, 2, ..., M - 1$.
Then

$$y - \sum_{k=1}^{M-1} c_k 10^{-k} = y - \sum_{k=1}^{M} c_k 10^{-k} + c_M 10^{-M} = (1 + c_M)10^{-M} \qquad (6.11)$$

from which we see that (by (6.10)) $1 + c_M < 10$, that is, $c_M < 9$. In the case where $M = 1$ $y = c_1 10^{-1} + 10^{-1} = (c_1 + 1)10^{-1}$ so since we know $y < 1$ it follows that $c_M < 9$ in this case too. Now because $c_M < 9$, $c_M + 1 \in \{0, 1, ..., 9\}$ and if we let $d_k = c_k$ (for $k < M$) and $d_M = c_M + 1$, we have from (6.11) that $y = \sum_{k=1}^{M} d_k 10^{-k}$ $0 \leq y - \sum_{k=1}^{M} d_k 10^{-k} < 10^{-M}$. Therefore $d_M = b_M$, that is $c_M = b_M - 1$. Also since $y = \sum_{k=1}^{M} b_k 10^{-k}$ we see that $b_k = 0$ for all $k > M$.

Finally, it follows since $y - \sum_{k=1}^{M-1} c_k 10^{-k} = 10^{-M}$ that $\sum_{k=M+1}^{\infty} c_k 10^{-k} = 10^{-M}$. From this it follows that $c_k = 9$ for all $k > M$, for

$$\sum_{k=M+1}^{\infty} c_k 10^{-k} \leq \sum_{k=M+1}^{\infty} 9.10^{-k} = 10^{-M}$$

and if c_N, say, were less than 9, then for all $n \geq N$

$$\sum_{k=M+1}^{\infty} c_k 10^{-k} \; \le \; \sum_{k=M+1}^{\infty} 9.10^{-k} - 10^{-N}$$

$$= \; \frac{9}{10^{M+1}} \frac{1}{1-10^{-1}} - 10^{-N} \; \le \; 10^{-M} - 10^{-N} .$$

$$< \; 10^{-M} . \qquad \square$$

(Notice in passing that we have assumed that if $c_k \le 9$ for all k, then $\sum_{k=M+1}^{\infty} c_k 10^{-k} \le \sum_{k=M+1}^{\infty} 9.10^{-k}$; this presumes a point about the behaviour of limits to which we shall return later.)

5.3 Some Interesting Consequences

There are some interesting consequences of all this. By looking at the way in which $b_1, b_2, b_3,...$ are defined, it is not too hard to show that if y is rational, the decimal for y eventually *repeats*, that is, there is a number k such that for all $n \ge N$ $a_{n+k} = a_n$ (where N and k are appropriately chosen). This arises essentially because if $x = p/q$ (for integers p, q) and we carry out "long division" by q, with remainders in $\{0, 1, ..., q\text{-}1\}$ there are only q possible remainders, so we eventually reach one we have had before. Notice that $1/4 = 0 \cdot 25 = 0 \cdot 2500000....$ has a repeating decimal (repeating 0). By working a little harder we can see the converse, that every repeating decimal represents a rational number.

The most interesting consequence is rather different. We have seen that we can write the rational numbers out in a sequence. For the sequence

$$\frac{1}{1}, \frac{2}{1}, \frac{1}{2}, \frac{3}{1}, \frac{2}{2}, \frac{1}{3}, \frac{4}{1}, \frac{3}{2}, \frac{2}{3}, \frac{1}{4}, \frac{5}{1}, \frac{4}{2}, ...$$

contains all positive rational numbers, so it follows that

$$0, \frac{1}{1}, -\frac{1}{1}, \frac{2}{1}, -\frac{2}{1}, \frac{1}{2}, -\frac{1}{2}, \frac{3}{1}, -\frac{3}{1}, \frac{2}{2}, -\frac{2}{2}, ...$$

contains all rational numbers. The surprising thing is that we *cannot* do this with the real numbers, not even with those between 0 and 1. We show this by contradiction.

Suppose that we write out a sequence $x_1, x_2, x_3, ...$ of numbers all satisfying $0 \le x_i \le 1$. Then each of these numbers has a decimal expansion. Should x_i have two decimal expansions we agree to choose the one ending in 0s rather than the one ending in 9s. Now we can write this as

$$x_1 \; = \; \cdot a_1^{(1)} \; a_2^{(1)} \; a_3^{(1)} \; ....$$

$$x_2 \; = \; \cdot a_1^{(2)} \; a_2^{(2)} \; a_3^{(2)} \; ,...$$

$$\vdots \qquad \qquad \vdots$$

Now suppose that we have done this so that $x_1, x_2, ...$ contains all real numbers for which $0 \le x_i < 1$. Define a new real number y by $y = 0 \cdot b_1 \, b_2 \, b_3 ...$, (or in terms of a series $\sum_{k=1}^{\infty} b_k 10^{-k}$), where the coefficients b_n are defined by

$$b_n \; = \; 4 \; \text{if} \; a_n^{(n)} \ne 4$$

$$b_n \; = \; 5 \; \text{if} \; a_n^{(n)} = 4 .$$

Then $\cdot b_1 b_2 b_3 \ldots$ certainly does not end in infinitely many 0s or 9s, so the number $y = \cdot b_1 b_2 b_3$ has only one decimal expansion. Therefore $y \neq x_1$, for the first digit in the decimal expansion of y is b_1 which is not equal to $a_1^{(1)}$, the first decimal digit of x_1. Since y has only one decimal expansion there is no possibility that it could be expressed as a number with another decimal expansion. In the same way $y \neq x_2$ (they differ in the second place of their decimal expansions) and for each $n \in \mathbb{N}$ $y \neq x_n$ (the nth decimal place is different). That is , y is different from all of $x_1, x_2, x_3, \ldots$.   But this is a contradiction, for we chose the sequence $x_1, x_2, x_3, \ldots$, to include *all* real numbers between 0 and 1.

It follows that we cannot arrange all real numbers between 0 and 1 in a sequence. Putting this in another way, no matter how cleverly we construct the sequence $x_1, x_2, x_3, \ldots$, there are still real numbers which are omitted from the list; there are "too many" real numbers to arrange them in a sequence.

This shows the surprising result that infinite sets come in different sizes: the set of rational numbers seems to be "smaller" than the set of real numbers, yet both are infinite.

Problems

1. Suppose that $x = \Sigma_{k=1}^n a_k 10^k$ where the numbers $a_1, \ldots, a_n$ all belong to $\{0, 1, \ldots, 9\}$.   Check that $10x = \Sigma_{j=1}^{n+1} b_j 10^j$ where for $j = 2, 3, \ldots, n+1$, $b_j = a_{j-1}$ and $b_1 = 0$. (That is, if the digits for x , starting with the highest power are $a_n a_{n-1} \ldots a_1$ then those for $10x$ are $a_n a_{n-1} \ldots a_1 0$ (there being one more of them).

2. A *terminating decimal* is one which ends in repeated zeros, e.g. $0\cdot250000 \ldots$ (or $0\cdot25$). It is clear (is it?) that if a number can be expressed as a terminating decimal then it is of the form of m/n where m is an integer and n is a power of 10. Deduce that if a number has a terminating decimal then it is rational and is of the form p/q where p and q are integers with no common factor and the only prime factors of q are 2 or 5 (or both), that is, $q = 2^a 5^b$ for some non-negative integers a and b).

    Check the converse, that if $q = 2^a 5^b$ for non-negative integers a and b , and if p is an integer, then p/q has a terminating decimal.

3*. Let p and q be natural numbers with $p < q$, so that $y = p/q$ lies between 0 and 1. Let $y = \cdot b_1 b_2 b_3 \ldots$ as in Lemma 6.2. By that lemma $b_1 = [10y]$.

    Show that $10p = b_1 q + r_1$ where r_1 is the remainder on long division of $10p$ by q , and b_1 is the quotient.

    Now, from Lemma 6.2 $b_2 = [10^2(y - b_1 10^{-1})]$. Use the last paragraph to show that $10^2(y - b_1 10^{-1}) = \frac{10r_1}{q}$ and hence that $b_2 = [\frac{10r_1}{q}]$. Deduce from this that $10r_1 = b_2 q + r_2$ where $r_2 \in \{0, 1, \ldots, q-1\}$. In this case b_2 is the quotient on long division of $10r_1$ by q and r_2 the remainder.

    The sequence $b_1, b_2, b_3, \ldots$ proceeds this way so that if r_n is the remainder at the nth stage , b_{n+1} is the quotient and r_{n+1} the remainder on long division of $10r_n$ by q. Notice that as there are only q possible remainders, at some stage r_n will equal a previous remainder and the process will repeat itself from then onwards.

4.   (i) Consider the number $\cdot 181818...$. (This can be written as $\sum_{n=1}^{\infty} a_n 10^{-n}$ where $a_n = 1$ if n is odd and $a_n = 8$ if n is even.) Show that this equals

$$\left(\tfrac{18}{100}\right)\left(1 + \tfrac{1}{100} + \tfrac{1}{(100)^2} + \tfrac{1}{(100)^3} + \cdots\right) \ .$$

By summing the geometric series to infinity show that
$$\cdot 181818... \ = \ \tfrac{18}{99} \ \left(=\tfrac{2}{11}\right) \ .$$

(ii) Use the same ideas to show that
$$\cdot 027\,027\,027 ... \ = \ \tfrac{1}{37} \ .$$

5*.   Suppose that for all $n \geq N$, $a_{n+k} = a_n$. (So the sequence $a_N, a_{N+1}, ..., a_{N+k-1}, a_{N+k}, a_{N+k+1} ...$ is $a_N, a_{N+1}, ..., a_{N+k-1}, a_N, a_{N+1}, ...$.) By noticing that

$$a_{N+jk+r} \ 10^{-(N+jk+r)} \ = \ (a_{N+r} \ 10^{-(N+r)})10^{-jk}$$

show that

$$\sum_{n=N}^{N+mk-1} a_n 10^{-n} = \left(\sum_{n=N}^{N+k-1} a_n 10^{-n}\right)\left(1 + 10^{-k} + 10^{-2k} + ... + 10^{-(m-1)k}\right) \ .$$

Hence show that

$$\sum_{n=N}^{\infty} a_n 10^{-n} = \left(\sum_{n=N}^{N+k-1} a_n 10^{-n}\right)\frac{10^k}{10^k - 1} = \frac{\sum_{n=N}^{N+k-1} a_n 10^{N+k-1-n}}{10^{N-1}(10^k - 1)} \ .$$

By noticing that the last numerator is an integer if $a_n \in \{0, 1, ..., 9\}$ for all n , deduce that $\sum_{n=N}^{\infty} a_n 10^{-n}$ is a rational number.

Hence show that if a number is represented by a repeating decimal, then it is rational.

6*.   Suppose that p is a prime number, and $p \neq 2$ or 5. Use Problem 5 to show that $\dfrac{1}{p} = \dfrac{m}{10^{N-1}(10^k - 1)}$ for some positive integers m, N and k . By noticing that p does not divide exactly into 10^N (because $p \neq 2$ or 5) deduce that p does divide exactly into $10^k - 1$.

Notice this odd result: we know $1/p$ has a repeating decimal so for some k , $10^k - 1$ (= 999 ... 9 with k nines) is a multiple of p. Every prime number other than 2 or 5 divides into some number of the form 99 ... 9 with the appropriate number of 9s.

7

Limits

"... May we not call them the ghosts of departed quantities?"   G. Berkeley (1734)

7.1 The Idea of a Limit

The limit is one of the fundamental ideas of mathematics - and one of its most subtle.   The history of the idea is long and fairly turbulent and it is only in the last century or so that a satisfactory theory has been established, mainly by Weierstrass and his pupils in the 1860s and 1870s.

Consider first the idea of a sequence, that is, an infinite progression of numbers $a_1, a_2, a_3 \dots$.   We have already encountered sequences, and have used them in forming decimals and infinite series, but this is a good point to make the ideas formal.

Definition **A sequence** (a_n) of real numbers is an ordered progression of real numbers, indexed by the natural numbers.   That is, for each natural number n there is an associated real number (a_n).   (More formally, we may consider a sequence to be a function from $\mathbb{N}$ into the real numbers.)

Since the natural numbers have an obvious ordering, it is natural to think of a sequence in terms of an infinite list of its terms in their natural order: $a_1, a_2, a_3 \dots$. The notation (a_n) denotes the sequence as a whole, while a_n will mean the number which is its nth term.   The n in (a_n) is a "dummy" variable and another symbol could be substituted.   A sequence may be defined explicitly by a formula (e.g. $a_n = 2^n$)   or by a recurrence relation such as $a_1 = 1, a_{n+1} = (\frac{1}{2})(a_n + 2/a_n)$ (for all $n \in \mathbb{N}$).   For the second type of definition we can determine a_{n+1} once we know a_n provided, in this case, $a_n \neq 0$ so we need to check, by induction, that a_n is defined for each $n \in \mathbb{N}$, that is, that the definition does not fail at some point.   In the quoted example it is easy to see that $a_n > 0 \Rightarrow a_{n+1} > 0$ and that $a_1 > 0$, so, by induction, all terms are positive and no difficulty arises.

Consider the sequence $(1/n)$, whose nth term is $1/n$.   Each term is positive but as n becomes larger, $1/n$ becomes smaller and 'approaches zero'.   Indeed, $1/n$ can be made as small as we wish by choosing n sufficiently large and we think informally that 'in the limit' $1/n$ 'becomes zero'.   It is this idea that we must make precise.

Definition Let (a_n) be a sequence of real numbers and let a be a real number. Then we say that $a_n \to a$ as $n \to \infty$ if and only if for all $\varepsilon > 0$, there is a natural number N such that

$$\text{if} \quad n \geq N \quad \text{then} \quad |a_n - a| < \varepsilon \ .$$

In this case we say a is the **limit** of (a_n) and write $a = \lim_{n \to \infty} a_n$. The sign $\to$ is pronounced 'tends to'.

Remarks: 1. The definition demands that no matter what predetermined positive 'tolerance' ε we are given, there must be an integer N such that, for every $n \geq N$, the distance from a_n to a is less than ε. We may expect the suitable values of N to depend on ε.

2. ∞ is not a number. The word 'infinity' and the symbol ∞ are convenient terms to use to refer to this definition, but we have not given them a separate meaning.

3. Notice how we could show that $a_n \to a$ as $n \to \infty$. Choose ε to be positive and show that a suitable N exists, then observe that since we use only the fact that ε is positive, this must hold for all positive ε. The easiest way to show a value of N exists is to find one.

4. We use the symbols $\forall$ and $\exists$ to mean 'for all' and 'there exists' respectively. So $a_n \to a$ as $n \to \infty$ may be written, using s.t. to denote 'such that',

$$\forall \varepsilon > 0 \ \exists N \in \mathbb{N} \ \text{s.t.} \ \forall n \geq N \ |a_n - a| < \varepsilon \ . \tag{*}$$

In this statement, ε, N and n are 'dummy' variables so that (*) says something about the sequence (a_n) and the number a but not about ε, N or n. Notice that the order in which the symbols $\forall$ and $\exists$ appear is important; (*) says that, no matter how $\varepsilon > 0$ is selected, once it has been selected, a suitable N exists. This is not at all the same as saying that a value of N exists which is suitable for all $\varepsilon > 0$. Despite the need to learn yet more symbols, using $\forall$ and $\exists$ shortens the statements involved to the extent that we can usually write them in a single line, which can make it easier to see the general trend of an argument.

5. There is nothing magic about the symbol ε. We may equally well rewrite (*) as
$$\forall \delta > 0 \ \exists N \in \mathbb{N} \ \text{s.t.} \ \forall n \geq N \ |a_n - a| < \delta.$$
Now choose $\varepsilon > 0$. Then, since $\varepsilon/2 > 0$, (*) tells us that
$$\exists N_1 \ \text{s.t.} \ \forall n \geq N_1 \ |a_n - a| < \varepsilon/2$$
$$\text{and} \ \exists N_2 \ \text{s.t.} \ \forall n \geq N_2 \ |a_n - a| < \varepsilon.$$
(Put $\delta = \varepsilon/2, \varepsilon$ in turn.) We must be careful not to presume that the numbers N_1 and N_2, which we know exist, are equal, which is why we remind ourselves of this by using different suffices.

6. Notice that $|a_n - a| < \varepsilon \Leftrightarrow -\varepsilon < a_n - a < \varepsilon \Leftrightarrow a - \varepsilon < a_n < a + \varepsilon$. Therefore, $a_n \to a$ as $n \to \infty$ can also be written
$$\forall \varepsilon > 0 \ \exists N \in \mathbb{N} \ \text{s.t.} \ \forall n \geq N \ a - \varepsilon < a_n < a + \varepsilon \ .$$
This form is very often useful, especially when order properties are involved.

Example 7.1 Let $a_n = 1/n$ for $n = 1, 2, 3, \ldots$. Then $a_n \to 0$ as $n \to \infty$. This, of course, should be no surprise, but we have made a technical definition of limit so we ought to check that it has the properties we might expect.

Solution: Let $\varepsilon > 0$. The problem is to show that there is a natural number N such that $\forall n \geq N \ |a_n - 0| < \varepsilon$. First, the rough work: $|a_n - 0| = 1/n$ and $1/n < \varepsilon$ if and only if $n > 1/\varepsilon$. Therefore all we have to do is choose $N > 1/\varepsilon$, then $n \geq N \Rightarrow n > 1/\varepsilon \Rightarrow |a_n - 0| < \varepsilon$. So, for the formal proof we have:

Let N be an integer greater than $1/\varepsilon$. Then:

$$\forall n \geq N \ |a_n - 0| = 1/n \leq 1/N < \varepsilon \ .$$

Since ε was a typical positive number we have shown that this is true for all $\varepsilon > 0$, that is, $\forall \varepsilon > 0 \ \exists N \in \mathbb{N}$ such that $\forall n \geq N \ |a_n - 0| < \varepsilon$. $\square$

The proof just given raises an issue that it is easy to overlook: given a typical real number x, *is there* an integer $N > x$? (We used this tacitly above by choosing $N > 1/\varepsilon$ and we need such an N to exist.) Now, our picture of the real number system as lying along a line strongly suggests this, but we have not yet proved it. For the time being, we shall set this aside by stating the property we wish and making it one of our assumptions; we shall prove it when we come to consider the structure of the real numbers in more detail in Chapter 9.

Archimedean Property of the Real Numbers
If x is a real number, there is a natural number n for which $n > x$.

Example 7.2 For all $n \in \mathbb{N}$, let $a_n = a$. Then $\forall \varepsilon > 0 \ \exists N \in \mathbb{N}$ (e.g. $N = 1$) such that $\forall n \geq N \ |a_n - a| = 0 < \varepsilon$. Thus $a_n \to a$ as $n \to \infty$. $\square$

In other words, if (a_n) is the constant sequence, $a, a, a, \ldots$, consisting of the same number repeated indefinitely, then (a_n) tends to that number a as $n \to \infty$.

Example 7.3 Let $0 < x < 1$ and, for all $n \in \mathbb{N}$, set $a_n = x^n$. Prove that $a_n \to 0$ as $n \to \infty$.

Solution: We need to connect x^n to something simpler and for this we recall Bernoulli's Inequality (Theorem 5.1). This applied to a number $y > 1$ to tell us that $y^n \geq 1 + n(y - 1)$ and hence $y^n > n(y - 1)$. In this case, because $0 < x < 1$ we use Example 5.11 to observe that because $\frac{1}{x} > 1$, for all $n \in \mathbb{N}$

$$\left(\tfrac{1}{x}\right)^n > n\left(\tfrac{1}{x} - 1\right) = n\,\tfrac{1-x}{x} \text{ and therefore } x^n < \tfrac{1}{n}\tfrac{x}{1-x} .$$

Now $\frac{1}{n}\frac{x}{1-x} \leq \varepsilon \Leftrightarrow n \geq \frac{1}{\varepsilon}\frac{x}{1-x}$ so we proceed as follows:

Let $\varepsilon > 0$. Choose N to be an integer with $N \geq x/((1 - x)\varepsilon)$.

$$\text{Then } \forall n \geq N, \ 0 < x^n < \tfrac{1}{n}\tfrac{x}{1-x} \leq \tfrac{1}{N}\tfrac{x}{1-x} \leq \varepsilon \ .$$

$$\text{Therefore } \forall n \geq N \ |a_n - 0| = x^n < \varepsilon .$$

Since ε was arbitrary we have shown that

$$\forall \varepsilon > 0 \ \exists N \in \mathbb{N} \text{ such that } \forall n \geq N \ |a_n - 0| < \varepsilon ,$$

thus $a_n \to 0$ as $n \to \infty.$ $\square$

Problem Prove that if $-1 < x < 1$, then $x^n \to 0$ as $n \to \infty.$

7.2 Manipulating Limits

Armed with our formal definition, we need some rules for manipulating limits. The most plausible results are:

Theorem 7.1 Suppose that (a_n) and (b_n) are sequences of real numbers, that $a_n \to a$ and $b_n \to b$ as $n \to \infty$ and that λ is a real number. Then

    (i) $\lambda a_n \to \lambda a$ as $n \to \infty$;

    (ii) $a_n + b_n \to a + b$ as $n \to \infty$;

    (iii) $a_n b_n \to ab$ as $n \to \infty$.

(The statement $a_n + b_n \to a + b$ means that the sequence whose n^{th} term is $a_n + b_n$ tends to $a + b$ as $n \to \infty$.)

Proof. In all cases we have to show that for all $\varepsilon > 0$ there exists an N with certain properties. We shall do this by choosing an arbitrary positive ε and, using the information given, finding a suitable N. The procedure for thinking out these proofs is less formal than the proof itself, so we shall separate this from the proof. The left-hand column below would normally be omitted.

Rough Working

(i) Let $\varepsilon > 0$. We need to find $N \in \mathbb{N}$ so that

$$n \geq N \Rightarrow |\lambda a_n - \lambda a| = |\lambda||a_n - a| < \varepsilon .$$

It would be enough to ensure that if $n \geq N$ then $|a_n - a| < \varepsilon/|\lambda|$, which is possible for $\lambda \neq 0$. If $\lambda = 0$, $|\lambda a_n - \lambda a| = 0 < \varepsilon$.

Formal Proof

(i) Suppose $\lambda \neq 0$. Let $\varepsilon > 0$. Then, since $\varepsilon/|\lambda| > 0$ and $a_n \to a$, there is an N such that $\forall n \geq N \ |a_n - a| < \varepsilon/|\lambda|$. For this N,

$$\forall n \geq N \ |\lambda a_n - \lambda a| = |\lambda||a_n - a| < \varepsilon .$$

Since $\varepsilon > 0$ was arbitrary, this holds for all such ε, so

$\forall \varepsilon > 0 \ \exists N$ such that

$$\forall n \geq N \ |\lambda a_n - \lambda a| < \varepsilon ..$$

Therefore $\lambda a_n \to \lambda a$ as $n \to \infty$.

If $\lambda = 0$, then $\forall n \in \mathbb{N} \ \lambda a_n = 0 = \lambda a$ so Example 7.2 shows $\lambda a_n \to \lambda a$ as $n \to \infty$.

(ii) Let $\varepsilon > 0$. We must find N such that

$$\forall n \geq N \ |(a_n + b_n) - (a + b)| < \varepsilon \ .$$

We can deduce something about $|a_n - a|$ and $|b_n - b|$, so we notice that

$$|(a_n + b_n) - (a + b)|$$
$$= |(a_n - a) + (b_n - b)|$$
$$\leq |a_n - a| + |b_n - b|$$

(by the triangle inequality). It would now be enough to find an N for which $\forall n \geq N$ both

$$|a_n - a| < \varepsilon/2 \text{ and } |b_n - b| < \varepsilon/2 \ .$$

Since $\varepsilon/2 > 0$, and $a_n \to a$ as $n \to \infty$ $\exists N$ such that $\forall n \geq N \mid a_n - a \mid < \varepsilon/2$. Choose such a value of N and call it N_1, and, similarly, choose an N_2 such that $\forall n \geq N_2 \mid b_n - b \mid < \varepsilon/2 \ .$

(iii) Let $\varepsilon > 0$ We need to find an N such that $\forall n \geq N \mid a_n b_n - ab \mid < \varepsilon$. The information we have concerns $|a_n - a|$ and $|b_n - b|$, so we notice that

$$|a_n b_n - ab| = |a_n(b_n - b) + (a_n - a)b|$$
$$\leq |a_n||b_n - b| + |a_n - a||b| \ . \ (*)$$

Since the right-hand side consists of more than one part we shall need care to ensure each part is small enough that their sum is less than ε.

At the moment we cannot see how small will be sufficient, so we introduce a positive constant k which we shall determine later once things become clearer. As in part (ii) we can find $N_0 \in \mathbb{N}$ such that

$$\forall n \geq N_0 \ |a_n - a| < k\varepsilon \text{ and } |b_n - b| < k\varepsilon \ .$$

$\therefore \forall n \geq N_0$, using (*)

$$|a_n b_n - ab| < |a_n|k\varepsilon + |b|k\varepsilon$$
$$= (|a_n| + |b|)k\varepsilon \ . \quad (\dagger)$$

Again, there is a hitch. The right-hand of ($\dagger$) depends on n, so we cannot

(ii) Let $\varepsilon > 0$. Since $\varepsilon/2 > 0$ and both $a_n \to a$ and $b_n \to b$ as $n \to \infty$, we can choose $N_1 \in \mathbb{N}$ and $N_2 \in \mathbb{N}$ such that

$$\forall n \geq N_1 \ |a_n - a| < \varepsilon/2 \text{ and}$$
$$\forall n \geq N_2 \ |b_n - b| < \varepsilon/2 \ .$$

Now set $N = \max(N_1, N_2)$. Then $N \in \mathbb{N}$ and $n \geq N \Rightarrow n \geq N_1$ and $n \geq N_2$, so

$$\forall n \geq N \ |(a_n + b_n) - (a + b)|$$
$$\leq |(a_n - a) + (b_n - b)|$$
$$< \varepsilon/2 + \varepsilon/2 = \varepsilon.$$

Since $\varepsilon > 0$ was arbitrary, we have shown that

$$\forall \varepsilon > 0 \ \exists N \in \mathbb{N} \text{ such that}$$
$$\forall n \geq N \ |(a_n + b_n) - (a + b)| < \varepsilon \ .$$

(iii) Let $\varepsilon > 0$ Since $a_n \to a$ and $b_n \to b$ as $n \to \infty$, we can choose $N_1, N_2,$ and N_3 all in $\mathbb{N}$ such that

$$\forall n \geq N_1 \quad |a_n - a| < \varepsilon/(1 + |a| + |b|) \quad (1)$$
$$\forall n \geq N_2 \quad |b_n - b| < \varepsilon/(1 + |a| + |b|) \quad (2)$$
$$\text{and } \forall n \geq N_3 \quad |a_n - a| < 1 \ . \quad (3)$$

Let $N = \max(N_1, N_2, N_3)$, so $N \in \mathbb{N}$ and (since $n \geq N \Rightarrow n \geq N_1$ and $n \geq N_2$ and $n \geq N_3$)

$$\forall n \geq N \ |a_n b_n - ab|$$
$$= |a_n(b_n - b) + (a_n - a)b|$$
$$\leq |a_n||b_n - b| + |a_n - a||b|$$
$$< (|a_n| + |b|)\varepsilon/(1 + |a| + |b|)$$
$$\leq \frac{(|a_n - a| + |a| + |b|)\varepsilon}{1 + |a| + |b|}$$
$$< (1 + |a| + |b|)\varepsilon/(1 + |a| + |b|)$$
$$= \varepsilon \ .$$

(The first and third inequalities by the triangle inequality, the second by (1) and

just let k be the reciprocal of $|a_n| + |b|$, since k is to be constant. Now,

$$|a_n| = |(a_n - a) + a| \le |a_n - a| + |a|$$
$$< k\varepsilon + |a| .$$

We could use this, but comparing (†) with the inequalities in Chapter 5, it is simpler to obtain $|a_n| \le C$ where C is a constant independent of ε. Now choose N_0' such that

$$\forall n \ge N_0' \quad |a_n - a| < 1 .$$

(Possible since $a_n \to a$ and $1 > 0$.) Then if $n \ge N_0$ and $n \ge N_0'$,

$$|a_n b_n - ab| < (|a_n| + |b|)k\varepsilon$$
$$< (1 + |a| + |b|)k\varepsilon .$$

We now see that we should have chosen $k = 1/(1 + |a| + |b|)$.

(2) and the fourth by (3).) Since $\varepsilon > 0$ was arbitrary we have shown that

$$\forall \varepsilon > 0 \ \exists N \in \mathbb{N} \text{ such that}$$
$$\forall n \ge N \ |a_n b_n - ab| < \varepsilon . \square$$

The ideas used in constructing the proofs above are more intuitive than the final proof - and for this reason it is essential to write out the final version neatly and clearly to check that the ideas in the outline do come together properly. (Not always, but that's life!) The most important points are the choice of the number $\varepsilon/(1 + |a| + |b|)$, which is made by working through the problem to see what is needed, and the choice of N_3 such that $\forall n \ge N_3 \ |a_n - a| < 1$. As with the sort of inequality problem in Chapter 5, there are many suitable alternatives to the last; we could, for example, have chosen N_1 to satisfy $\forall n \ge N_1 \ |a_n - a| < \min(1, \varepsilon)$.

In constructing proofs of limit results, notice that $a_n \to a$ as $n \to \infty$ if and only if $a_n - a \to 0$ as $n \to \infty$. This emphasises the nature of the information we are given by the statement $a_n \to a$ as $n \to \infty$; it is an elaborate statement about the 'smallness' of $a_n - a$. Thus to show $a_n b_n \to ab$ we must express $a_n b_n - ab$ in terms of 'small' quantities.

We now know that $1/n \to 0$ as $n \to \infty$. By Theorem 7.1(i), $a/n \to 1$ as $n \to \infty$ if a is a constant. By setting $a_n = b_n = 1/n$ for all $n \in \mathbb{N}$, (iii) shows us that $1/n^2 \to 0$ as $n \to \infty$. Using part (iii) again we could deduce $1/n^3 \to 0$ as $n \to \infty$ and so on. We shall leave it to the reader to prove that for all natural numbers k, $1/n^k \to 0$ as $n \to \infty$. (Induction?)

Theorem 7.2 Let (a_n) and (b_n) be two sequences of real numbers with $a_n \to a$ and $b_n \to b$ as $n \to \infty$. Then:

    (i) $|a_n| \to |a|$ as $n \to \infty$;

    (ii)   If $a \ne 0$, $1/a_n \to 1/a$ as $n \to \infty$;

    (iii)   If $\forall n \in \mathbb{N} \ a_n \le b_n$ then $a \le b$.

Helpful Graffiti:

(i) We use the inequality from Lemma 5.3, $||x|-|y||\le|x-y|$, so that if $|a_n - a| < \varepsilon$ then also $||a_n| - |a|| < \varepsilon$.

Proof:

(i) Let $\varepsilon > 0$. Then $\exists n \in \mathbb{N}$ such that $\forall n \ge N$ $|a_n - a| < \varepsilon$. Using the alternative form of the triangle inequality we see that

$$\forall n \ge N \quad ||a_n|-|a|| < |a_n - a| < \varepsilon .$$

Since $\varepsilon > 0$ was arbitrary, we have proved that $|a_n| \to |a|$ as $n \to \infty$.

(ii) Let $\varepsilon > 0$. Let k denote a positive constant whose value we shall fix once we see what we need. $\exists N_1$ such that $\forall n \ge N_1$ $|a_n - a| < k\varepsilon$. Then

$$\forall n \ge N_1 \quad \left|\frac{1}{a_n} - \frac{1}{a}\right| = \frac{|a-a_n|}{|a_n||a|}$$

$$< \frac{k\varepsilon}{|a_n||a|} .$$

The $|a|$ factor on the right could have been eliminated by a suitable choice of k, but the $|a_n|$ term is dependent on n. We want the right-hand side to be less than ε so it would be enough to ensure that $1/|a_n|$ is less than a constant for all sufficiently large n. This would be true if we ensured that $|a_n|$ was not close to zero. Since $a_n \to a$ as $n \to \infty$ $|a_n| \to |a| \ne 0$, so if we ensure a_n is 'close' to a, $|a_n|$ should be 'close' to $|a|$ and thus 'far' from 0. Choose N_2, such that

$$\forall n \ge N_2 \quad |a_n - a| < |a|/2.$$

Then,

$$\forall n \ge N_2 \quad ||a_n|-|a|| \le |a_n - a| < |a|/2$$

so $|a| - |a|/2 < |a_n| < |a| + |a|/2$
thus $|a_n| > |a|/2$ and $1/|a_n| < 2/|a|$.

Therefore, for $n \ge \max(N_1, N_2)$ we have

$$\left|\frac{1}{a_n} - \frac{1}{a}\right| = \frac{|a-a_n|}{|a||a_n|} < \frac{k\varepsilon}{|a|} \cdot \frac{2}{|a|}$$

and so we see we should have chosen $k = |a|^2/2$.

(ii) Let $\varepsilon > 0$. Since $a_n \to a$ as $n \to \infty$, we can choose N_1 and N_2 in $\mathbb{N}$ such that

$$\forall n \ge N_1 \quad |a_n - a| < (|a|^2/2)\varepsilon \qquad (1)$$

and $\forall n \ge N_2$ $|a_n - a| < |a|/2$. $\qquad (2)$

Let $N = \max(N_1, N_2)$. Then, by (2),

$$\forall n \ge N \quad ||a_n|-|a|| \le |a_n - a|$$
$$< |a|/2$$

so $\quad |a|/2 = |a|-|a|/2 < |a_n| . \quad (3)$

Therefore, $\forall n \ge N$

$$\left|\frac{1}{a_n} - \frac{1}{a}\right| = \frac{|a-a_n|}{|a_n||a|}$$

$$< \frac{2}{|a|^2}|a - a_n| \quad (\text{by (3)})$$

$$< \frac{2}{|a|^2}\frac{|a|^2}{2}\varepsilon = \varepsilon \quad (\text{by (1)}) .$$

Since ε was arbitrary,

$$1/a_n \to 1/a \text{ as } n \to \infty .$$

Note: We have used the information $a \ne 0$ in three places: we divided by a, and in (1) and (2) the numbers on the right-hand side, $(|a|^2/2)\varepsilon$ and $|a|/2$, must be positive.

(iii) We prove this by contradiction. Suppose $b < a$. Then in Fig. 7.1 for

(iii) Suppose the result is false, that is, $a > b$. Then let $\varepsilon = (a - b)/2 > 0$ so,

large enough n, b_n is close to b and a_n close to a. This should ensure $a_n > b_n$ which we know to be false. We need to choose $\varepsilon > 0$ small enough that no number can simultaneously be greater than $b + \varepsilon$ and less than $a - \varepsilon$.

since $a_n \to a$ and $b_n \to b$ as $n \to \infty$, we can choose N_1 and N_2 in $\mathbb{N}$ such that

$$\forall n \geq N_1 \quad a - \varepsilon < a_n < a + \varepsilon \text{ and}$$
$$\forall n \geq N_2 \quad b - \varepsilon < b_n < b + \varepsilon .$$

Therefore letting $N = \max(N_1, N_2)$,
$$b_N < b + \varepsilon = (a + b)/2 = a - \varepsilon < a_N .$$

But we know that $\forall n \in \mathbb{N} \; a_n \leq b_n$ so this is a contradiction, hence our supposition is wrong and $a \leq b$. □

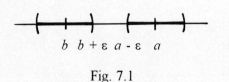

$$b \quad b + \varepsilon \quad a - \varepsilon \quad a$$

Fig. 7.1

The proof in (iii) relies on our being able to find a natural number N for which $N \geq N_1$ and $N \geq N_2$. If, as in this case, N_1 and N_2 are natural numbers this is easy, but had they been real numbers we would have had to rely on the Archimedean property.

There is an issue here which we have avoided. If $a_n \to a$ as $n \to \infty$ and $a \neq 0$, then we showed in part (ii) above that there is a number N_2 with the property that $\forall n \geq N_2 \; |a_n| > |a|/2$ and hence, in particular, $\forall n \geq N_2 \; a_n \neq 0$. It may happen that there are values of n less than N_2 for which $a_n = 0$ and thus $1/a_n$ is meaningless. Strictly speaking, this means that the sequence $(1/a_n)$ is not properly defined. However, if we set $b_n = 1/a_n$ whenever $a_n \neq 0$ and $b_n = 0$ whenever $a_n = 0$, then by the proof above, $b_n \to 1/a$ as $n \to \infty$, since $\forall n \geq N_2 \; b_n = 1/a_n$. We shall presume this interpretation if the limit of $1/a_n$ is to be found.

Example 7.4 Find $\lim\limits_{n \to \infty} \dfrac{n^2 + 2n}{n^2 + 1}$.

Solution: We need to express this as a quotient of terms which tend to a limit, which we do by dividing numerator and denominator by the dominant term, n^2 :

$$\frac{n^2 + 2n}{n^2 + 1} = \frac{1 + 2/n}{1 + 1/n^2} .$$

Since $1/n \to 0$, by Theorem 7.1(i) $2/n \to 0$ and 7.1(ii) tells us that $1 + 2/n \to 1$ as $n \to \infty$. Parts (ii) and (iii) show that $1 + 1/n^2 \to 1$, hence $1/(1 + 1/n^2) \to 1$ by 7.2(ii), so using 7.1(iii) again, $\dfrac{1 + 2/n}{1 + 1/n^2} \to \dfrac{1}{1} = 1$ as $n \to \infty$. Therefore

$$\lim_{n \to \infty} \frac{n^2 + 2n}{n^2 + 1} = \lim_{n \to \infty} \frac{1 + 2/n}{1 + 1/n^2} = 1 . \quad \square$$

Example 7.5 Find the limit as $n \to \infty$ of $\dfrac{n^3 + n + 1}{n^4 + 1}$.

Solution: Using the theorems several times,

$$\frac{n^3+n+1}{n^4+1} = \frac{1/n+1/n^3+1/n^4}{1+1/n^4} \to \frac{0+0+0}{1+0} = 0 \quad \text{as} \quad n \to \infty \,. \quad \square$$

If the denominator in an expression tends to 0 as $n \to \infty$, the theorems are no help, since the result about $\lim(1/a_n)$ requires a_n to tend to a non-zero limit. Such cases are dealt with individually.

Example 7.6 Show that $(n^4 + 1)/(n^3 + 1)$ does not tend to a limit as $n \to \infty$.

Solution: $\dfrac{n^4+1}{n^3+1} = \dfrac{1+1/n^4}{1/n+1/n^4}$, so as $n \to \infty$ the numerator tends to 1, but the denominator tends to 0, so Theorem 7.2(ii) does not help. Now, for $n \in \mathbb{N}$,

$$\frac{n^4+1}{n^3+1} > \frac{n^4}{n^3+1} \geq \frac{n^4}{n^3+n^3} = \frac{n}{2}$$

(since $n \geq 1$, so $n^3 + n^3 \geq n^3 + 1$). Now let $a_n = (n^4 + 1)/(n^3 + 1)$ and suppose $a_n \to a$ as $n \to \infty$. Then, since $1 > 0$, $\exists N \in \mathbb{N}$ such that $\forall n \geq N$ $a - 1 < a_n < a + 1$.

Therefore $\forall n \geq N$ $n/2 < a + 1$, and in turn $\forall n \geq N$ $n < 2(a + 1)$, which is nonsense.

This contradiction shows that (a_n) does not tend to a limit. $\square$

7.3 Developments

Definition A sequence (a_n) of real numbers is said to be **bounded above** if there is a constant U such that $\forall n \in \mathbb{N}$ $a_n \leq U$, and **bounded below** if there is a constant L such that $\forall n \in \mathbb{N}$ $a_n \geq L$. U and L are said to be **upper** and **lower bounds** respectively (a_n) is said to be **bounded** if it is both bounded above and below.

This allows us a general result including the last example as a special case.

Theorem 7.3 If (a_n) is a sequence of real numbers which tends to a limit, then (a_n) is bounded.

Proof: Suppose that $a_n \to a$ as $n \to \infty$. Since $1 > 0$, we can choose N such that $\forall n \geq N$ $a - 1 < a_n < a + 1$. Therefore if we let $M = \max(a + 1, a_1, ..., a_{N-1})$ and $m = \min(a - 1, a_1, ..., a_{N-1})$, we have $\forall n \in \mathbb{N}$ $m \leq a_n \leq M$. Thus (a_n) is bounded. $\square$

Remark: Notice the important issue here: the existence of the limit gives us information about all but finitely many values of n, so that in forming m and M we take the minimum or maximum of a *finite number* of quantities.

The converse of Theorem 7.3 is false, as we shall see in a moment. Before tackling this, however, we must consider how to negate statements involving 'for all' or 'there exists'. Let $P(x)$ be some statement about x, and A be some set from which x is chosen: in our examples so far, A could be the set of natural numbers, or the set of natural numbers greater than N, etc. Then to say that 'For all x in A, $P(x)$ is true' is false, is equivalent to saying that 'There is an x in A for which $P(x)$ is false'; in symbols, the negation of '$\forall x \in A \; P(x)$' is '$\exists x \in A$ such that not $P(x)$'. Similarly the negation of '$\exists x \in A$ such that $P(x)$' is '$\forall x \in A$ not $P(x)$'. For example, the negation of '$\forall x \in \mathbb{N} \; x \geq 1$' is '$\exists x \in \mathbb{N}$ such that $x < 1$'.

To negate a statement involving more than one $\forall$ or $\exists$ statement, we proceed in steps. The statement $a_n \not\to a$ as $n \to \infty$, meaning that a_n does not tend to a as $n \to \infty$, may be written in each of the following ways, the last being the most revealing:

$$\text{not}\big(\forall \varepsilon > 0 \qquad \exists N \in \mathbb{N} \text{ such that } \forall n \geq N \qquad |a_n - a| < \varepsilon\big)$$

$$\exists \varepsilon > 0 \text{ such that } \quad \text{not}\big(\exists N \in \mathbb{N} \text{ such that } \forall n \geq N \qquad |a_n - a| < \varepsilon\big)$$

$$\exists \varepsilon > 0 \text{ such that } \quad \forall N \in \mathbb{N} \text{ not} \big(\forall n \geq N \qquad |a_n - a| < \varepsilon\big).$$

$$\exists \varepsilon > 0 \text{ such that } \quad \forall N \in \mathbb{N} \; \exists n \geq N \text{ such that } \qquad |a_n - a| < \varepsilon .$$

Let us now use this to show that the sequence (a_n), where $a_n = (-1)^n$, does not tend to a limit. Let a be a real number; we show that $a_n \not\to a$ as $n \to \infty$. If $a \geq 0$, set $\varepsilon = 1$ and we see that $\forall N \in \mathbb{N} \; \exists n \geq N$ (e.g. $n = 2N + 1$) such that $|a_n - a| = |-1 - a| = a + 1 \geq \varepsilon$ so we have shown that $\exists \varepsilon > 0$ such that $\forall N \in \mathbb{N}$ $\exists n \geq N$ such that $|a_n - a| \geq \varepsilon$. If $a < 0$, let $\varepsilon = 1$ again and $\forall N \in \mathbb{N}$ $\exists n \geq N$ (e.g. $n = 2N$) such that $|a_n - a| = 1 - a \geq \varepsilon$. So in either case, a is not the limit of (a_n), so (a_n), has no limit. Notice, though, that (a_n) is bounded, since $\forall n \in \mathbb{N} \; -1 \leq a_n \leq 1$, so the converse of Theorem 7.3 is false.

There is, however, something we wish to say in the way of a converse to Theorem 7.3. Consider a sequence (a_n) which is bounded above by U and in which each term is no smaller than the previous term, so $\forall n \in \mathbb{N} \quad a_{n+1} \geq a_n$ and $a_n \leq U$. Since the terms are increasing in size but never exceed U, Fig. 7.2

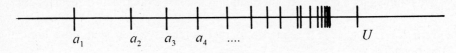

Fig. 7.2

suggests that they must 'bunch up' in some sense and tend to a limit as $n \to \infty$. (Since U is just *an* upper bound and the sequence will have many upper bounds, there is no reason to expect the limit to be U.) This is quite a substantial statement since it can be thought of as saying that there are sufficiently many real numbers that there is one which is the limit of the sequence. In fact, this is a property that distinguishes the system of real numbers from the system of rational numbers, since an increasing sequence of rational numbers need not have a rational limit; the sequence 1, 1.4, 1.41, 1.414, ... of approximations to $\sqrt{2}$ has no rational limit.

We have not yet assumed enough about the real number system to prove the result we wish here, so we shall state it now and prove it once we have made a more detailed consideration of the real number system.

Definition A sequence (a_n) of real numbers is said to be **increasing** if $\forall n \in \mathbb{N}$ $a_{n+1} \geq a_n$, and **strictly increasing** if $\forall n \in \mathbb{N}$ $a_{n+1} > a_n$. If $\forall n \in \mathbb{N}$ $a_{n+1} \leq a_n$ $(a_{n+1} < a_n)$ we say (a_n) is **decreasing (strictly decreasing)**.

Theorem 7.4 If (a_n) is an increasing sequence of real numbers which is bounded above, then (a_n) tends to a limit as $n \to \infty$.
    The proof will appear in Chapter 9. $\square$

Corollary If (a_n) is a decreasing sequence of real numbers which is bounded below, then (a_n) tends to a limit as $n \to \infty$.

Proof. Suppose that (a_n) is decreasing and $\forall n \in \mathbb{N}$ $a_n \geq L$. Let $b_n = -a_n$. Then $\forall n \in \mathbb{N}$ $b_{n+1} = -a_{n+1} \geq -a_n = b_n$, and $\forall n \in \mathbb{N}$ $b_n = -a_n \leq -L$ so (b_n) is increasing and bounded above. By the Theorem there is a b such that $b_n \to b$ as $n \to \infty$, so $a_n \to -b$ as $n \to \infty$. $\square$

    Notice that the example $a_n = (-1)^n/n$ shows that a sequence may tend to a limit yet be neither increasing nor decreasing.

Example 7.7 Define a sequence (a_n) by

$$a_1 = 2, \ \forall n \in \mathbb{N} \ a_{n+1} = \tfrac{1}{2}(a_n + \tfrac{2}{a_n})$$

and discuss whether or not it tends to a limit.

Solution: We first need to check that the definition is valid, that is, that a_n is not zero for any n or else the definition of a_{n+1} would fail. It is, however, clear that $a_n > 0 \Rightarrow a_{n+1} > 0$ and, since $a_1 > 0$, induction shows that $\forall n \in \mathbb{N}$ $a_n > 0$.
    To discover whether (a_n) is increasing or not, consider

$$a_{n+1} - a_n = \tfrac{1}{2}(a_n + \tfrac{2}{a_n}) - a_n = \frac{2 - a_n^2}{2a_n} . \qquad (*)$$

To use $(*)$ we need to know the sign of $2 - a_n^2$. Now

$$2 - a_{n+1}^2 = 2 - \tfrac{1}{4}(a_n^2 + 4 + \tfrac{4}{a_n^2}) = -\tfrac{1}{4}(a_n^2 - 4 + \tfrac{4}{a_n^2}) = -\tfrac{1}{4}(a_n - \tfrac{2}{a_n})^2 \leq 0 ,$$

so that $\forall n \in \mathbb{N}$ $a_{n+1}^2 \geq 2$. Hence $\forall n \geq 2$ $a_n^2 \geq 2$ and since we know $a_1^2 \geq 2$ we deduce that $\forall n \in \mathbb{N}$ $a_n^2 \geq 2$. From $(*)$ we now see that $\forall n \in \mathbb{N}$ $a_{n+1} - a_n \leq 0$, that is, (a_n) is decreasing. Since (a_n) is bounded below (all the terms are positive) the Corollary above assures us that (a_n) tends to some real number, say a, as $n \to \infty$.
    We must find a. Since $a_n \to a$ as $n \to \infty$, $a_{n+1} \to a$ as $n \to \infty$ (see Problem 2). Also $(a_n^2 > 2$ and $a_n > 0) \Rightarrow a_n > 1$, so $\forall n$ $a_n \geq 1$, hence $a \geq 1$ (by Theorem 7.2(iii) with $a_n = 1$ for all $n \in \mathbb{N}$). In particular $a \neq 0$, and

$$a = \lim_{n \to \infty} a_{n+1} = \lim_{n \to \infty} \frac{a_n^2 + 2}{2a_n} = \frac{a^2 + 2}{2a} .$$

From this we deduce that $a^2 = 2$, so $a_n \to a$, where a is a positive number whose square is 2, i.e. $a = \sqrt{2}$. $\square$

Notice, in passing, that this result provides a proof that there is a real number whose square is 2; the existence of a was deduced from Theorem 7.4.   We shall return to this later.

When dealing with sequences defined by an equation of the form $a_{n+1} = f(a_n)$ it is often very helpful to draw a diagram.   By plotting the two curves $y = f(x)$ and $y = x$ on the same graph we can gauge the behaviour of (a_n) as follows.   Mark a_n on the x-axis, so that the point where the line $x = a_n$ meets $y = f(x)$ has y-co-ordinate $f(a_n)$, that is, a_{n+1}.   By drawing the line through $(a_n, f(a_n))$ parallel to the x-axis until it meets the line $y = x$,   we obtain a point whose x-co-ordinate is a_{n+1}, and the process may be repeated; see Fig. 7.3.

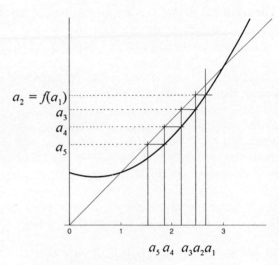

Fig 7.3.   Plotting a_n where $a_{n+1} = f(a_n)$ and $f(x) = \frac{1}{3}(x^3 - x) + 1$.

When using diagrams in this way, care must be taken in the drawing and in the correct placing of turning points and the like.   Diagrams do, however, often suggest clearly how a suitable analysis proof is to be made by suggesting whether or not the sequence decreases, what suitable upper or lower bounds are, etc.   At this stage in your career you must then convert the ideas suggested by the diagram into a formal proof, which will show up any significant shortcomings in the sketch made.

In some examples we shall find that the terms of a sequence (a_n) become large as n increases and are eventually greater than any given number.   Since (a_n) is not bounded in this case it cannot tend to a limit as $n \to \infty$, but we can describe its behaviour.

Definition We say that $a_n \to \infty$ as $n \to \infty$ if, for all real R, $\exists N \in \mathbb{N}$ such that $\forall n \geq N \ a_n > R$.   We say that $a_n \to -\infty$ if $-a_n \to \infty$ as $n \to \infty$.

Notice that '$a_n \to \infty$ as $n \to \infty$' is a useful notation but we do not regard (a_n) as tending to a limit, nor do most of the results about limits apply. ∞ is not a real number and must not be treated as one.

Example 7.8 Show that $n(2 + (-1)^n) \to \infty$ as $n \to \infty$.

Solution: Let R be a real number.   Choose N to be a natural number greater than R.   Then

$$\forall n \geq N \ n(2 + (-1)^n) \geq n(2 - 1) = n \geq N > R .$$

Since R was arbitrary, this holds for all $R \in \mathbb{R}$, and we have shown that $n(2 + (-1)^n) \to \infty$ as $n \to \infty$. $\square$

The next two results, although 'elementary', require a certain amount of low cunning.

Example 7.9 Let $a > 0$. $a^{1/n} \to 1$ as $n \to \infty$.

Solution: We shall use Bernoulli's Inequality, since that will relate a to $a^{1/n}$, so suppose first that $a \geq 1$. Then $a^{1/n} \geq 1$ so, by Bernoulli's Inequality, $a = (a^{1/n})^n > n(a^{1/n} - 1)$ whence $0 \leq a^{1/n} - 1 < a/n$, this being true for all $n \in \mathbb{N}$.
Let $\varepsilon > 0$. Choose $N \geq a/\varepsilon$. Then

$$\forall n \geq N \; |a^{1/n} - 1| < a/n \leq a/N \leq \varepsilon.$$

Since $\varepsilon > 0$ was arbitrary we have proved that $a^{1/n} \to 1$ as $n \to \infty$ (if $a \geq 1$).
For $0 < a < 1$, notice that $1/a > 1$ so $(1/a)^{1/n} \to 1$ as $n \to \infty$. Now $a^{1/n} = 1/(1/a)^{1/n} \to 1/1$ as $n \to \infty$. $\square$

Example 7.10 $n^{1/n} \to 1$ as $n \to \infty$.

Solution: This one is more awkward, but illustrates how progress can be made. The same use of Bernoulli's Inequality as above would yield $n > n(n^{1/n} - 1)$ from which we deduce $n^{1/n} - 1 < 1$, which, unfortunately, does not imply that $n^{1/n} - 1$ is small for large n. We use the trick of using square roots, but in a different guise; apply Bernoulli's Inequality to $n^{1/2n}$ to obtain

$$\sqrt{n} = (n^{1/(2n)})^n > n(n^{1/(2n)} - 1),$$

whence $0 < n^{1/2n} - 1 < 1/\sqrt{n}$.
Let $\varepsilon > 0$. Then $\forall n \geq N \geq 1/\varepsilon^2$,

$$|n^{1/2n} - 1| = n^{1/2n} - 1 < 1/\sqrt{n} \leq 1/\sqrt{N} \leq \varepsilon.$$

Thus $n^{1/2n} \to 1$ as $n \to \infty$, whence $n^{1/n} = (n^{1/2n})^2 \to 1$ as $n \to \infty$. $\square$

We have seen that the definition of limit we have introduced allows us to prove the results we would expect and, with the aid of the assumed properties of the real number system, also allows us to prove, without the aid of hand waving, results like the existence of $\sqrt{2}$, which holds the promise of the technique's proving powerful. On the debit side there is no doubt that the definition and the basic results using it are technically complicated. This is necessary, as the next two examples illustrate, in the nature of warnings.
The definition of a limit demands that, as $n \to \infty$, a_n tends to a number, that is, something independent of n. It is tempting to write such things as '$a_n \to b_n$' to mean that a_n and b_n become arbitrarily close for large n, but the temptation should be resisted. Do we mean the difference becomes small, $a_n - b_n \to 0$? (True if $a_n = 1/n$, $b_n = 1/n^2$, for example.) Or do we mean that the ratio tends to 1, $a_n/b_n \to 1$? (True, for example, if $a_n = n^2 + 1$, $b_n = n^2 + n$.) Even worse, one is tempted to think that '$a_n \to b_n$' should imply '$a_n^2 \to b_n^2$' which is false in some interpretations. Consider $a_n = n + 1/n$, $b_n = n$ (so that $a_n - b_n \to 0$ and

$a_n/b_n \to 1$ as $n \to \infty$), but $a_n^2 = n^2 + 2 + 1/n^2$, $b_n^2 = n^2$ so $a_n^2 - b_n^2 \nrightarrow 0$. Of course, if we notice $a_n - b_n \to 0$ we deduce, correctly, that $(a_n - b_n)^2 \to 0$, a different result. This is not profound, but it illustrates the sort of sloppiness that has to be avoided; indeed, in many problems the difficulty is in deciding precisely what the correct question is, the solution being easy after that.

A second warning is more serious. Consider an argument along these lines: Let $f(m, n)$ depend on m and n and suppose $g(m) = \lim_{n \to \infty} f(m,n)$ exists, and that $g(m) \to G$ as $m \to \infty$. Thus G is close to $f(m, n)$ for sufficiently large m and n. If also $h(n) = \lim_{m \to \infty} f(m,n)$ and $h(n) \to H$ as $n \to \infty$, then $f(m, n)$ will be close to H for sufficiently large m and n. Does this mean $G = H$? It seems that we could show that G and H are as close together as we wish, for each is within ε of $f(m, n)$ for sufficiently large m and n. Unfortunately this is wrong, as the following example shows.

$$\lim_{m \to \infty} \frac{m+n^2}{mn+n^2} = \lim_{m \to \infty} \frac{1+n^2/m}{n+n^2/m} = \frac{1}{n} \quad \text{and} \quad \lim_{n \to \infty} \frac{1}{n} = 0 \ .$$

also

$$\lim_{n \to \infty} \frac{m+n^2}{mn+n^2} = \lim_{n \to \infty} \frac{m/n^2+1}{m/n+1} = 1 \quad \text{and} \quad \lim_{m \to \infty} 1 = 1 \ ,$$

so

$$0 = \lim_{n \to \infty} \left\{ \lim_{m \to \infty} \frac{m+n^2}{mn+n^2} \right\} \neq \lim_{m \to \infty} \left\{ \lim_{n \to \infty} \frac{m+n^2}{mn+n^2} \right\} = 1 \ .$$

This result shows us that the order in which limits are taken cannot be changed without justification. This is an important limitation, with considerable nuisance value, as it happens from time to time that we wish to reverse the order of taking two limiting operations, and we shall have to circumvent this difficulty.

Readers may wish to try proving $G = H$ above, to see exactly where the argument fails. Loosely speaking, the difficulty is that if we wish $|g(m) - G| < \varepsilon$, m needs to be sufficiently large. Then to ensure that $|f(m, n) - g(m)| < \varepsilon$, n needs to be sufficiently large, but *how large is required depends on the value of m chosen.* In simpler words, to make $|f(m, n) - G| < 2\varepsilon$ we need to have n related in some way to m. Similarly to make $|f(m, n) - H| < 2\varepsilon$ we need to have m determined by n. It may not be possible to find a pair of values m and n which satisfies both criteria.

Problems

1. Prove directly from the definition that:
    (i)   if $a_n = (n - 1)/(n + 1)$ then $a_n \to 1$ as $n \to \infty$,
    (ii)   if $\forall n \in \mathbb{N}$ $1 - (2/n^2) \le a_n \le 1 + 1/n$ then $a_n \to 1$ as $n \to \infty$,
    (iii)   if $a_n = \begin{cases} 1/n & (n \text{ a perfect square}), \\ -1/n^2 & (\text{otherwise}), \end{cases}$ then $a_n \to 0$ as $n \to \infty$.

2, Suppose that (a_n) is a sequence of real numbers and that (b_n) is defined by $b_n = a_{n+1}$. Prove that $b_n \to L$ as $n \to \infty$ if and only if $a_n \to L$ as $n \to \infty$.

(Hint: How large does the suffix on a need to be so that we are certain that $|a_n - L| < \varepsilon$?)

3. Let (a_n), (b_n) and (c_n) be three sequences of real numbers such that $\forall n \in \mathbb{N}$ $a_n \le b_n \le c_n$. Prove that if $a_n \to L$ and $c_n \to L$ as $n \to \infty$, then $b_n \to L$ as $n \to \infty$. (Sometimes called the 'sandwich rule' or 'squeeze rule'.)

4. Find the limit as $n \to \infty$, when one exists, or prove it does not exist, for the sequences whose nth term is:

$$\frac{2n^3+n}{n^3+4n^2}, \quad \frac{n^2-n+1}{2n^2+n}, \quad \frac{n^4+n^2-1}{3n^5+2}, \quad \frac{2^n+3^n}{1+5^n}, \quad \frac{n^3+n}{n^2-2}, \quad \frac{n^3-3n^2+2n-1}{2n^3+3n-1}.$$

5. Prove that an increasing sequence is bounded if and only if it is bounded above.

6. Prove that (a_n) is bounded if and only if $(|a_n|)$ is bounded above.

7. Let (a_n) be an increasing sequence which is not bounded above. Prove that $a_n \to \infty$ as $n \to \infty$. Deduce that if (b_n) is increasing then either (b_n) tends to a limit or $b_n \to \infty$ as $n \to \infty$.

8. Define the sequence (a_n) by $a_1 = \alpha$, and $\forall n \in \mathbb{N}$ $a_{n+1} = a_n^2 - 2a_n + 2$, where $1 < \alpha < 2$. Show that (i) $\forall n \in \mathbb{N}$ $1 < a_n < 2$, (ii) (a_n) is decreasing, (iii) (a_n) tends to a limit, say a, as $n \to \infty$ and $a = 1$ or 2, (iv) $\forall n \in \mathbb{N}$ $a_n \le \alpha$ and $a \le \alpha$ and (v) $a_n \to 1$ as $n \to \infty$.

9. The sequence (a_n) is defined by $a_1 = \alpha$, $a_{n+1} = (a_n^2 - 2a_n + 3)/2$. Prove that if $1 \le \alpha < 3$ then $a_n \to 1$ as $n \to \infty$, while if $\alpha > 3$ then (a_n) is increasing and does not tend to a limit, hence $a_n \to \infty$ as $n \to \infty$ (see Question 7). By considering a_2, discuss the behaviour of (a_n) in the cases in which $\alpha < 1$. (A diagram should help here.)

10. Define (a_n) by $a_1 = \alpha$, $a_{n+1} = (a_n^2 + a_n)/2$. Discuss the behaviour of (a_n) as $n \to \infty$ in the case where $0 \le \alpha < 1$. Discuss also the cases $\alpha = 1$, $\alpha > 1$, $-1 \le \alpha < 0$, $-2 < \alpha < -1$, $\alpha = -2$ and $\alpha < -2$. (A diagram may help see why these are the 'cases' to be considered.)

11. Let $a_1 = \alpha$, $a_{n+1} = 2a_n/(a_n + 1)$ and show that $a_n \to 1$ as $n \to \infty$ if $\alpha > 0$. How does the sequence behave if $\alpha < -1$? (Harder: Show that if $-1 < \alpha < 0$ and α is not of the form $-1/(2^m - 1)$ for some $m \in \mathbb{N}$, then $\forall n \in \mathbb{N}$ $a_n \ne -1$ so the sequence is well-defined and $a_n \to 1$ as $n \to \infty$.

12. Let $a_1 = \alpha$, $a_{n+1} = 2a_n^2 - a_n^3$. Show that $a_n \to 0$ as $n \to \infty$ if $0 \le \alpha < 1$ and that $a_n \to 1$ as $n \to \infty$ if $1 \le \alpha \le (1 + \sqrt{5})/2$. (Hint: To show that $1 \le a_n \le (1 + \sqrt{5})/2 \Rightarrow 1 \le a_{n+1}$, consider $a_{n+1} - 1$ and factorise. It should help to draw a diagram!)

13. Suppose that $a_1 = \alpha$ and $a_{n+1} = 2a_n/(1 + 2|a_n|)$. Find the behaviour of (a_n) as $n \to \infty$, depending on the value of α. (Sort out the cases yourself.)

14. Show that if $a_n \to \infty$ as $n \to \infty$ then $1/a_n \to 0$ as $n \to \infty$. Show also that if $\forall n \in \mathbb{N}$ $a_n > 0$ and $a_n \to 0$ as $n \to \infty$, then $1/a_n \to \infty$.

15. From Question 14 it is easily seen that if $a_n \to \infty$ or $a_n \to -\infty$ as $n \to \infty$ then $1/a_n \to 0$ $n \to \infty$. The converse is false; show this by finding a sequence (a_n) of non-zero numbers such that $1/a_n \to 0$ as $n \to \infty$ but a_n tends to neither ∞ nor $-\infty$. (Hint: Think of $|a_n|$.)

16. Let $a_n \to a$ as $n \to \infty$, and suppose that $\forall n \in \mathbb{N}$ $|b_n - a_n| < 1/n$. Prove that $b_n \to a$ as $n \to \infty$.

17. Suppose that $a_n \to a$ as $n \to \infty$ and $a_n \to b$ as $n \to \infty$; show that $a = b$, that is, show that if a sequence has a limit, then it has only one. (You may wish to ponder on whether you have tacitly assumed that to be true, without even asking the question!)

18.* Let $a_n \to a$ as $n \to \infty$ and define b_n by $b_{2n} = b_{2n-1} = a_n$ $\forall n \in \mathbb{N}$, so (b_n) is the sequence $(a_1, a_1, a_2, a_2, a_3, a_3, ...)$. Prove that $b_n \to a$ as $n \to \infty$. If (c_n) is the sequence $(a_1, a_2, a_2, a_3, a_3, a_3, ...)$ in which a_n occurs n times, show that $c_n \to a$ as $n \to \infty$.

19.* Let $a_n \to a$ and $b_n \to b$ as $n \to \infty$ and suppose that $\forall n \in \mathbb{N}$ $a_n < b_n$. Show by example that this information alone does not imply that $a < b$. Given that there is a constant $A > 0$ such that $\forall n \geq N$ $a_n + A \leq b_n$, prove that $a < b$.

20.* Theory tells us that if $a_n \to a$ then $a_n^2 \to a^2$ and $a_n^3 \to a^3$ as $n \to \infty$. In this question we investigate the converse. All properties of square or cube roots used here must be proved true.
  (i)   Show that if $a_n^2 \to 0$ as $n \to \infty$ then $a_n \to 0$ as $n \to \infty$.
  (ii)   Show that if $a_n^2 \to a^2$ then $|a_n| \to |a|$ as $n \to \infty$. You may find the equation $x - y = (x^2 - y^2)/(x + y)$ helpful for the case $a \neq 0$.
  (iii) Give an example to show that it is possible that (a_n^2) tends to a limit while (a_n) does not.
  (iv)   Prove that if $a_n^3 \to a^3$ as $n \to \infty$ then $a_n \to a$. (Hint: A suitable expression to use with the idea in (ii) is $(x - y) = (x^3 - y^3)/(x^2 + xy + y^2)$.)
  (v)   Prove that if $a_n^2 \to a^2$ and $a_{n+1} - a_n \to 0$ as $n \to \infty$ then either $a_n \to a$ or $a_n \to -a$ as $n \to \infty$. (Hint: For $a \neq 0$ show that $\exists N$ such that $\forall n \geq N$ a_n has constant sign.)

8

Infinite Series

"And whatever the Ordinary Analysis performs using equations of a finite number of terms (when it be possible) this always performs using infinite equations; I have not hesitated to give it, too, the name of Analysis. Indeed, the arguments in it are no less certain than in the other, nor are the equations less exact, though we Mortals of finite intelligence can neither express nor conceive all their terms to know thence exactly the desired quantities."　　　　　　　Sir Isaac Newton

8.1 Introduction

In Chapter 5 we met infinite series, where we noticed that in considering the sums $1 + 1/2$, $1 + 1/2 + 1/4$, $1 + 1/2 + 1/4 + 1/8$, ... it is not hard to guess that $1 + 1/2 + ... + 1/2^k = 2 - 1/2^k$, which can then be proved for all $k \in \mathbb{N}$ by induction. Since $1/2^k \to 0$ as $k \to \infty$ we are inclined to think that if we added up all the numbers of the form $1/2^k$ for $k = 0, 1, 2, ...$ the sum should be 2. As there are infinitely many terms we cannot actually perform all the additions, but it is still clear what the answer 'should be'. Notice, however, that there are sequences whose terms cannot be added in this way; if we consider the sums $1, 1 + 1, 1 + 1 + 1, ...$ then it is clear that the sum of the first k terms does not tend to a limit as $k \to \infty$. The definition we made in Chapter 6 was in terms of limits, which we have now looked at in detail.

Definition Let (a_n) be a sequence of real numbers and define s_n by $s_n = a_1 + a_2 + ... + a_n$ $(n \in \mathbb{N})$. If s_n tends to a limit as $n \to \infty$ we say that the **series** Σa_n **converges**, while if s_n does not tend to a limit we say that Σa_n **diverges**. When $s_n \to s$ as $n \to \infty$ we say that s is the **sum** of the series and write

$$s = \sum_{n=1}^{\infty} a_n .$$

　　　The Σ sign denotes summation, so that when we talk of 'the series' Σa_n we are referring to the process of summing the terms a_n, that is, we are interested in the behaviour of the associated sequence s_n. The statement Σa_n converges' refers to the behaviour of (s_n) and does not directly say anything about (a_n). The notation $\sum_{n=1}^{\infty} a_n$ is used for a number, and its use implicitly requires the series Σa_n to converge since otherwise no such number exists. a_n is called the nth **term** of the series, and s_n is sometimes referred to as the nth **partial sum**.

Examples 8.1　　(i) If $|x| < 1$ then Σx^n converges and $\sum_{n=1}^{\infty} x^n = x/(1-x)$.

　　　　　(ii) $\Sigma 1/(n(n+1))$ converges and $\sum_{n=1}^{\infty} 1/(n(n+1)) = 1$.

Solution: Part (i) was done in Chapter 6. Do it (without looking at that Chapter) as a problem.

(ii) Let $s_n = \sum_{k=1}^{n} 1/(k(k+1)) = \sum_{k=1}^{n} \{1/k - 1/(k+1)\} = 1 - 1/(n+1)$. Then $s_n \to 1$ as $n \to \infty$, by inspection. $\square$

8.2 Convergence Tests

The two examples just quoted and those used in Chapter 6 are unusual in that we can find a convenient expression for s_n. Normally we have to prove that s_n tends to a limit by indirect means. The simplest result about series is:

Theorem 8.1 If Σa_n converges, then $a_n \to 0$ as $n \to \infty$.

Proof: Let $s_n = a_1 + a_2 + \dots + a_n$, and suppose Σa_n converges. Then for some real number s, $s_n \to s$ as $n \to \infty$. Therefore $s_{n-1} \to s$ as $n \to \infty$ so that $a_n = s_n - s_{n-1} \to s - s = 0$ as $n \to \infty$. $\square$

Caution: The converse of Theorem 8.1 is false. There are sequences (a_n) for which $a_n \to 0$ as $n \to \infty$ but Σa_n diverges, as we shall show in a moment. This is a feature of the study of infinite series which leads to the diversity of the subject: there is no simple criterion which determines the convergence of a general series, just a host of partial results. Notice that Theorem 8.1 shows us that not all series converge.

Example 8.2 Let $a_n = 1/n$ $\forall n \in \mathbb{N}$. Then $a_n \to 0$ as $n \to \infty$ but Σa_n diverges.

Solution: Let $s_n = 1 + 1/2 + \dots + 1/n$ and choose m to be a non-negative integer for which $n \ge 2^m$. Then

$$
\begin{aligned}
s_n &= 1 + \frac{1}{2} + \frac{1}{3} + \dots + \frac{1}{2^m} + \dots + \frac{1}{n} \\
&\ge 1 + \frac{1}{2} + \left(\frac{1}{3} + \frac{1}{4}\right) + \dots + \left(\frac{1}{2^{m-1}+1} + \dots + \frac{1}{2^m}\right) \\
&\ge 1 + \frac{1}{2} + \frac{1}{2} + \dots + \frac{1}{2}
\end{aligned}
$$

(since the bracket ending in $1/2^k$ has 2^{k-1} terms each no smaller than $1/2^k$, for $k = 1, \dots, m$)

$$= 1 + m/2 .$$

Therefore $s_{2^m} \ge 1 + m/2$ for all $m \in \mathbb{N}$, so if (s_n) were bounded above, by S say, then $\forall m \in \mathbb{N}$ $1 + m/2 \le s_{2^m} \le S$ which is a contradiction. It follows that (s_n) is not a bounded sequence and therefore does not tend to a limit as $n \to \infty$. $\square$

Before we develop the main body of theory there are some obvious little results:

Lemma 8.2 Let (a_n) and (b_n) be sequences of real numbers and λ be a constant. If Σa_n and Σb_n converge, so do $\Sigma \lambda a_n$ and $\Sigma (a_n + b_n)$; moreover,

$$\sum_{n=1}^{\infty} \lambda a_n = \lambda \sum_{n=1}^{\infty} a_n, \text{ and } \sum_{n=1}^{\infty} (a_n + b_n) = \sum_{n=1}^{\infty} a_n + \sum_{n=1}^{\infty} b_n.$$

Proof: Let $s_n = a_1 + a_2 + \ldots + a_n$ and $t_n = b_1 + b_2 + \ldots + b_n$. Then if $s_n \to s$ and $t_n \to t$ as $n \to \infty$,

$$\lambda a_1 + \lambda a_2 + \ldots + \lambda a_n = \lambda s_n \to \lambda s \text{ as } n \to \infty$$

and $(a_1 + b_1) + (a_2 + b_2) + \ldots + (a_n + b_n) = s_n + t_n \to s + t$ as $n \to \infty$.) $\square$

Lemma 8.3 Suppose that there is a natural number N for which $\forall n \geq N \ a_n = b_n$. Then Σa_n and Σb_n either both converge or both diverge.

Proof: Let $s_n = a_1 + a_2 + \ldots + a_n$ and $t_n = b_1 + b_2 + \ldots + b_n$. Then $\forall n \geq N$ $s_n = t_n + C$ where $C = a_1 + a_2 + \ldots + a_{N-1} - b_1 - b_2 - \ldots - b_{N-1}$ is independent of n. (s_n) tends to a limit if and only if (t_n) tends to a limit (though the two limits will differ by C). $\square$

Remark: Changing the first few terms of a series does not affect the convergence - but it will normally affect the sum.

We now start with series of non-negative terms, whose behaviour is simpler than the general case.

Lemma 8.4 Let Σa_n be a series with non-negative terms. Then Σa_n is convergent if and only if the associated sequence (s_n) is bounded, that is, if and only if there is a constant K for which $\forall N \in \mathbb{N}$ $\sum_{n=1}^{N} a_n \leq K$.

Proof: $\forall n \in \mathbb{N}$ $s_{n+1} = s_n + a_{n+1} \geq s_n$ (since $\forall k \ a_k \geq 0$), so (s_n) is increasing. Therefore from Theorem 7.4, (s_n) tends to a limit if and only if it is bounded above. $\square$

This leads to our first standard test for convergence. Notice that, very roughly, the test says that a series of non-negative terms converges if $a_n \to 0$ 'sufficiently rapidly'.

Lemma 8.5 The Comparison Test. Suppose that $\forall n \in \mathbb{N}$ $0 \leq a_n \leq b_n$ and that Σb_n converges. Then Σa_n converges.

Proof: Let $s_n = a_1 + a_2 + \ldots + a_n$ and $t_n = b_1 + b_2 + \ldots + b_n$. Then, since Σb_n converges, (t_n) is bounded. Let $t_n \leq T$ $\forall n \in \mathbb{N}$. Therefore, $\forall n \in \mathbb{N}$,

$$s_n = a_1 + a_2 + \ldots + a_n \leq b_1 + b_2 + \ldots + b_n = t_n \leq T$$

so (s_n) is bounded above. Since (s_n) is increasing $(a_n \geq 0)$, (s_n) tends to a limit as $n \to \infty$, and Σa_n converges. $\square$

Corollary 1 Suppose that $N \in \mathbb{N}$, $\forall n \geq N$ $0 \leq a_n \leq b_n$ and Σb_n converges. Then Σa_n converges.

Proof: Let $a'_n = b'_n = 0$ $(n = 1, 2, \ldots, N - 1)$ and $a'_n = a_n$ $\forall n \geq N$, $b'_n = b_n$

$\forall n \geq N$. By Lemma 8.3, $\sum b'_n$ converges, so by the comparison test $\sum a'_n$ converges and Lemma 8.3, once more, tells us that $\sum a_n$ converges. $\square$

Corollary 2 Suppose that $N \in \mathbb{N}$, that $\forall n \geq N$ $0 \leq c_n \leq d_n$ and that $\sum c_n$ diverges. Then $\sum d_n$ diverges.

Proof: If $\sum d_n$ were convergent, Corollary 1 would show $\sum c_n$ convergent. $\square$

Remarks: These results rely on the terms being non-negative, at least for large n. We can use them to show that $\sum 1/n^2$ converges as follows: $\sum 1/n(n+1)$ we know converges, hence $\sum 2/(n(n+1))$ converges, from which, noticing that $\forall n \in \mathbb{N}$ $1/n^2 \leq 2/(n(n+1))$, the comparison test shows $\sum 1/n^2$ converges. By noticing that $\forall n \in \mathbb{N}$ $0 \leq 1/n \leq 1/\sqrt{n}$ we see that $\sum 1/\sqrt{n}$ diverges, since $\sum 1/n$ does.

Example 8.3 Suppose that, for all $n \in \mathbb{N}$, $a_n \in \{0, 1, ..., 9\}$. Then the series $\sum a_n 10^{-n}$ converges. This is easily checked by the comparison test, for $\forall n \in \mathbb{N}$ $0 \leq a_n \leq 9$ and $\sum 9.10^{-n}$ converges because it is a geometric series whose common ratio is $\frac{1}{10}$ and $0 < \frac{1}{10} < 1$.

This result shows us that every decimal represents a real number, whereas Chapter 6's main thrust was to show that every number had a decimal expansion. The results of Chapter 6 show us that some real numbers (those of the form of an integer divided by a power of 10) have two decimal expansions. $\square$

Corollary 3 Suppose that $\forall n \in \mathbb{N}$ $a_n \geq 0$ and $b_n > 0$ and that $a_n / b_n \to \lambda$ as $n \to \infty$. Then if $\sum b_n$ is convergent, so is $\sum a_n$.

Proof: Since $a_n / b_n \to \lambda$, $\exists N$ such that $\forall n \geq N$ $\lambda - 1 < a_n / b_n < \lambda + 1$. Therefore, $\forall n \geq N$ $a_n < (\lambda + 1)b_n$ and, if $\sum b_n$ is convergent, $\sum (\lambda + 1)b_n$ is also, whence Corollary 1 shows that $\sum a_n$ converges. $\square$

Remarks: This Corollary is not quite symmetric in (a_n) and (b_n). If $a_n / b_n \to \lambda$ as $n \to \infty$, then, provided $\lambda \neq 0$, $b_n / a_n \to 1/\lambda$ as $n \to \infty$ so the two series will either both converge or both diverge in the case $\lambda \neq 0$. If $\lambda = 0$ it may happen that $\sum a_n$ converges and $\sum b_n$ diverges.

The drawback of all these comparison tests is that we need to have a stock of series at our disposal to compare new series with. To establish some we need one rather specialised result.

Theorem 8.6 Cauchy's Condensation Test. Suppose that (a_n) is a decreasing sequence of positive terms. Then $\sum a_n$ converges if and only if $\sum 2^n a_{2^n}$ converges.

Proof: Let $s_n = a_1 + a_2 + ... + a_n$ and $t_n = 2^1 a_2 + 2^2 a_4 + ... + 2^n a_{2^n}$. Since (a_n) is decreasing and $a_1 \geq 0$

$$s_{2^n} = a_1 + a_2 + (a_3 + a_4) + \ldots + (a_{2^{n-1}+1} + \ldots + a_{2^n})$$

$$\geq 0 + (a_2) + (a_4 + a_4) + \ldots + (a_{2^n} + \ldots + a_{2^n})$$

where the kth bracket contains 2^{k-1} occurrences of a_{2^k}

$$\geq a_2 + 2a_4 + \ldots + 2^{k-1}a_{2^k} + \ldots + 2^{n-1}a_{2^n}$$

$$= \tfrac{1}{2}t_n .\tag{1}$$

Also, $s_{2^n} = a_1 + (a_2 + a_3) + (a_4 + \ldots + a_8) + \ldots + (a_{2^{n-1}} + \ldots + a_{2^n-1}) + a_{2^n}$

$$\leq a_1 + (a_2 + a_2) + (a_4 + \ldots + a_4) + \ldots + (a_{2^{n-1}} + \ldots + a_{2^{n-1}}) + a_{2^n}$$

where the kth bracket contains 2^k occurrences of a_{2^k}

$$= a_1 + 2a_2 + 4a_4 + \ldots + 2^{n-1}a_{2^{n-1}} + a_{2^n}$$

$$\leq a_1 + t_n .\tag{2}$$

If the sequence (s_k) is bounded, then (s_{2^n}) is bounded (since its terms are a selection of those of (s_k)) and so by (1), (t_n) is bounded above. Also if (t_n) is bounded above, by T say, then (2) shows that

$$\forall n \in \mathbb{N} \quad s_n \leq s_{2^n} \leq a_1 + t_n \leq a_1 + T \;,$$

(the first inequality because $n < 2^n$ and the extra terms in (s_{2^n}) are positive) whence (s_n) is bounded above. Thus (s_n) is bounded above if and only if (t_n) is bounded above, and since all terms in both series are positive, Lemma 8.4 completes the proof. $\square$

Example 8.4 $\sum 1/n^\alpha$ converges if $\alpha > 1$ and diverges if $\alpha \leq 1$.

Solution: We need only consider $\alpha > 0$, since if $\alpha \leq 0$ then $1/n^\alpha \not\to 0$, so the series certainly diverges. Let $a_n = 1/n^\alpha$ and $\alpha > 0$. Then $\forall n \in \mathbb{N}$ $a_n > 0$ and $a_{n+1} < a_n$ so we may use the condensation test.

$$2^n a_{2^n} = 2^n (1/2^n)^\alpha = 2^n 2^{-\alpha n} = (2^{1-\alpha})^n .$$

Therefore $\sum 2^n a_{2^n}$ is a geometric series with common ratio $2^{1-\alpha}$, and this converges if and only if $|2^{1-\alpha}| < 1$, that is, if and only if $1 - \alpha < 0$, which is equivalent to $\alpha > 1$. Therefore by the condensation test, $\sum 1/n^\alpha$ converges if and only if $\alpha > 1$. $\square$

The series $\sum 1/n^\alpha$ are useful as yardsticks in the comparison test. To be truthful, the condensation test is useful for few other results, though some series involving powers and logarithms are susceptible to this attack.

Example 8.5 Decide whether $\sum (n + 1)/(n^3 + 2n)$ is convergent or divergent.

Solution: For large n we expect that $(n + 1)/(n^3 + 2n)$ ought to behave like $1/n^2$ so, since $\sum 1/n^2$ converges, we show the series converges by comparison. Let $a_n =$

$(n + 1)/(n^3 + 2n)$ and $b_n = 1/n^2$. Σb_n converges and $a_n/b_n \to 1$ as $n \to \infty$ so Corollary 3 to the comparison test shows that Σa_n converges. $\square$

Although we have established enough theory to be able to tackle a range of examples, this is not wide enough to give useful experience at this stage. This is partly because we have yet to develop routine techniques and partly because the experience of judging which test to try first is best gained when the main tests are available. So we shall march on and expand our repertoire to allow series with terms of varying sign.

Theorem 8.7 Let Σa_n be a series of real numbers (positive, negative or zero). Then if $\Sigma |a_n|$ converges so does Σa_n.

Proof: Let
$$b_n = \begin{cases} a_n & \text{if } a_n \geq 0, \\ 0 & \text{if } a_n < 0, \end{cases} \qquad c_n = \begin{cases} 0 & \text{if } a_n \geq 0, \\ -a_n & \text{if } a_n < 0. \end{cases}$$
Then $\forall n \in \mathbb{N}$ $b_n \geq 0$ and $c_n \geq 0$ and $a_n = b_n - c_n$. . Also, $0 \leq b_n \leq |a_n|$ and $0 \leq c_n \leq |a_n|$, so if $\Sigma |a_n|$ is convergent the comparison test proves that Σb_n and Σc_n are convergent, and thus so is $\Sigma (b_n - c_n)$. $\square$

Definition A series for which $\Sigma |a_n|$ is convergent is said to be **absolutely convergent**. We have just proved that an absolutely convergent series is convergent in the ordinary sense.

Corollary If Σa_n is a series of real numbers, Σb_n is a convergent series of non-negative numbers and for some integer N we have $\forall n \geq N$ $|a_n| \leq b_n$, , then Σa_n is (absolutely) convergent.

Proof: By the earlier forms of comparison test, $\Sigma |a_n|$ converges. $\square$

In practice we refer to the foregoing result together with the comparison test and all its Corollaries as 'the comparison test', without distinguishing which component we are referring to. The last form is the most useful, largely because the majority of important series are absolutely convergent. However, all forms of the comparison test suffer from the disadvantage that they require the user to invent a series with which to compare the given series. We now develop some more routine tests, the main two, the ratio and root tests, being based on comparison with a geometric series.

Theorem 8.8 The ratio test. Let Σa_n be a series of real numbers for which $\left| \dfrac{a_{n+1}}{a_n} \right| \to L$ as $n \to \infty$. Then

                if $L < 1$ the series converges (absolutely);

                if $L > 1$ the series diverges;

and   if $L = 1$ we obtain no information.

Proof: Suppose first that $L < 1$. Let $\varepsilon = (1 - L)/2$ so that $\varepsilon > 0$ and $L + \varepsilon < 1$. Then $\exists N$ such that $\forall n \geq N \ |a_{n+1}/a_n| < L + \varepsilon$. Therefore,

$$\exists N \text{ such that } \forall n \geq N \ |a_{n+1}| < (L + \varepsilon)|a_n| ,$$

and, keeping N fixed, we see that

$$\forall n > N , \ |a_n| < (L + \varepsilon)|a_{n-1}| < (L + \varepsilon)^2|a_{n-2}| < ... < (L + \varepsilon)^{n-N}|a_N| .$$

Since $|a_N|$ and $L + \varepsilon$ are independent of n and $\sum(L + \varepsilon)^n$ converges (for $0 < L + \varepsilon < 1$), the comparison test now shows that $|\sum a_n|$ converges.

Next, let $L > 1$. Then let $\varepsilon = (L - 1)$ so that $\varepsilon > 0$ and $\exists N$ such that $\forall n \geq N \ |a_{n+1}/a_n| > L - \varepsilon = 1$. Therefore

$$\exists N \text{ such that } \forall n \geq N \ |a_{n+1}| > |a_n| ,$$

and, keeping N fixed, we have

$$\forall n \geq N \ |a_n| > |a_{n-1}| > ... > |a_N| .$$

Since $|a_N| \neq 0$ (or else a_{N+1}/a_N would not exist) the last line shows that $a_n \nrightarrow 0$ so, by our first result on series, $\sum a_n$ diverges.

To see that the case $L = 1$ really gives no information, consider $\sum 1/n$ and $\sum 1/n^2$. In both cases, $|a_{n+1}/a_n| \rightarrow 1$ as $n \rightarrow \infty$, but one of the series converges and the other diverges. $\square$

Example 8.6 For all real x, $\sum x^n/n!$ converges.

Solution: If $x = 0$ the result is obvious, for $x^n = 0$ $(n \geq 1)$, so suppose $x \neq 0$ and let $a_n = x^n/n!$ $(\neq 0)$. Then $|a_{n+1}/a_n| = |x|/(n + 1) \rightarrow 0$ as $n \rightarrow \infty$ and since $0 < 1$ the series converges by the ratio test. $\square$

Theorem 8.9 The Root Test. Suppose that $\sum a_n$ is a series of real numbers and that $|a_n|^{1/n} \rightarrow L$ as $n \rightarrow \infty$. Then

$$\text{if } L < 1 \quad \text{the series converges (absolutely);}$$
$$\text{if } L > 1 \quad \text{the series diverges;}$$
$$\text{and} \qquad \text{if } L = 1 \quad \text{we obtain no information.}$$

Proof: Let $L < 1$. Set $\varepsilon = (1 - L)/2 > 0$ so $\exists N$ such that $\forall n \geq N \ |a_n|^{1/n} < L + \varepsilon$ whence $|a_n| < (L + \varepsilon)^n$. Since $L + \varepsilon < 1$, the series $\sum(L + \varepsilon)^n$ converges, so $\sum|a_n|$ converges by comparison.

Let $L > 1$. This time set $\varepsilon = L - 1 > 0$ so $\exists N$ such that $\forall n \geq N$ $|a_n|^{1/n} > L - \varepsilon = 1$ whence $\forall n \geq N \ |a_n| > 1$. Thus $a_n \nrightarrow 0$ and $\sum a_n$ diverges.

As with the ratio test, the two series $\sum 1/n$ and $\sum 1/n^2$ show that the case $L = 1$ can occur with a convergent or a divergent series. $\square$

It is worth noticing that both these tests, in the form in which we have given them, only give information if the terms of $\sum a_n$ are sufficiently regular that the limit exists. In practice this covers most of the cases which arise but notice that the proof of the ratio test shows that $\sum a_n$ is absolutely convergent if there is a constant $k < 1$ and $N \in \mathbb{N}$ such that $\forall n \geq N \ |a_{n+1}/a_n| \leq k$, since the first step of the proof is devoted to using the limit condition to establish this (with $L + \varepsilon$ in place of k). It is, however, easy to mistake this result, since k must be *independent of* n and less that 1; for various plausible but wrong results, see the problems at the end of the chapter. The root test will, in principle, give information in every case which the ratio test copes with and some additional ones and is thus theoretically better, though in practice the nth roots are more awkward to handle. Again, the details are in the problems.

All of the tests so far will only indicate that a series converges absolutely, and are not capable of distinguishing a series which is convergent but not absolutely convergent. The only test we shall give for this is:

Theorem 8.10 The alternating series theorem. Suppose that (a_n) is a decreasing sequence of positive numbers and that $a_n \to 0$ as $n \to \infty$. Then $\Sigma(-1)^{n-1}a_n$ converges.

Proof: Let s_n denote the sum of the first n terms of $\Sigma(-1)^{n-1}a_n$., so $s_n = a_1 - a_2 + a_3 - a_4 \ldots + (-1)^{n-1}a_n$. For $m \in \mathbb{N}$ and $n \geq 2m$,

$$s_n \;=\; s_{2m} + a_{2m+1} - a_{2m+2} + \cdots + (-1)^{n-1}a_n$$

$$= \begin{cases} s_{2m} + (a_{2m+1} - a_{2m+2}) + \cdots + (a_{n-1} - a_n) & (n \text{ even}), \\ s_{2m} + (a_{2m+1} - a_{2m+2}) + \cdots + (a_{n-2} - a_{n-1}) + a_n & (n \text{ odd}). \end{cases}$$

Since each bracket is non-negative and $a_n \geq 0$, we see that $s_n \geq s_{2m}$. Similarly if $n \geq 2m + 1$

$$s_n = \begin{cases} s_{2m+1} - (a_{2m+2} - a_{2m+3}) - \cdots - (a_{n-2} - a_{n-1}) - a_n & (n \text{ even}), \\ s_{2m+1} - (a_{2m+2} - a_{2m+3}) - \cdots - (a_{n-1} - a_n) & (n \text{ odd}), \end{cases}$$

so, since the brackets are non-negative, $s_n \leq s_{2m+1}$. Therefore

$$\forall n \geq 2m \quad s_{2m} \leq s_n \leq s_{2m+1} \qquad (*)$$

This shows that (s_n) is bounded, but we still need care. The sequence $(s_2, s_4, s_6, \ldots) = (s_{2n})$ is increasing, since $\forall m \in \mathbb{N}$ $s_{2m} \leq s_{2m+2}$, and (s_{2n}) is bounded above, by s_3 for example, hence (s_{2n}) tends to a limit as $n \to \infty$. Let $s_{2n} \to s$ as $n \to \infty$. Also $(s_1, s_3, s_5, \ldots) = (s_{2n+1})$ is decreasing, since $\forall m \in \mathbb{N}$ $s_{2(m+1)+1} \leq s_{2m+1}$, and bounded below by s_2 so (s_{2n+1}) also tends to a limit as $n \to \infty$. Call its limit t. Then

$$t - s = \lim_{n \to \infty}(s_{2n+1} - s_{2n}) = \lim_{n \to \infty} a_{2n+1} = 0, \text{ so } s = t.$$

Finally, $s_n \to s$ as $n \to \infty$. To see this, let $\varepsilon > 0$. Then:

$$\exists N_1 \text{ such that } \forall m \geq N_1 \quad |s_{2m} - s| < \varepsilon \quad (s_{2m} \to s)$$

and $\qquad \exists N_2$ such that $\forall m \geq N_2 \quad |s_{2m+1} - s| < \varepsilon \quad (s_{2m+1} \to s)$.

Let $N = \max(2N_1, 2N_2 + 1)$. Then if $n \geq N$, either $n = 2m$ or $n = 2m + 1$, where m is an integer; in the first case $m \geq N_1$, in the second $m \geq N_2$ so in both cases $|s_n - s| \leq \varepsilon$. Therefore $\forall n \geq N$ $|s_n - s| \leq \varepsilon$.

Since $\varepsilon > 0$ was arbitrary we have proved that (s_n) tends to a limit as $n \to \infty$, that is, that $\Sigma(-1)^{n-1}a_n$ is convergent. $\square$

Notice the inequality (*) in the alternating series theorem. Keeping m fixed and letting $n \to \infty$ allows us to deduce that $s_{2m} \le s \le s_{2m+1}$, giving a simple estimate for $\sum_{n=1}^{\infty} (-1)^{n-1} a_n$.

Example 8.7 By the alternating series theorem, the series $1 - \frac{1}{2} + \frac{1}{3} - \frac{1}{4} + \cdots$, that is, $\sum (-1)^{n-1}/n$, converges. Since $\sum 1/n$ diverges, we have discovered a series which converges but is not absolutely convergent. □

Infinite series are not just sums of infinite collections of numbers, or, at least, that is not how the sum to infinity was defined. We must, therefore, be careful not to presume without proof that all of the properties of finite sums hold for sums of series. In fact, although the expected properties do hold under suitable conditions, some may fail to be true in complete generality.

Example 8.8 Consider the series $\sum (-1)^{n-1}/n$. Let s_n be the sum of the first n terms, i.e. $s_n = 1 - \frac{1}{2} + \frac{1}{3} + \cdots + (-1)^{n-1}/n$. We know that for some s, $s_n \to s$ as $n \to \infty$ and the estimate in the proof of the alternating series theorem allows us to deduce that $s_2 \le s \le s_3$, that is, $\frac{1}{2} \le s \le \frac{5}{6}$. In particular, $s > 0$.

Now rearrange the order of the terms so that we add in turn the first positive term, the first two negative ones, the next positive term, the next two negative ones, ... The resulting series is

$$1 - \frac{1}{2} - \frac{1}{4} + \frac{1}{3} - \frac{1}{6} - \frac{1}{8} + \frac{1}{5} - \cdots,$$

and if we let t_n denote the sum of the first n terms of this series, then for $n \in \mathbb{N}$

$$
\begin{aligned}
t_{3n} &= 1 - \frac{1}{2} - \frac{1}{4} + \frac{1}{3} - \cdots + \frac{1}{2n-1} - \frac{1}{4n-2} - \frac{1}{4n} \\
&= s_{2n} - \left(\frac{1}{2n+2} + \frac{1}{2n+4} + \cdots + \frac{1}{4n} \right) \\
&= s_{2n} - \frac{1}{2} \left(\frac{1}{n+1} + \frac{1}{n+2} + \cdots + \frac{1}{2n} \right).
\end{aligned}
$$

Now let $u_n = 1 + \frac{1}{2} + \frac{1}{3} + \cdots + \frac{1}{n}$, so that

$$
\begin{aligned}
\frac{1}{n+1} + \frac{1}{n+2} + \cdots + \frac{1}{2n} &= u_{2n} - u_n = u_{2n} - 2(\tfrac{1}{2} u_n) \\
&= (1 + \tfrac{1}{2} + \cdots + \tfrac{1}{2n}) - 2(\tfrac{1}{2} + \tfrac{1}{4} + \cdots + \tfrac{1}{2n}) = s_{2n}.
\end{aligned}
$$

Therefore $t_{3n} = s_{2n} - \frac{1}{2} s_{2n} = \frac{1}{2} s_{2n}$, so $t_{3n} \to \frac{1}{2} s \ne s$ as $n \to \infty$. By considering t_{3n+1} and t_{3n+2} it is not hard to show that $t_n \to s/2$ as $n \to \infty$, that is, *the rearranged series converges to a different sum from the original one.* □

A simpler point, easily missed, concerns combining terms. The series $1 - 1 + 1 - 1 + \cdots$ diverges, but if we insert brackets to combine the terms, we produce $(1 - 1) + (1 - 1) + (1 - 1) + \ldots$ and all of the terms in the new series are 0, so combining or splitting terms may alter convergence. If s_n denotes the sum of the first n terms of the original series then $s_n = 1$ or 0 according as n is odd or even. By putting in the brackets we obtain a series, the sum of whose first n terms is s_{2n}, and $s_{2n} \to 0$ as $n \to \infty$..

These pathologies should not be a source of dismay. The seventeenth century view of series as an 'infinite sum' would regard these examples as paradoxes, since the less rigorous demonstrations of mathematical results in use then tend to contradict them. Our more precise logical approach is capable of coping with these subtleties, and one by-product is these pathological results. These should not be given too much emphasis, except to counsel caution. In a sense, these apparently odd results indicate that our intuitive idea of an 'infinite sum' corresponds to something better behaved than the mere convergence of a series. The intuitive idea is closer to absolute convergence; although we shall not prove it here, a rearrangement of an absolutely convergent series converges to the same sum as the original series.

Example 8.9 Discuss the convergence of $\sum nx^n$ and $\sum x^{3n}/(2^n n)$ for all real values of x.

Solution: The easiest tests are the ratio and root tests, so we try these first. For $x \neq 0$, since $|(n + 1)x^{n+1}/(nx^n)| = \{(n + 1)/n\}|x| \to |x|$ as $n \to \infty$, the ratio test shows that the first series is convergent for $|x| < 1$ and divergent for $|x| > 1$. It is obvious that the series converges for $x = 0$ so this leaves only the case $|x| = 1$ undecided. However, since $|x| = 1$ implies $|nx^n| = n \not\to 0$ as $n \to \infty$ we see that the series diverges for $x = \pm 1$.

Applying the ratio test to the second series, we notice that if $a_n = x^{3n}/(2^n n)$ then (for $x \neq 0$) $|a_{n+1}/a_n| = (n/(n+1))(|x|^3/2) \to |x|^3/2$ as $n \to \infty$. By the ratio test, then, the series converges if $|x|^3 < 2$ (including the trivial case when $x = 0$) and diverges if $|x|^3 > 2$. The ratio (and root) tests fail us if $|x| = 2^{1/3}$, so we proceed separately. Let $x = 2^{1/3}$; then $a_n = 1/n$ and the series is $\sum 1/n$ which we know diverges. Also, if $x = -2^{1/3}$, $a_n = (-1)^n 1/n$, which gives the series $\sum(-1)^n 1/n$, convergent by the alternating series theorem. (Strictly speaking, the theorem shows that $\sum(-1)^{n-1} 1/n$ converges, but multiplying each term by -1 does not affect the convergence.) In this case the series converges for $-2^{1/3} \leq x \leq 2^{1/3}$ and diverges for other real x. $\square$

These examples are fairly typical in that the ratio or root tests may be brought to bear on most cases (or most values of a parameter) and are thus worth trying first, before the others are tried. Experience will bear out this rule of thumb and show the type of example for which the ratio test yields no information.

8.3 Power Series
Definition A series of the form $\sum a_n x^n$ where, for each n, the number a_n does not depend on x, is called a **power series**. It is usual to allow the index of summation of a power series to start at $n = 0$ giving a term $a_0 x^0$ which we interpret as a_0 and which is independent of x.

Lemma 8.11 Suppose that the power series $\sum a_n w^n$ converges. Then the series $\sum a_n x^n$ converges absolutely for all x with $|x| < |w|$.

Proof: Since $\sum a_n w^n$ converges, $a_n w^n \to 0$ as $n \to \infty$ and so $(a_n w^n)$ is a bounded sequence. Choose M such that $\forall n \geq 0 \ |a_n w^n| \leq M$. Then for $|x| < |w|$ and $n \geq 0$,

$$|a_n x^n| = |a_n w^n||x/w|^n \leq M|x/w|^n .$$

Since $|x/w|$ is independent of n and $|x/w| < 1$, $\sum |x/w|^n$ converges and so, by comparison, $\sum |a_n x^n|$ converges. $\square$

Corollary Suppose $\sum a_n w^n$ diverges. Then $\sum a_n x^n$ diverges for all x with $|x| > |w|$.

Proof: If $\sum a_n x^n$ converged and $|x| > |w|$, then by the Theorem, $\sum a_n w^n$ would converge. $\square$

Lemma 8.11 and its Corollary tell us that a power series $\sum a_n x^n$ will converge if $|x|$ is 'small enough' and diverge if $|x|$ is 'large enough'. This may be put more precisely.

Definition Let $\sum a_n x^n$ be a power series. We say that the number $R \geq 0$ is the **radius of convergence** of the series if $\sum a_n x^n$ converges for all x with $|x| < R$ and diverges for all x with $|x| > R$.

Considering a few of the examples that have arisen (or will arise when you do the problems!) we see that $\sum x^n/n$ converges if $|x| < 1$, diverges if $|x| > 1$, while for $|x| = 1$ the series converges for $x = -1$ and diverges for $x = +1$. Here the radius of convergence is 1, but notice that this fact on its own does not indicate the behaviour for $|x| = 1$. $\sum n!x^n$ has radius of convergence 0 (the ratio test shows that the series diverges if $|x| \neq 0$) while $\sum x^n/n!$ does not have a radius of convergence; it converges for all real x. We shall prove later that a power series either converges for all real x or has a radius of convergence. The examples given are typical of many, in that the ratio test will determine the convergence for all x except for $|x|$ equal to the radius itself where other tests must be involved.

Example 8.10 Let $\sum a_n x^n$ be a power series, and suppose that $|a_{n+1}/a_n| \to L$ as $n \to \infty$. Then if $L > 0$ the series has radius of convergence $1/L$ while if $L = 0$ the series converges for all real x. If $|a_{n+1}/a_n| \to \infty$ as $n \to \infty$ the series converges only for $x = 0$. The proof is a straightforward use of the ratio test. $\square$

8.4 Decimals again

We saw in Chapter 6 that every real number has a decimal expansion, that is, it can be expressed in the form of an integer plus a number of the form $\sum_{n=1}^{\infty} a_n 10^{-n}$, where each of the coefficients a_n belongs to $\{0, 1, ..., 9\}$. We can absorb the integer part for neatness to give us an expression of the form of $\sum_{n=-N}^{\infty} a_n 10^{-n}$ if the number is non-negative.   That is, every non-negative real number has at least one infinite decimal expansion.   We also know that certain numbers have more than one such expansion.   By Example 8.3, every series of the form $\sum_{n=-N}^{\infty} a_n 10^{-n}$ with $a_n \in \{0, 1, ..., 9\}$ converges, and hence it represents a real number.   We therefore have a correspondence between infinite decimals and non-negative real numbers.   This is useful, although we need care occasionally to avoid difficulties with those numbers which have two decimal expansions.

Problems

1.   Prove that $\sum_{n=1}^{N} 1/(n(n+1)(n+2)) = 1/4 - 1/(2(N+1)(N+2))$ and deduce that $\sum_{n=1}^{\infty} 1/(n(n+1)(n+2)) = 1/4$.

2.   Let $a_n = 1/2^n$ if n is odd and $a_n = (3/4)^n$ if n is even.   Show that $\sum a_n$ converges.

3.   By comparing the following series with $\sum 1/n^{\alpha}$ for suitable α , decide whether each converges or diverges: $\sum(n + 1)/(n^2 + 1)$, $\sum(n^2 + 1)/(n + 1)$, $\sum(n + 1)/(n^3 + 1)$, $\sum(n + 1)/n^4$.

4.   Suppose that $\forall n \in \mathbb{N}$ $a_n \geq 0$ and that $\sum a_n$ converges.   Prove that $\sum a_n^2$ converges.   Give an example where $\sum b_n^2$ converges but $\sum b_n$ does not.

5.   Given that $\sum a_n$ is absolutely convergent, prove that $\left| \sum_{n=1}^{\infty} a_n \right| \leq \sum_{n=1}^{\infty} |a_n|$.

6.   Decide which of the following series converge: $\sum n/2^n$, $\sum 2^n/(n3^n)$, $\sum (n2^n/3^n)$, $\sum 1/n^n$, $\sum(n + 1)^2/2^n$, $\sum 2^n/n!$, $\sum(2n)!/(n!)^2$, $\sum(2n)!/(8^n n!n!)$, $\sum(-1)^{n-1}/\sqrt{n}$.

7.   For which positive values of α does the series $\sum(-1)^{n-1}/n^{\alpha}$ converge?

8.   Find for which real x the series $\sum x^n/(1 + x^{2n})$ converges; make sure you test all x .

9.   Find the radius of convergence of the following series, or, if there is none, show that the series converges for all real x: $\sum x^n/n^2$, $\sum 2^n x^n$, $\sum x^n/(n^2 + 1)$, $\sum x^n/n$, $\sum x^{2n}/n$, $\sum x^{2n}/2^n$, $\sum x^n/(n2^n)$, $\sum(1 + 1/n)^n x^n$, $\sum n!x^n$, $\sum \frac{n!}{(2n)!} x^n$, $\sum \frac{(2n)!}{n!n!} x^n$, $\sum \frac{(3n)!}{n!(2n)!} x^n$, $\sum \frac{x^n}{n^n}$.

10.   In all but the last three parts of Question 9, consider those x for which $|x|$ equals the radius of convergence and decide whether or not the series converges for those x .

11. By observing that if $n \geq k$, $n! \geq (k+1)^{n-k} k!$, show that $\sum_{n=k}^{\infty} 1/n! \leq (k+1)/(k.k!)$. Let e denote the number $1+\sum_{n=1}^{\infty} 1/n!$, and show that if $s_m = 1+\sum_{n=1}^{m-1} 1/n!$, then $s_m \leq e \leq s_m + (m+1)/(m \cdot m!)$. Deduce that $2 \leq e \leq 3$.

12. Choose a sequence (a_n) which is strictly decreasing and for which $a_n \to 1$ as $n \to \infty$. Notice that for this sequence $\forall n \in \mathbb{N}$ $|a_{n+1}/a_n| < 1$ but $\sum a_n$ diverges since (a_n) does not even tend to 0.

13. Suppose that $|a_{n+1}/a_n| \to L$ as $n \to \infty$. Given $\varepsilon > 0$ show that for some N, $n \geq N \Rightarrow (L-\varepsilon)^{n-N} |a_N| < |a_n| < (L+\varepsilon)^{n-N} |a_N|$ and deduce that $|a_n|^{1/n} \to L$ as $n \to \infty$. Notice that this result shows that the root test gives a result whenever the ratio test does and is thus 'stronger'. In practice, the roots may be awkward to evaluate, and this result may help - for example, it may be used to show $(1/n!)^{1/n} \to 0$ as $n \to \infty$.

14. By showing that $\sum n^{\alpha} x^n$ converges, or otherwise, show that if α is fixed and $|x| < 1$, $n^{\alpha} x^n \to 0$ as $n \to \infty$.

15. Suppose that $\sum a_n$ and $\sum b_n$ are both convergent and that $\forall n \in \mathbb{N}$ $a_n \leq b_n$. Show that $\sum_{n=1}^{\infty} a_n \leq \sum_{n=1}^{\infty} b_n$. Given the additional information that there is an $N \in \mathbb{N}$ for which $a_N < b_N$, show that $\sum_{n=1}^{\infty} a_n < \sum_{n=1}^{\infty} b_n$. [Hint: Let (s_n) and (t_n) be the corresponding sequences of partial sums. It is not enough merely to notice that $\forall n \geq N$ $s_n < t_n$ since this guarantees only that $\lim s_n \leq \lim t_n$. Notice that $\forall n \geq N$ $s_n \leq t_n - (b_N - a_N)$ where $b_N - a_N$ is positive and independent of n.]

16. Suppose that $\sum a_n$ is convergent and $\sum b_n$ is divergent. Prove that $\sum (a_n + b_n)$ is divergent.

17. Suppose that $\sum a_n$ is convergent, but not absolutely convergent, and let $b_n = \max(a_n, 0)$ and $c_n = \min(a_n, 0)$ so that $a_n = b_n + c_n$ and $|a_n| = b_n - c_n$. Prove that $\sum b_n$ and $\sum c_n$ both diverge.

18.* Let (a_n) be a sequence such that $a_n = 0$ if n is odd and let $b_n = a_{2n}$. Show that $\sum b_n$ converges if and only if $\sum a_n$ does and that $\sum_{n=1}^{\infty} b_n = \sum_{n=1}^{\infty} a_n$. Can you generalise this?

19.* Let $\sum a_n$ and $\sum b_n$ be convergent series and define the sequence (c_n) by $c_{2n} = b_n$, $c_{2n-1} = a_n$, so (c_n) is $(a_1, b_1, a_2, b_2, ...)$. Prove that $\sum c_n$ converges and $\sum_{n=1}^{\infty} c_n = \sum_{n=1}^{\infty} a_n + \sum_{n=1}^{\infty} b_n$.

20.* Using the notation of Question 11 show that e is irrational by observing that if $e = p/q$ where p and q are positive integers,

$$|p/q - s_{q+1}| \leq (q + 2)/((q + 1)(q + 1)!)$$

and that $q!|p/q - s_{q+1}|$ is an integer. Obtain a contradiction.

21. Show that for integers k, n satisfying $0 \leq k \leq n$, $0 \leq \binom{n}{k}\frac{1}{n^k} \leq \frac{1}{k!}$ and hence that $\forall n \in \mathbb{N}\ 2 \leq (1 + 1/n)^n \leq e$. Notice, in particular, that, however the sequence $(1 + 1/n)^n$ behaves, it does not tend either to 1 (which is $\lim_{n \to \infty}(1 + 1/n)^k$) or to ∞ (which is $\lim_{n \to \infty}(1 + 1/k)^n$) where k here denotes a fixed natural number.

9

The Structure of the Real Number System

"The integers God made; all the rest is the work of man."   L. Kronecker.

The invention of the calculus presented the mathematical world with a severe difficulty. No one was in doubt of its usefulness, nor that it could produce many correct results that could not be obtained by other methods, yet the foundations of the calculus were unsatisfactory. There was some doubt as to the exact nature of a limit, and proofs of various results left much to be desired: indeed, even from the earliest times of the calculus, mathematicians had produced 'proofs' of results which were visibly false. The calculus became a matter of some controversy, well exemplified in Berkeley's book, *The Analyst, or A Discourse Addressed to an Infidel Mathematician*, published in 1734; this is still worth reading.

As is human nature, mathematicians, when faced with the apparently insoluble problem of putting calculus on a rigorous foundation, tried to ignore the matter, and until the end of the eighteenth century calculus was accepted largely as a matter of faith. By that time, however, the issue had become pressing, and major progress was made in the early nineteenth century, with the rigorous development of the idea of limit and its consequences. The stumbling block at last became apparent; there was no precise statement of the properties of the real number system. This was rectified around 1870 and the modern subject of analysis assumed increasing importance thereafter.

To an extent, we have followed a similar path. We introduced the idea of limit and deduced its consequences, but we noticed that there were results we needed to assume. For example, we did not prove the theorem that a bounded increasing sequence must tend to some real limit, but deferred this issue. The important thing is to appreciate that there *is* an issue here; no matter how plausible the statement may seem, it does require justification unless it is to be accepted as a basic premise. If we accept the property as true, for the moment, then it provides a distinction between the set of real numbers and the set of rational numbers. To see this, notice that if $s_n = 1 + 1/1! + 1/2! + .. + 1/n!$, then (s_n) is increasing and bounded above by 3 but the limit, the number usually called e , is not rational even though all the terms s_n are rational. (These assertions are the substance of Problems 8.11 and 8.20.)

To cope with this task we need to state clearly what we are assuming about $\mathbb{R}$. In fact the result we wish (that an increasing sequence of real numbers which is bounded above tends to some real limit) cannot be proved from the assumptions we have so far made, either explicitly or tacitly. These are what we rather glibly described in Chapter 3 as "the normal properties of the number system" and the four order properties at the beginning of Chapter 5, which for convenience we shall re-state formally here. These we shall accept as given; the formal word is **axioms**.

AXIOMS

A1   For all real x, y, and z, $x + (y + z) = (x + y) + z$ and $x(yz) = (xy)z$.

A2   For all real x and y, $x + y = y + x$ and $xy = yx$.

A3   There is a real number 0 such that, for all real x, $x + 0 = 0 + x = x$.

A4   For all real numbers x, there is a real number y for which
       $x + y = y + x = 0$.

A5   There is a real number 1 such that, for all real x, $x1 = 1x = x$.

A6   For all real $x \neq 0$, there is a real y for which $xy = yx = 1$.

A7   For all real x, y, and z, $x(y + z) = (xy) + (xz)$.

A8 $1 \neq 0$.

A9   For all real x, y, and z, exactly one of the three statements $x < y$, $x = y$,
       $x > y$ holds.

A10   For all real x, y, and z, if $x < y$ and $y < z$ then $x < z$.

A11   For all real x, y, and z, if $x < y$ then $x + z < y + z$.

A12   For all real x, y, and z, if $x < y$ and $z > 0$ then $xz < yz$.

These axioms, however, are also true of the system of rational numbers, so any result which can be deduced from them alone must be true of the system of rational numbers and, indeed, of any other system satisfying A1 - A12. Since we know that a bounded increasing sequence in the system of rational numbers need not have a limit within that system, the property we wish is not true of $\mathbb{Q}$ and therefore not deducible from A1 - A12 alone.

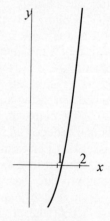

       We have already stated that $\mathbb{R}$ is a set with various arithmetical properties and our picture of the real number system is that it corresponds to the points of a line. We therefore wish a property which would allow us to replace the following with something more precise (see Fig. 9.1): The function given by $y = x^3 - 2$ has a negative value at $x = 1$ and a positive one at $x = 2$, so, since the curve has no discontinuities, it "must" cross the axis at some point. This assertion that the curve "must" cross the axis is not so much a proof as an act of defiance ("you tell me how else it could change from being negative to being positive").

       The feature we shall make precise is the idea that there are no gaps in the set of real numbers, as we think of there being no gaps in a line. (There certainly are gaps in the set of rational numbers and $\sqrt[3]{2}$ lies in one of

Fig. 9.1

them.) Looking more closely, we see that we can isolate two subsets of $\mathbb{R}$, $\{x \in \mathbb{R}: x^3 - 2 < 0\}$ and $\{x \in \mathbb{R}: x^3 - 2 > 0\}$, with the property that every member of the first set is less than every member of the second set. The existence of the number $\sqrt[3]{2}$ is equivalent to the existence of a real number which marks the division

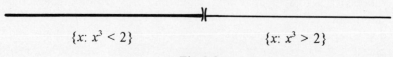

$\{x: x^3 < 2\}$ $\{x: x^3 > 2\}$

Fig 9.2

between these two sets. This corresponds to a picture of dividing a line into two: if we divide the line into two parts, one lying entirely to the left of the other, then there must be a point which marks the division, and this point must lie in one portion or the other (see Fig. 9.2). This is what motivates the definition below. If A and B are two sets, we define their **union** , $A \cup B$, to be the set $\{x: x \in A \text{ or } x \in B\}$ and their **intersection** , $A \cap B$, to be $\{x: x \in A \text{ and } x \in B\}$.

Axiom of Continuity (Dedekind's Axiom)
The set $\mathbb{R}$ has the property that if $\mathbb{R} = A \cup B$ where A and B are both non-empty and for all $x \in A$, and all $y \in B$, $x < y$, then either A has a greatest element or B has a smallest.

Notice that the conditions ensure that A and B have no elements in common, and that the axiom guarantees that real certain numbers exist.

Notation: Let $a, b \in \mathbb{R}$ and $a \leq b$. We define the **closed interval** $[a, b]$ to be the set $\{x \in \mathbb{R} : a \leq x \leq b\}$ and the **open interval** (a, b) to be $\{x \in \mathbb{R} : a < x < b\}$. $(a, b]$ and $[a, b)$ have their obvious analogous meanings. For convenience we use a similar notation for $\{x \in \mathbb{R} : a \leq x\}$, denoting this by $(-\infty, a]$ and $\{x \in \mathbb{R} : x > a\}$ by (a, ∞), with the corresponding interpretation of (a, ∞) and $[a, \infty)$. As before , ∞ is used as a notation, not a number.

Theorem 9.1 Let $\mathbb{R} = A \cup B$ where A and B are both non-empty and for all $x \in A$, and all $y \in B$, $x < y$. Then there is a real number a such that *either* $A = (-\infty, a]$ and $B = (a, \infty)$ *or* $A = (-\infty, a)$ and $B = [a, \infty)$.

Proof: By the axiom of continuity, either A has a greatest element or B has a smallest.
    First, suppose A has a greatest element , a . Then $a \in A$ and $\forall x \in A$, $x \leq a$ so A contains only numbers belonging to $(-\infty, a]$; we need to show A contains all such numbers. Let $y \in \mathbb{R}$, $y \leq a$. Since $y \in \mathbb{R}$, $y \in A$ or $y \in B$. But $y \in B \Rightarrow a < y$ (since $a \in A$, by the condition given connecting A and B), so this is impossible. Thus $y \in A$ so $A = (-\infty, a]$, since y was a typical number in $(-\infty, a]$. Since $B = \{x \in \mathbb{R} : x \notin A\} = \{x \in \mathbb{R} : x \not\leq a\}$, $B = (a, \infty)$. (That $B = \{x \in \mathbb{R} : x \notin A\}$ uses our knowledge that $A \cup B = \mathbb{R}$ and $A \cap B = \varnothing$.)
    The case when B has a smallest element is similar. $\square$

In practice, the Axiom of Continuity is inconvenient to use, so we must develop a more readily applicable tool.

Definitions A subset S of $\mathbb{R}$ is said to be **bounded above** if there is a number u with the property that $\forall x \in S$, $x \leq u$. Such a number u is called an **upper bound** for S . Notice that if u is an upper bound for S so are $u + 1$, $u + 2$, $u + \frac{1}{2}$, ... so a set which is bounded above will have many upper bounds.

Example 9.1 Show that the number u is *not* an upper bound for the set S if and only if $\exists x \in S$ such that $x > u$.

Solution: x is not an upper bound for S if and only if it is false that $\forall x \in S$, $x \leq u$, that is, if and only if $\exists x \in S$ such that $x > u$. In other words, u satisfies the negation of the statement that it be an upper bound for S.

This is a useful property, well worth remembering, for it gives us a positive property of numbers which are not upper bounds of a given set. By similar reasoning the number l is not a lower bound of the set S if and only if $\exists x \in S$ such that $x < l$. □

Theorem 9.2 Let S be a non-empty subset of $\mathbb{R}$ which is bounded above. Then among all of the upper bounds of S there is a smallest upper bound, i.e. the set of upper bounds for S has a smallest element.

Proof: Let B be the set of all upper bounds for S; we know $B \neq \varnothing$. Let A be the set of all real numbers which are *not* upper bounds for S. From the definitions it is obvious that $A \cup B = \mathbb{R}$. Since $S \neq \varnothing$, there is an element $s \in S$, whence $s - 1$ is not an upper bound for S and $A \neq \varnothing$.

To apply the Axiom of Continuity we need to show that every element of A is less than every element of B. Let $x \in A$ and $y \in B$. Since x is not an upper bound for S, there is an element, say s, of S such that $x < s$. Since y is an upper bound for S, $s \leq y$. Thus $x < y$. Because x and y were typical elements of A and B respectively., we have shown that $\forall x \in A$ and $\forall y \in B$ $x < y$.

By the Axiom of Continuity, either A has a largest element or B has a smallest. In fact, A cannot have a largest element. Let $a \in A$. Then a is not an upper bound for S so there is an element of S, say s, such that $a < s$. Then also $a < (a + s)/2 < s$ so $(a + s)/2$ is not an upper bound for S, (because it is less than the element s of S), and therefore $(a + s)/2 \in A$. This shows that a is not the greatest element of A. Since a was a typical element of A, A can have no greatest element. By the Axiom of Continuity B has a smallest element. This is what we require. □

Comments: 1. This result does tell us something significant. The set $(0,1)$ has no smallest element even though all its members are greater than 0, so not all bounded sets have a smallest element. The Theorem tells us, however, that those sets which are the set of all upper bounds of a non-empty set are special in that they must have a smallest element.

2. We do not claim that S has a largest element, but if it does, this element is the smallest upper bound. (For the largest element of a set is an upper bound and no smaller number can be an upper bound by Example 9.1.)

Example 9.2 The smallest upper bound for $[0,1]$ is 1, and the smallest upper bound for $(0,1)$ is also 1. Notice that 1 belongs to one of the sets but not the other. The smallest upper bound of the set $\{\frac{1}{2}, \frac{2}{3}, \frac{3}{4}, ...\}$ is 1.

Solution: In all three cases 1 is an upper bound for the set. Because $1 \in [0, 1]$

no smaller number can be an upper bound, so in the first case we see that 1 is the least upper bound. In the second case if we choose a number $x < 1$ then there is an element of $(0, 1)$ which is greater than x (for example 1/2 if $x < 1/2$, or $(x + 1)/2$ if $x \geq 1/2$), so gain we see that no number smaller than 1 can be an upper bound for $(0, 1)$. In the last case again if we choose $x < 1$ then there is an element of the set, in this case a number of the form $n/(n + 1) = 1 - 1/(n + 1)$ for large enough n which is greater than x, showing again that x is not an upper bound for the set. (This uses the Archimedean property and will be dealt with more fully in the next few pages.) □

Definition Let S be a non-empty set which is bounded above. We call the smallest upper bound for S the **supremum** of S and denote it by sup S.

Lemma 9.3 Let S be a subset of $\mathbb{R}$. Then $s = \sup S$ if and only if

(i) $\forall x \in S$, $x \leq s$, and (ii) $\forall s' < s$ $\exists x \in S$ such that $s' < x$.

Proof: Statement (i) says that s is an upper bound for S and (ii) that no smaller number is an upper bound. A moment's thought will show that a number possesses both properties if and only if it is the smallest upper bound for S. □

The two conditions of Lemma 6.3 offer a practical method for establishing that a particular number is the supremum of a given set. Notice that there are *two* conditions to be checked and that this complication is intrinsic. The supremum need not be a member of the set concerned; matters would be greatly simplified if this were so, for then every set with an upper bound would have a greatest element, but it just is not true. Let us use these ideas to prove one of our deferred results.

Theorem 9.4 Let (a_n) be an increasing sequence of real numbers which is bounded above. Then there is a real number a such that $a_n \rightarrow a$ as $n \rightarrow \infty$.

Proof: We know that $\{a_n : n \in \mathbb{N}\}$ is bounded above; it is non-empty since it contains a_1. Let $a = \sup \{a_n : n \in \mathbb{N}\}$.

Let $\varepsilon > 0$. Since $a - \varepsilon < a$, by Lemma 9.3 there is an element of $\{a_n : n \in \mathbb{N}\}$ which is greater than $a - \varepsilon$; let this element be a_N.. Then, since (a_n) is increasing we see that $\forall n > N$ $a - \varepsilon < a_N \leq a_n \leq a$, the last inequality being true since $a = \sup \{a_n : n \in \mathbb{N}\}$. Therefore $\forall n > N$ $|a_n - a| < \varepsilon$. Since $\varepsilon > 0$ was not further specified, this holds for all $\varepsilon > 0$ and we have shown that $a_n \rightarrow a$ as $n \rightarrow \infty$. □

Notice that Lemma 9.3 allows us to deduce the existence of at least one element of the set $\{a_n : n \in \mathbb{N}\}$ greater than $a - \varepsilon$; we do not know whether or not $a - \varepsilon$ itself belongs to the set.

This leaves us in a position to tackle the detailed structure of $\mathbb{R}$. We base all our deductions on the assumptions we have made about $\mathbb{R}$, which are that $\mathbb{R}$ satisfies the arithmetical and order axioms Al-A12 stated above and that $\mathbb{R}$ obeys the axiom of continuity. By the arithmetical properties, we can prove that $\mathbb{R}$ contains all rational numbers (e.g. 2/3=(1 + 1)/(1 + 1 + 1)), whose construction uses the arithmetical properties). To show that $\mathbb{R}$ contains irrational numbers we need the axiom of continuity, which we use indirectly. Define a sequence (a_n) by $a_1 = 2$, $a_{n+1}, = (a_n + 2/a_n)/2$. Then, following earlier work in Example 7.7, we see that (a_n) is decreasing and bounded below, so there is a real number a such that $a_n \to a$ as $n \to \infty$. (This uses Theorem 9.4 above, which we have now proved.) As before, $a^2 = 2$ so a is not rational. This guarantees the existence of a real number whose square is 2. That is, our assumptions are enough to prove formally, and without appealing to anything other than these assumptions, that irrational numbers exist.

The rational numbers are distributed throughout $\mathbb{R}$ and are thoroughly intermingled with the irrational numbers. Before showing this we prove the Archimedean property, which we previously had to assume.

Theorem 9.5 The set of natural numbers is not bounded above. In other words, for every real number x, there is a natural number n for which $n > x$.

Remarks: This result may seem obvious. It certainly is obvious that $\mathbb{N}$ has no greatest element, though that on its own does not preclude the existence of an upper bound. Theorem 9.5 can be viewed as showing that $\mathbb{R}$ contains no "infinitely large" numbers and is a form of confirmation that there are no further additional properties which we shall have to assume. Though it need not concern us here, there do exist systems satisfying A1-A12 for which the set corresponding to $\mathbb{N}$ is bounded above.

Proof: Suppose $\mathbb{N}$ were bounded above. Then, since $\mathbb{N} \neq \emptyset$, there must be a real number which is the supremum of $\mathbb{N}$; call it s. Then, by the properties of the supremum, we see that since $s - \frac{1}{2} < s$, there is a natural number n with $s - \frac{1}{2} < n$. Then $n + 1 > s + \frac{1}{2} > s$, which is a contradiction since $n + 1 \in \mathbb{N}$, but $n + 1 > \sup \mathbb{N}$. This contradiction shows $\mathbb{N}$ cannot be bounded above.

The second part of the statement is just the observation that if $x \in \mathbb{R}$ then x is not an upper bound for $\mathbb{N}$. $\square$

Corollary If $x \in \mathbb{R}$ and $x > 0$ then there is a natural number n with $1/n < x$.

Proof: Since $x > 0$, $1/x > 0$ so by Theorem 9.5 $\exists n \in \mathbb{N}$ such that $n > 1/x$. For this n, $1/n < x$. $\square$

This result shows us that if x is positive, no matter how small, then if we add x to itself sufficiently often, the result is greater than 1: this just notices that $\exists n \in \mathbb{N}$ such that $nx > 1$ and $nx = x + x + ... + x$. This could be interpreted as showing that there are no 'infinitesimally small' elements of $\mathbb{R}$.

Lemma 9.6 If x and y are real numbers and $x < y$, there is a rational number q with $x < q < y$. (Loosely, between every two reals there is a rational number.)

Proof: Since $y - x > 0$ there is $N \in \mathbb{N}$ with $1/N < y - x$. We now show that for some integer m, m/N is between x and y.

Since Nx is real, there is an $m_1 \in \mathbb{N}$ for which $m_1 > Nx$ so that $x < m_1/N$. Since $-Nx \in \mathbb{R}$, there is an $m_0 \in \mathbb{N}$ for which $m_0 > -Nx$, hence $x > -m_0/N$. The set $A = \{m/N : m \in \mathbb{Z}, -m_0 \le m \le m_1\}$ is a finite set which contains an element greater than x and another less than x. If m_2/N is the smallest member of A which is greater than x then $(m_2 - 1)/N \le x$. (This number exists since every *finite* set of real numbers has a minimum.) Let $q = m_2/N$ so $q \in \mathbb{Q}$, $x < q$ and $q - 1/N < x$. Thus $q < x + 1/N < x + (y - x)$ whence $x < q < y$. $\square$

Obviously, we can repeat the process just carried out to produce a rational number q_1 with $x < q_1 < q < y$ and repeat again to produce as many rationals as we wish between x and y. There are, therefore, infinitely many rationals between x and y, since if this were not so there would be some maximum number of rationals which could be found between x and y.

Corollary 1 If $x \in \mathbb{R}$ then $x = \sup \{q \in \mathbb{Q} : q < x\}$.

Proof: Let $A = \{q \in \mathbb{Q} : q < x\}$. We show the result by using Lemma 9.3. Obviously, $\forall y \in A$, $y \le x$. Also if $y' < x$ then by Lemma 9.6 there is a $q \in \mathbb{Q}$ with $y' < q < x$, so $\exists q \in A$ such that $y' < q$. Therefore y' is not an upper bound for A, and by Lemma 9.3 $x = \sup A$. $\square$

Corollary 2 Every real number is the limit of a sequence of rational numbers.

Proof: Let $a \in \mathbb{R}$. For each $n \in \mathbb{N}$, there is a rational number a_n with $a - 1/n < a_n < a$, by the Lemma. Then $\forall n \in \mathbb{N}$ $|a_n - a| < 1/n$ so that $a_n \to a$ as $n \to \infty$. $\square$

In the foregoing work we have introduced and used the idea of the supremum of a set. This is an important tool, so we need to obtain some skill in using it. Notice that the definition of the supremum as the smallest upper bound is somewhat complicated, and the equivalent statement in Lemma 9.3 has two parts. It is usual to have to establish these two properties of the supremum separately.

Example 9.3 Let A be a non-empty subset of $\mathbb{R}$ which is bounded above, and let $B = \{2x : x \in A\} = \{y : y/2 \in A\}$. Then $\sup B = 2\sup A$.

Solution: Let $a = \sup A$. We need to show that $2a$ has the two properties required by Lemma 9.3.

Let $y \in B$. Then $y = 2x$ for some $x \in A$, so $y/2 \in A$. Thus $y/2 \le a$ (as $a = \sup A$), and $y \le 2a$. Therefore $\forall y \in B$, $y \le 2a$.

Let $b' < 2a$. Then $b'/2 < a$. Since $a = \sup A$, from the second part of Lemma 9.3, $\exists x \in A$ such that $b'/2 < x$ and so $b' < 2x$. Thus $\exists y \in B$ ($y = 2x$ before) such that $b' < y$. This establishes the condition $\forall b' < 2a$ $\exists y \in B$ such that $b' < y$.

We have established both conditions of Lemma 9.3, so $2a = \sup B$. $\square$

The two-part nature of the proof above is typical.

Definitions A real number l is said to be a **lower bound** for the set A if $\forall x \in A$, $x \geq l$. A set is **bounded below** if it possesses a lower bound. The real number m is called the **infimum** of the set A and denoted by $\inf A$ if (and only if) it is the greatest lower bound of A. A set is said to be **bounded** if it is both bounded above and bounded below.

To save repeating our work to produce the properties of inf, we notice the following:

Lemma 9.7 Let A be a non-empty set of real numbers which is bounded above and let $B = \{x: -x \in A\}$. Then B is bounded below and $\inf B = -\sup A$.

Proof: Let $a = \sup A$. Then $x \in B \Rightarrow -x \in A \Rightarrow -x \leq a \Rightarrow x \geq -a$. Thus $-a$ is a lower bound for B. In passing notice that this shows B is bounded below.

Suppose that b is a lower bound for B; we need to relate this to A. $\forall x \in A$, $-x \in B$ so $-x \geq b$ and $x \leq -b$. Thus $\forall x \in A$, $x \leq -b$. Therefore $-b$ is an upper bound for A, hence $a \leq -b$ since $a = \sup A$ is the smallest upper bound for A. Hence $b \leq -a$. We have shown that if b is a lower bound for B, $b \leq -a$, so since $-a$ is a lower bound for B, it is the greatest lower bound, that is, $\inf B$. $\square$

Lemma 9.8 Every non-empty set of real numbers which is bounded below has a (real) infimum. If A is a subset of $\mathbb{R}$, then $t = \inf A$ if and only if

(i) $\forall x \in A$ $x \geq t$ and (ii) $\forall t' > t$ $\exists x \in A$ such that $t' > x$.

Proof: Let A be non-empty and bounded below. Let $C = \{x: -x \in A\}$. Then C is non-empty and bounded above and thus C has a supremum. (If l is a lower bound for A, $-l$ is an upper bound for C, as is easily checked.) By Lemma 9.7, then, $\inf A = -\sup C$ so $\inf A$ exists. The proof that the criteria are correct can either be obtained by relating them to $-t = \sup C$ or by analogy with the proof of Lemma 6.3. $\square$

Lemma 9.9 If A is a bounded non-empty subset of $\mathbb{R}$, $\inf A \leq \sup A$.

Proof: Since $A \neq \emptyset$, we can choose $x \in A$. Then $\inf A \leq x$ and $x \leq \sup A$ so $\inf A \leq \sup A$. $\square$

Example 9.4 Let A and B be two non-empty sets which are bounded above and let $C = \{x + y: x \in A$ and $y \in B\}$. Show that $\sup C = \sup A + \sup B$.

Solution: Let $a = \sup A$ and $b = \sup B$. Then if $z \in C$ there are $x \in A$ and $y \in B$ such that $z = x + y \leq a + b$ (since $x \leq a$ and $y \leq b$). Thus $\forall z \in C$ $z \leq a + b$.

Now let $c' < a + b$. Then $c' - b < a$, so, since $a = \sup A$, $\exists x \in A$ with $c' - b < x$. For this value of x, $c' - x < b = \sup B$, so $\exists y \in B$ with $c' - x < y$.

Thus letting $z = x + y$, we have $z \in C$ with $c' < z$. This gives the second property of $\sup C$, so $a + b = \sup C$ (by Lemma 9.3). $\square$

Example 9.5 Suppose that A and B are two non-empty sets in $\mathbb{R}$ with the property that $\forall x \in A$ and $\forall y \in B$ $x \leq y$. Show that $\sup A \leq \inf B$.

Solution: Let $x \in A$. Then $\forall y \in B$ $x \leq y$ so that x is a lower bound for B. Therefore $x \leq \inf B$. (inf B is the greatest lower bound.)

The conclusion of the above paragraph was reached only on knowing $x \in A$ so it must hold for all such x, hence $\forall x \in A$ $x \leq \inf B$. Thus $\inf B$ is an upper bound for A, so $\sup A \leq \inf B$. $\square$

Experience will show that in most examples where a supremum or infimum is to be shown equal to a given number, the proof is in two steps. Where an inequality involving sup or inf is concerned, one step may suffice and the two parts of Lemma 9.3 can be viewed as establishing two opposite inequalities (e.g. if a is an upper bound for A then $\sup A \leq a$).

Notice that the supremum of a set need not belong to the set (consider $(0,1)$); life would be much simpler if this were true! Also notice that if $a' < \sup A$ $\exists x \in A$ such that $a' < x$. This guarantees the existence of some $x \in A$ with $a' < x$ but *not* that a' itself is in A. (Consider $A = \{\frac{1}{2}, \frac{2}{3}, \frac{3}{4}, \ldots\}$.)

We give closed intervals their own name because they have useful properties, one of which is that the closed interval $[a, b]$ contains both its supremum and infimum. An open interval contains neither its supremum nor its infimum. In the course of the next chapter we shall encounter several situations where the distinction is crucial.

There is one common type of set which is simpler in this respect. Subsets of $\mathbb{N}$, other than the empty set, necessarily have a smallest element so in this case the set contains its infimum. This was proved in Lemma 4.1.

Problems

1.   The following sets are bounded above; identify the set of upper bounds in each case: $(0, 1)$, $\{x: 1/x > 2\}$, $\{x \in \mathbb{Q}: x < 4\}$, $\{x \in \mathbb{Q}: x \leq 4\}$, $\{\frac{1}{2}, \frac{2}{3}, \frac{3}{4}, \ldots\}$, $\{\frac{1}{2}, \frac{1}{3}, \frac{1}{4}, \ldots\}$, $\{\frac{1}{2}, \frac{1}{4}, \frac{1}{8}, \frac{1}{16}, \ldots\}$,

   $\{\frac{m}{n}: m, n, \in \mathbb{N}, 0 < m < n$ and n is a power of 2$\}$.

2.   Let A be a non-empty subset of $\mathbb{R}$ such that $\forall x \in A$, $x < 2$. Prove that $\sup A \leq 2$. Give two examples of sets A satisfying this property, one with $\sup A = 2$, one with $\sup A < 2$.

3.   Find the supremum of each of the sets in Q1 and check that the properties demanded by Lemma 9.3 are satisfied.

4.   Recall that if r is an irrational number and q is a *non-zero* rational number, rq is irrational.   Use this to show that if x and y are real numbers and $x < y$ then there is an irrational number s with $x < s < y$. (Hint: Consider $x/\sqrt{2}$ and $y/\sqrt{2}$.)

5.   Show that for each $n \in \mathbb{N}$ there is a rational number a_n satisfying $\sqrt{2} - 1/n < a_n < \sqrt{2}$, and deduce that there is a sequence of rational numbers whose limit is $\sqrt{2}$.

6*   Modify the construction in Q5 to produce an *increasing* sequence (b_n) of rational numbers whose limit is $\sqrt{2}$.

7.   Let A be a non-empty subset of $\mathbb{R}$ which is bounded above, and define B and C by $B = \{x \in \mathbb{R}: x - 1 \in A\}$ and $C = \{x \in \mathbb{R}: (x + 1)/2 \in A\}$. Prove that $\sup B = 1 + \sup A$, $\sup C = 2\sup A - 1$.

8.   Let A be a non-empty set of *positive* numbers, which is bounded above, and set $B = \{x \in \mathbb{R}: 1/x \in A\}$.   Prove that B is bounded below and $\inf B = 1/\sup A$.

9.   Suppose that A and B are non-empty bounded subsets of $\mathbb{R}$.   Prove that
$$\sup (A \cup B) = \max(\sup A, \sup B),$$
$$\inf \{x + y: x \in A \text{ and } y \in B) = \inf A + \inf B,$$
$$\sup \{x - y: x \in A \text{ and } y \in B\} = \sup A - \inf B.$$

10.   Let A and B be non-empty, bounded sets of *positive* numbers and define C by $C = \{xy: x \in A$ and $y \in B\}$.   Prove that C is bounded and that $\sup C = \sup A \sup B$, $\inf C = \inf A \inf B$.

10

Continuity

10.1 Introduction

The idea of continuity of a function f is simple - we wish $f(x)$ to become close to $f(a)$ if x is close to a. Put a little more precisely, we want $f(x)$ to tend to $f(a)$ as x tends to a. This suggests that all we have to do is modify our definition of the limit of a sequence, where the variable is an integer, to cover the situation where the variable is a real number. Assuming we can do this, then we would expect to be able to modify the results of Chapter 7 fairly easily by changing the old terms into the new ones.

A function of a real variable, however, may tend to a limit as x tends to a number a in its domain, and it may do this for some values of a and not for others. That is, there are infinitely many points at which f may, or may not, behave conveniently. In the case of a sequence we had to worry only about the behaviour as $n \to \infty$. The results in this chapter will include a second group of properties obtained by observing that a function f tends to a limit at every point of its domain, that is, we have to collate the properties relating to the behaviour at each individual point of the domain.

Continuity is essentially the formalisation of the notion that a small change in the "input" x results in a small change in the "output" $f(x)$. This is a natural sort of assumption to make about real-life processes. Although continuity is not usually the final idea of interest in mathematical processes, it turns out that once we have a firm grasp of the treatment of continuity the other, apparently more complicated, ideas can be dealt with. In that sense continuity is the heart of analysis.

We have so far assumed that the idea of a function is understood. This may well be so, but we had better be clear and make a formal definition.

Definition

Let A and B be two sets. A **function** from A to B is a rule which associates with each member x of A a unique member $f(x)$ of B. A is called the **domain** of f and B the **codomain**. The notation $f: A \to B$ means that f is a function with domain A and codomain B.

Comments: The definition just given is the way one thinks of a function in practice. It is possible, and in courses on mathematical logic necessary, to make this more formal but we shall not require to do so. The main point is that for every $x \in A$ there is *one* "image" point, or **value**, $f(x)$. This prohibits defining a function by the rule $\pm\sqrt{x}$, for example, for we have not specified which of $\sqrt{x}$ or $-\sqrt{x}$ is the image of x. Notice that the definition associates a unique $f(x)$ with each x, but it does not demand that the images associated with distinct members x and y of A should be different. Notice that there is a difference between the function f and its value at a particular point, $f(x)$. For example, we may define the function $f: \mathbb{R} \to \mathbb{R}$ by the rule $f(x) = x^2$. In this case the function f is the rule, or the process of associating $f(x)$ with x , while $f(x)$ denotes the number x^2 , the value of f at the point x .

Definition The definition of a function may be abbreviated. For example we may specify a function by naming its domain and codomain and the rule for the process. We might consider the function $x \mapsto x^2$ from $\mathbb{R}$ to $\mathbb{R}$, that is the function with domain and codomain $\mathbb{R}$ which maps each x into x^2 . When the domain and codomain are understood we abbreviate this to the notation $x \mapsto x^2$.

10.2 The Limit of a Function of a Real Variable
We begin our discussion of continuity by adapting our definition of the limit of a sequence to functions depending on a real variable.

Definition Suppose that, for some $h > 0$, $(c - h, c) \cup (c, c + h)$ belongs to the domain of the function f. We say that $f(x) \to L$ as $x \to c$ if and only if

$$\forall \varepsilon > 0 \ \exists \delta > 0 \ \text{such that} \ \forall x \in \mathbb{R} \ 0 < |x - c| < \delta \Rightarrow |f(x) - L| < \varepsilon.$$

Notice a few points about this definition: as with the definition of the limit of a sequence, once we have decided on our 'tolerance' ε, so long as this is positive, there is a positive δ so that if x is near enough c ('near enough' being less than δ from c) then $f(x)$ differs from L by less than ε. We do not consider $x = c$, since we are interested in the behaviour of $f(x)$ as x approaches c; the value of $f(c)$ itself is not relevant - indeed $f(c)$ need not be defined. If $f(x) \to L$ as $x \to c$ we write $L = \lim_{x \to c} f(x)$.

Example 10.1 We establish the existence of the limit of a function in much the same way as for sequences. Let $f(x) = x^2$ for all $x \in \mathbb{R}$ so $f: \mathbb{R} \to \mathbb{R}$. Let $c \in \mathbb{R}$. We prove that $\lim_{x \to c} f(x) = c^2$.

Solution: Let $\varepsilon > 0$. Then let $\delta = \min(1, \varepsilon/(2|c| + 1)) > 0$. Then
$$\forall x \in \mathbb{R} \ |x - c| < \delta \Rightarrow |x^2 - c^2| = |x - c||x + c|$$
$$\leq \delta(|x| + |c|) < \delta(|c| + 1 + |c|) \leq \varepsilon$$
(where we have used the result
$$|x - c| < \delta \leq 1 \Rightarrow |x| \leq |x - c| + |c| < \delta + |c| \leq 1 + |c|) \ .$$
Since $\varepsilon > 0$ was arbitrary, the above is true for all $\varepsilon > 0$, hence
$$\forall \varepsilon > 0 \ \exists \delta > 0 \ \text{such that} \ \forall x \in \mathbb{R} \ |x - c| < \delta \Rightarrow |f(x) - f(c)| < \varepsilon . \ \square$$

It should now be natural to expect the following Lemma to be true, by analogy with what we know of sequences:

Lemma 10.1 Let $f(x) \to L$ and $g(x) \to M$ as $x \to c$. Then if λ is a constant,
(i) $f(x) + g(x) \to L + M$, (ii) $\lambda f(x) \to \lambda L$, (iii) $f(x)g(x) \to LM$, (iv) $|f(x)| \to |L|$
and, (v) provided $L \neq 0$ $1/f(x) \to 1/L$, all as $x \to c$.

Proof: We adapt the results of Theorems 7.1 and 7.2 to the new situation.

Preliminary thoughts

(i) Let $\varepsilon > 0$. Then if k is a constant there are δ_1 and δ_2, both positive, for which

$$0 < |x - c| < \delta_1 \Rightarrow |f(x) - L| < k\varepsilon$$

and

$$0 < |x - c| < \delta_2 \Rightarrow |g(x) - L| < k\varepsilon.$$

So if $|x - c| < \delta_1$ and $|x - c| < \delta_2$ then
$$|f(x) + g(x) - (L + M)|$$
$$\leq |f(x) - L| + |g(x) - M|$$
$$< k\varepsilon + k\varepsilon = 2k\varepsilon.$$
We should have chosen $k = \frac{1}{2}$.

Proof:

(i) Let $\varepsilon > 0$. Then since $\frac{1}{2}\varepsilon > 0$ and $f(x) \to L$ as $x \to c$,
$$\exists \delta_1 > 0 \text{ s.t. } 0 < |x - c| < \delta_1$$
$$\Rightarrow |f(x) - L| < \frac{1}{2}\varepsilon$$
and similarly
$$\exists \delta_2 > 0 \text{ s.t. } 0 < |x - c| < \delta_2$$
$$\Rightarrow |f(x) - L| < \frac{1}{2}\varepsilon.$$
Then $\exists \delta = \min(\delta_1, \delta_2) > 0$ such that
$$0 < |x - c| < \delta \Rightarrow |f(x) + g(x) - (L + M)|$$
$$< |f(x) \to L| + |g(x) \to M|$$
$$< \frac{1}{2}\varepsilon + \frac{1}{2}\varepsilon = \varepsilon.$$
Since $\varepsilon > 0$ was arbitrary we have shown that $f(x) + g(x) \to L + M$ as $x \to c$.

(ii) Let $\varepsilon > 0$ and k be a constant. Then $\exists \delta > 0$ s.t.
$$0 < |x - c| < \delta \Rightarrow |f(x) - L| < k\varepsilon.$$
Further
$$0 < |x - c| < \delta \Rightarrow |\lambda f(x) - \lambda L| < |\lambda| k\varepsilon.$$
We should have chosen $k|\lambda| \leq 1$.

(ii) Let $\varepsilon > 0$. Then if $\lambda \neq 0$, $\varepsilon/|\lambda| > 0$ so $\exists \delta > 0$ s.t.
$$0 < |x - c| < \delta \Rightarrow |f(x) - L| < \varepsilon/|\lambda|$$
$$\Rightarrow |\lambda f(x) - \lambda L| < \varepsilon.$$
Since $\varepsilon > 0$ was arbitrary this proves the case when $\lambda \neq 0$.
For $\lambda = 0$ $|\lambda f(x) - \lambda L| = 0 < \varepsilon$
so
$$\forall \varepsilon > 0 \; \exists \delta > 0 \text{ (e.g. } \delta = 1\text{) such that}$$
$$0 < |x - c| < \delta \Rightarrow |\lambda f(x) - \lambda L| = 0 < \varepsilon.$$

(iii) Let $\varepsilon > 0$ and k be constant. Then we can arrange δ_1 and δ_2 so that
$$0 < |x - c| < \delta_1 \Rightarrow |f(x) - L| < k\varepsilon$$
and $0 < |x - c| < \delta_2 \Rightarrow |g(x) - M| < k\varepsilon$.

Then if $0 < |x - c| < \delta_1$ and $0 < |x - c| < \delta_2$ we split $f(x)g(x) - LM$ into parts we know are "small", that is

(iii) Let $\varepsilon > 0$. Then $\varepsilon/(1 + |L| + |M|) > 0$ so $\exists \delta_1, \delta_2$ and δ_3, all positive, such that
$$0 < |x - c| < \delta_1 \Rightarrow |f(x) - L| < \varepsilon/(1 + |L| + |M|)$$
$$0 < |x - c| < \delta_2 \Rightarrow |g(x) - M| < \varepsilon/(1 + |L| + |M|)$$
$$0 < |x - c| < \delta_3 \Rightarrow |f(x) - L| < 1.$$

$f(x) - L$ and $g(x) - M$.

$|f(x)g(x) - LM|$ =

$\quad |f(x)(g(x) - M) + (f(x) - L)M|$

$\quad \leq |f(x)||g(x) - M| + |f(x) - L||M|$

$\quad \leq (|f(x)| + |M|)k\varepsilon$

Now $|f(x)|$ depends on x, so we notice that

$$|f(x)| \leq |f(x) - L| + |L| < \varepsilon + |L|$$

and if we ensure that $|f(x) - L| < 1$ (or any other known constant) we obtain

$|f(x)g(x) - LM| \quad < (\varepsilon + |L| + |M|)k\varepsilon$

$\qquad\qquad\qquad < (1 + |L| + |M|)k\varepsilon$

We should have chosen $k = 1(1 + |L| + |M|)$.

(iv) This one is as easy as with sequences because, by the triangle inequality in its alternative form

$$\left\| |f(x)| - |L| \right\| \leq |f(x) - L| \quad .$$

(v) If we arrange things so that $|f(x) - L| < k\varepsilon$ then

$$\left| \frac{1}{f(x)} - \frac{1}{L} \right| = \frac{|L - f(x)|}{|f(x)L|}$$

$$< \frac{k\varepsilon}{|f(x)||L|} \quad .$$

We need to deal with the $|f(x)|$ on the denominator, and it would be enough to know that $1/|f(x)|$ is less than some constant. To do this we need $|f(x)|$ greater than some positive constant. Now $|f(x)|$ can be forced to be close to $|L|$ (because by (iv) $|f(x)| \to |L|$ as $x \to c$) and hence greater than $|L| - |L|/2$ if x is close enough to c. (Here we use $|L|/2$ in place of ε.) Then if $|f(x)| > |L| - |L|/2 = |L|/2$, we have

Then if $\delta = \min(\delta_1, \delta_2, \delta_3)$, $\delta > 0$ and $0 < |x - c| < \delta \implies$

$\quad |f(x)g(x) - LM|$

$\quad\quad \leq |f(x)||g(x) - M| + |f(x) - L||M|$

$$\leq \frac{(|f(x)| + |M|)\varepsilon}{(1 + |L| + |M|)}$$

$$\leq \frac{(|f(x) - L| + |L| + |M|)\varepsilon}{(1 + |L| + |M|)}$$

$\quad\quad < \varepsilon .$

Since $\varepsilon > 0$ was general, this holds for all such ε and (iii) is proved.

(iv) Let $\varepsilon > 0$. Then since $f(x) \to L$ as $x \to c$, $\exists \delta > 0$ such that

$$0 < |x - c| < \delta \implies |f(x) - L| < \varepsilon .$$

Then

$0 < |x - c| < \delta \implies \left\| |f(x)| - |L| \right\|$

$$\leq |f(x) - L| < \varepsilon .$$

This proves (iv).

(v) Let $\varepsilon > 0$. Then since $f(x) \to L$ as $x \to c$, $\exists \delta_1 > 0$ such that

$0 < |x - c| < \delta_1 \implies |f(x) - L| < \frac{1}{2}|L|^2\varepsilon$.

(Notice that we use the fact that $L \neq 0$ to ensure $\frac{1}{2}|L|^2\varepsilon > 0$.)

Then since $|f(x)| \to |L|$ as $x \to c$ by (iv), $\exists \delta_2 > 0$ such that

$0 < |x - c| < \delta_2 \implies \left\| |f(x)| - |L| \right\| < \frac{1}{2}|L|$

$$\implies |L| - \frac{1}{2}|L| < |f(x)|$$

$$\implies \frac{1}{|f(x)|} < \frac{2}{|L|} .$$

Therefore, if $\delta = \min(\delta_1, \delta_2)$, $\delta > 0$ and $0 < |x - c| < \delta \implies$

$\frac{1}{|f(x)|} < \frac{2}{|L|}$ so that $\left|\frac{1}{f(x)} - \frac{1}{L}\right| < \frac{2}{|L|^2} k\varepsilon$.

We should have chosen $k = \frac{1}{2}|L|^2$.

$\left|\frac{1}{f(x)} - \frac{1}{L}\right| = \frac{|L - f(x)|}{|f(x)||L|}$

$< \frac{1}{2}|L|^2 \varepsilon \frac{2}{l^2} = \varepsilon$.

Since $\varepsilon > 0$ was arbitrary, this is true for all such ε and $\frac{1}{f(x)} \to \frac{1}{L}$ as $x \to c$. $\square$

There are two further results about sequences whose analogues we must consider. If we know that $h > 0$ and $\forall x \in (c - h, c) \cup (c, c + h)$ $f(x) \le g(x)$, that $f(x) \to L$ and that $g(x) \to M$ as $x \to c$ then we would expect that $L \le M$. This result is true, and left as a problem. As with sequences the additional information that $\forall x \in (c - h, c) \cup (c, c + h)$ $f(x) < g(x)$ does not, of itself, guarantee $L < M$.

Notation: If A and B are two sets, we denote by $A \backslash B$ the set of elements of A which do not belong to B , that is, $A \backslash B = \{x : x \in A \text{ and } x \notin B\}$.

Problem Suppose that for all $x \in (c - \delta_0, c + \delta_0) \backslash \{c\}$ (that is, all x satisfying $0 < |x - c| < \delta_0$) $f(x) \le g(x) \le h(x)$. Show that if $f(x) \to L$ and $h(x) \to L$ as $x \to c$, then $g(x) \to L$ as $x \to c$ also. This is sometimes called the "sandwich rule" or "squeeze rule", because if $f(x)$ and $h(x)$ are tending to the same limit L $g(x)$ is "squeezed" between them. $\square$

The remaining result on sequences is that which claims that if $a_n \to a$ as $n \to \infty$ then (a_n) is bounded. The corresponding result is not true of functions as the following loose argument suggests: if $a_n \to a$ as $n \to \infty$, then $a - 1 < a_n < a + 1$ for all n 'near ∞ ', which turns out (when made precise) to be for all but finitely many n . We can then consider the maximum and minimum values of (a_n) for these finitely many values of n . If $f(x) \to L$ as $x \to c$ then $L - 1 < f(x) < L + 1$ for x 'near c ', that is, for $x \in (c - \delta, c) \cup (c, c + \delta)$, but, unlike the case of a sequence, this will usually leave infinitely many values of x for which we have no information about $f(x)$.   A function can have 'behaviour' at many points, and it is not hard to find one which tends to a limit as $x \to 1$ but which is nevertheless unbounded; consider $f : (0, \infty) \to \mathbb{R}$ given by $f(x) = 1/x$.

The existence of $\lim_{x \to c} f(x)$ is a "local" property of f , that is, it depends only on the values of $f(x)$ for x near c . To be precise, if $\lim_{x \to c} f(x) = L$ and if $\delta_0 > 0$ and $g(x) = f(x)$ for all $x \in (c - \delta_0, c) \cup (c, c + \delta_0)$, then $\lim_{x \to c} g(x) = L$ also, even if $f(x)$ and $g(x)$ differ for values of x further away from c .   To see this notice that if $\varepsilon > 0$, $\exists \delta > 0$ s.t. $0 < |x - c| < \delta \Rightarrow |f(x) - L| < \varepsilon$.   Then

$0 < |x - c| < \min(\delta, \delta_0) \Rightarrow f(x) = g(x)$ and $|g(x) - L| = |f(x) - L| < \varepsilon$.

The definition of limit we have given involves the behaviour of $f(x)$ as x

$x < 0$ and it is obvious that we should consider $x > 0$ and $x < 0$ separately. It is easily checked that $f(x)$ does not tend to a limit as $x \to 0$, since no matter how we choose $\delta > 0$, there are values of x satisfying $|x| < \delta$ for which $f(x) = +1$ and others for which $f(x) = -1$. We therefore define 'one-sided' limits.

Definition Let $f:(c, c + h) \to \mathbb{R}$ for some $h > 0$. Then we say $f(x)$ tends to L as x tends to c **from the right**, or $f(x) \to L$ as $x \to c+$, if and only if $\forall \varepsilon > 0$ $\exists \delta > 0$ such that $\forall x \in (c, c + \delta)$ $|f(x) - L| < \varepsilon$. When this holds we write $L = \lim_{x \to c+} f(x)$. If $f:(c - h, c) \to \mathbb{R}$ for some $h > 0$ then we define $\lim_{x \to c-} f(x)$, the limit as x tends to c **from the left**, in the analogous way.

Example 10.2 If $f(x) = x/|x|$ for $x \in \mathbb{R}\backslash\{0\}$, then
$$\lim_{x \to 0+} f(x) = 1 \quad \text{and} \quad \lim_{x \to 0-} f(x) = -1.$$

Solution: Let $\varepsilon > 0$. Then $\exists \delta > 0$ (e.g. $\delta = 1$) such that $\forall x \in (0, \delta)$ $|f(x) - 1| = 0 < \varepsilon$. Since $\varepsilon > 0$ was arbitrary we have proved that $f(x) \to 1$ as $x \to 0+$. The proof for the left-hand limit is similar. $\square$

Example 10.3 Let $f:(c - h, c) \cup (c, c + h) \to \mathbb{R}$, for some $h > 0$. Then $\lim_{x \to c} f(x)$ exists if and only if $\lim_{x \to c-} f(x)$ and $\lim_{x \to c+} f(x)$ both exist and are equal.

Solution: Suppose $\lim_{x \to c} f(x)$ exists and that its value is L. Then $\forall \varepsilon > 0$ $\exists \delta > 0$ such that $\forall x \in \mathbb{R}$ $0 < |x - c| < \delta \Rightarrow |f(x) - L| < \varepsilon$. Since $x \in (c, c + \delta) \Rightarrow$ $x \in \mathbb{R}$ and $0 < |x - c| < \delta$ it is easy to see that $\lim_{x \to c+} f(x) = L$ and similarly $\lim_{x \to c-} f(x) = L$.

Conversely, suppose that $\lim_{x \to c-} f(x)$ and $\lim_{x \to c+} f(x)$ both exist and have the common value L. Let $\varepsilon > 0$. Then
$$\exists \delta_1 > 0 \text{ such that } \forall x \in (c - \delta_1, c) \, |f(x) - L| < \varepsilon$$
and $\quad\quad\quad \exists \delta_2 > 0$ such that $\forall x \in (c, c + \delta_2)$ $|f(x) - L| < \varepsilon$
Therefore, letting $\delta = \min(\delta_1, \delta_2)$, $\delta > 0$ and
$$\forall x \in \mathbb{R} \,\, 0 < |x - c| < \delta \Rightarrow x \in (c - \delta_1, c) \text{ or } x \in (c, c + \delta_2),$$
and in either case $|f(x) - L| < \varepsilon$. Thus
$$\exists \delta > 0 \text{ such that } \forall x \in \mathbb{R} \,\, 0 < |x - c| < \delta \Rightarrow |f(x) - L| < \varepsilon.$$
Since $\varepsilon > 0$ was arbitrary we have proved that $\lim_{x \to c} f(x) = L$. $\square$

An easy modification to the proof shows that Lemma 10.1 remains true if left-hand limits are substituted throughout for 'two-sided' limits or if right-hand limits are substituted throughout.   The one result we gain is the analogue of the result guaranteeing the existence of a limit of an increasing sequence which is bounded above.

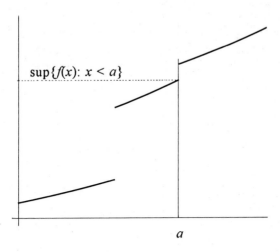

Fig. 10.1

Example 10.4 Suppose $f: \mathbb{R} \to \mathbb{R}$ is **increasing**, that is, $x \leq y \Rightarrow f(x) \leq f(y)$. Then for all $a \in \mathbb{R}$, $\lim_{x \to a-} f(x)$ exists and equals $\sup\{f(x): x < a\}$.   (See Fig. 10.1.)

Solution: Since f is increasing, $f(a)$ is an upper bound for $\{f(x): x < a\}$ so this set has a supremum; call the supremum α .

Let $\varepsilon > 0$.   Then since $\alpha - \varepsilon < \alpha$, from the properties of supremum $\exists x_0 < a$ for which $f(x_0) > \alpha - \varepsilon$.   Let $\delta = a - x_0$, so $\delta > 0$ and $x_0 = a - \delta$, and, using the increasing property of f, $\forall x \in (a - \delta, a)$ $\alpha - \varepsilon < f(x_0) \leq f(x) \leq \alpha$, whence $|f(x) - \alpha| < \varepsilon$.   Since $\varepsilon > 0$ was arbitrary this shows us that $\forall \varepsilon > 0$ $\exists \delta > 0$ such that $\forall x \in (a - \delta, a)$ $|f(x) - \alpha| < \varepsilon$, that is, that $\lim_{x \to a-} f(x) = \alpha$. $\square$

Problem Modify the proof of Example 10.4 to show that if $f: \mathbb{R} \to \mathbb{R}$ is increasing and bounded above, then $f(x) \to \alpha$ as $x \to \infty$ where $\alpha = \sup\{f(x): x \in \mathbb{R}\}$, and where by $f(x) \to \alpha$ as $x \to \infty$ we mean that $\forall \varepsilon > 0$ $\exists X$ such that $\forall x \geq X$ $|f(x) - \alpha| < \varepsilon$. $\square$

The fact that $f(x) \to L$ as $x \to c+$ gives us information about $f(x)$ for x "close enough" to and greater than c.   In some cases we can use this to extract simpler information from the limit.   For example, if $L > 0$ then $f(x)$ can be made as close as we wish to the positive number L if x is close to and greater than c.

This forces $f(x)$ to have the same sign as L for x close enough to c. Example 10.5 puts this in cold, precise terms.

Example 10.5 Let $f: (c, c + h) \to \mathbb{R}$ and $f(x) \to L$ as $x \to c+$. Then if $L > 0$, $\exists \delta > 0$ such that $c < x < c + \delta \Rightarrow f(x) > 0$.

Comment This is another example where we need to make a suitable choice of the ε which appears in the definition of limit. From the definition we see that we can choose $\delta > 0$ such that $c < x < c + \delta \Rightarrow L - \varepsilon < f(x) < L + \varepsilon$. It would be enough to choose ε so that $L - \varepsilon \geq 0$, and in this case $\varepsilon = L$ will do.

Solution: Let $\varepsilon = L$; since $L > 0$ this ensures that $\varepsilon > 0$. Then since $f(x) \to L$ as $x \to c+$, $\exists \delta > 0$ such that

$$c < x < c + \delta \Rightarrow L - \varepsilon < f(x) < L + \varepsilon$$
$$\Rightarrow 0 < f(x). \quad \square$$

The results corresponding to Example 10.5 for limits as $x \to c$ from the left, or as $x \to c$ as a two-sided limit are also true, and their proofs nearly identical to the one given. If $L < 0$ then $f(x) < 0$ for x "near" c in the appropriate sense. Notice, however, that if $L = 0$ then the proof above does not yield information (for knowing that $f(x) \in (0 - \varepsilon, 0 + \varepsilon)$ does not tell us the sign of $f(x)$). Problem 5 gives an example where $f(x)$ changes sign infinitely often near $x = 0$, yet $f(x) \to 0$ as $x \to 0$.

10.3 Continuity

The most important connection in which we use limits is where $\lim_{x \to a} f(x) = f(a)$.

A function with this property at all points a is one whose graph can be drawn without "jumps" and "without taking the pencil off the paper". The property is so common we give it a name.

Definition Let f be a function whose domain includes $(c - h, c + h)$ for some $h > 0$. Then we say f is **continuous at** c if and only if $\lim_{x \to c} f(x) = f(c)$. A

function is said to be **continuous** if and only if it is continuous at each point of its domain. This is only a matter of applying the definition just given if the domain is a set of one of the forms $\mathbb{R}$, $(-\infty, b)$, (a, ∞) or (a, b), but if $f:[a, b] \to \mathbb{R}$ we have to make a special case at the two endpoints (where, for all $h > 0$, f is not, for example, defined on $(a - h, a + h)$). A function $f:[a, b] \to \mathbb{R}$ is said to be **continuous** if and only if it is continuous at each point of (a, b) and $\lim_{x \to a+} f(x) = f(a)$ and $\lim_{x \to b-} f(x) = f(b)$.

We shall use continuity so much that it is worth while to restate it in terms of ε and δ: f is continuous at c if and only if

$$\forall \varepsilon > 0 \ \exists \delta > 0 \text{ such that } \forall x \in (c - \delta, c + \delta) \ |f(x) - f(c)| < \varepsilon$$

Notice that we do not need to exclude the case $x = c$ since the limit in this case is

$f(c)$ and $|f(c) - f(c)| = 0 < \varepsilon$. (See Fig. 10.2.) The continuity conditions at a and b in the case of $f:[a, b] \rightarrow \mathbb{R}$ are called **continuity on the right and left** respectively.

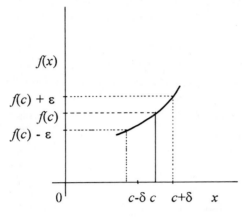

Fig. 10.2

Theorem 10.2 Suppose that f and g are continuous at c. Then if λ is a constant, the functions $\lambda f, f + g, fg$ and $|f|$ are continuous at c. If $f(c) \neq 0$ then $1/f$ is continuous at c. The analogous results follow for continuity on the left or the right.

Proof: Apply Lemma 10.1. □

There is, however, one operation on functions which has not arisen before, that of composition.

Definition Let $f: B \rightarrow C$ and $g: A \rightarrow B$ be two functions. The **composition** of f and g, $f \circ g$, is the function $f \circ g : A \rightarrow C$ defined by $f \circ g(x) = f(g(x))$.

Lemma 10.3 Let $f: B \rightarrow C$ and $g: A \rightarrow B$ be functions, where A, B and C are subsets of $\mathbb{R}$. If g is continuous at a and f is continuous at $b = g(a)$, then $f \circ g$ is continuous at a. If f and g are continuous, so is $f \circ g$.

Thoughts:
We need to ensure that for the appropriate x we have $|f(g(x)) - f(g(a))| < \varepsilon$.
Now $|f(y) - f(g(a))| < \varepsilon$ if y is close enough to $g(a)$, because f is continuous at $b = g(a)$. Then we have to fix it so that $g(x)$ satisfies the condition we have just required of y, that it be "close enough" to $g(a)$. By the continuity of g at a this will hold if x is close enough to a.

Proof:
Let $\varepsilon > 0$. Since f is continuous at $b = g(a)$, $\exists \delta_1 > 0$ such that
$$|y - b| < \delta_1 \Rightarrow |f(y) - f(b)| < \varepsilon$$
For this value of ε, since $\delta_1 > 0$ and g is continuous at a, $\exists \delta_2 > 0$ such that
$$|x - a| < \delta_2 \Rightarrow |g(x) - g(a)| < \delta_1.$$
Then

$$|x - a| < \delta_2 \quad \Rightarrow |g(x) - b| < \delta_1$$
$$\Rightarrow |f(g(x)) - f(b)|$$
$$= |f \circ g(x) - f \circ g(a)| < \varepsilon$$

Since ε was arbitrary we have shown that $f \circ g$ is continuous at a. $\square$

The results about continuous functions allow us to deduce immediately that many standard functions are continuous. We can see immediately that a constant function and the function $g: \mathbb{R} \to \mathbb{R}$ defined by $g(x) = x$ are continuous, so since any polynomial can be obtained from these by a finite number of operations of adding, multiplying by a constant or multiplying two functions, we see that polynomials are continuous by Theorem 10.2. Any function which is the quotient of two polynomials is continuous at all points where the denominator is non-zero.

The importance of continuous functions is rather more substantial than we might at this point expect, in that we can deduce 'global' properties of these functions which derive from the fact that they are continuous at all points of their domain. These results have an additional level of sophistication to the proof in which we bring the underlying properties of the real number system into play.

Theorem 10.4 The Intermediate Value Theorem Suppose that $f: [a, b] \to \mathbb{R}$ is continuous and that γ lies between $f(a)$ and $f(b)$, in the sense that either $f(a) \leq \gamma \leq f(b)$ or $f(a) \geq \gamma \geq f(b)$. Then there is a number $\xi \in [a, b]$ for which $f(\xi) = \gamma$.

Remarks: This result is 'obvious' in the sense that it is easily believed, so easily that one can scarcely conceive that it could be false. The fact that one cannot imagine how it could be wrong is, of course, not a proof. The interesting point is the way in which the proof has to use the properties of $\mathbb{R}$.

Proof: We can reduce this to a more particular case by noticing that if f is continuous, so is g, where $g(x) = f(x) - \gamma$, and then 0 lies between $g(a)$ and $g(b)$. By this device we need only prove the case when $\gamma = 0$. By considering $-g$ in place of g we could deduce the result for the case $g(a) \geq 0 \geq g(b)$ from that where $g(a) \leq 0 \leq g(b)$. Finally, if either $g(a)$ or $g(b)$ is zero, there is no trouble in finding a ξ for which $g(\xi) = 0$. We may therefore deduce all we wish if we can show that $f(a) < 0 < f(b)$ implies the existence of $\xi \in [a, b]$ for which $f(\xi) = 0$.

The way forward is to identify a candidate for ξ; from Fig.10.3 we choose $\xi = \sup S$, where $S = \{x \in [a, b]: f(x) < 0\}$. The set S is non-empty (it contains a) and bounded above (by b) so ξ exists and $a \leq \xi \leq b$. We have to show $f(\xi) = 0$.

Suppose $f(\xi) > 0$. Then, since f is continuous at ξ, setting $\varepsilon = f(\xi)$,

$$\exists \delta > 0 \text{ such that } \forall x \in (\xi - \delta, \xi + \delta) \; |f(x) - f(\xi)| < f(\xi).$$

Since $|f(x) - f(\xi)| < f(\xi) \Rightarrow f(\xi) - f(\xi) < f(x) < f(\xi) + f(\xi)$ we deduce that $\forall x \in (\xi - \delta, \xi + \delta) \; f(x) > 0$. Since $\xi - \delta < \sup S$, $\exists y > \xi - \delta$ such that $y \in S$,

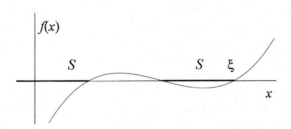

Fig 10.3

that is, such that $f(y) < 0$.   But this is a contradiction since $y \in (\xi - \delta, \xi + \delta) \Rightarrow$ $f(y) > 0$.   From this contradiction it follows that $f(\xi) \leq 0$.

Now suppose $f(\xi) < 0$.   Then by the continuity of f at ξ, and since $-f(\xi) > 0$, $\exists \delta > 0$ such that $\forall x \in (\xi - \delta, \xi + \delta)$ $|f(x) - f(\xi)| < -f(\xi)$ whence $\forall x \in (\xi - \delta, \xi + \delta)$ $f(x) < f(\xi) - f(\xi) = 0$.   Therefore $f(\xi + \frac{1}{2}\delta) < 0$ so $\xi + \frac{1}{2}\delta \in S$ and $\xi + \frac{1}{2}\delta > \sup S$, a contradiction.   This proves that $f(\xi) \geq 0$.

We are left only with the possibility $f(\xi) = 0$, which must be true. □

Example 10.6 This result allows us to show with virtually no effort that nth roots exist.   Let a be positive and $n \in \mathbb{N}$.   The function $f: [0, \infty) \to \mathbb{R}$ defined by $f(x) = x^n$ is continuous, and $f(0) < a$.   If $a \leq 1$ then $f(0) \leq a \leq f(1)$, while if $a > 1$ then $a^n \geq a$ so $f(0) \leq a \leq f(a)$.   Applying the Intermediate Value Theorem we see that there is a ξ for which $\xi^n = a$. □

The Intermediate Value Theorem may also be deployed to give information on the existence of solutions of more complicated equations.   One could show the existence of a solution to the equation $x^2 + 7x = x^3 + 1$ by showing that the (continuous) function f given by $f(x) = x^3 - x^2 - 7x + 1$ attains a positive value at $x = 0$ and a negative one at $x = -3$, and is thus zero at some intermediate point. There are obviously some practical matters to be attended to before this can be made greatly useful.

Definition Let $f: A \to \mathbb{R}$. f is said to be **bounded** if the set $\{f(x): x \in A\}$ is a bounded set.   For $S \subset A$, we say f is **bounded on** S if $\{f(x): x \in S\}$ is bounded.   The analogous definitions of **bounded above** and **bounded below** are also made.

Thus $f: \mathbb{R} \to \mathbb{R}$ defined by $f(x) = 1/(1 + x^2)$ is bounded, but $g: (0, 1) \to \mathbb{R}$ given by $g(x) = 1/x$ is not.

Experience suggests that if $f: [a, b] \to \mathbb{R}$ is continuous then f is bounded. This we shall prove shortly, but as a preliminary, let us isolate the difficulties.   For each $c \in [a, b]$, since f is continuous at c, $\exists \delta_c > 0$ such that $\forall x \in (c - \delta_c, c + \delta_c) \cap [a, b]$ $f(c) - 1 < f(x) < f(c) + 1$.   (We adopt this formulation to avoid exceptions at the end points.) Thus the function obtained by restricting f to the domain $(c - \delta_c, c + \delta_c) \cap [a, b]$ is bounded. One natural approach would be to choose a finite number of points $c_1, ..., c_n$ in such a way that every point of $[a, b]$ lies in one of the corresponding intervals $(c - \delta_c, c + \delta_c)$, where c equals one of the c_i. It is here that care is needed; δ_c may depend on c. Once we know the point c and that f is continuous at c, continuity guarantees the existence of $\delta_c > 0$, but the suitable values of δ may not be the same at another point c'. This prevents the simple approach that would be possible if δ_c were a constant, when it would be easy to find a finite number of intervals of fixed length covering $[a, b]$. The information we have at the moment does not exclude the possibility that for each $c \in [a, b]$ δ_c is half the distance between c and b, so that we seem to need infinitely many intervals to cover the whole of $[a, b]$. If we require infinitely many subintervals, the upper bounds for each of the various the various subintervals need not automatically form a bounded set.

Theorem 10.5 Let $f: [a, b] \to \mathbb{R}$ be continuous. Then f is bounded.

Proof: We introduce the set $S = \{x \in [a, b]: f$ is bounded on $[a, x]\}$. Our problem is to show that $b \in S$. S is bounded above (by b) and non-empty (it contains a) so S has a supremum. Let $\xi = \sup S$.

Suppose $\xi < b$. Then since f is continuous at ξ,
$$\exists \delta > 0 \text{ such that } \forall x \in (\xi - \delta, \xi + \delta) \cap [a, b] \ f(\xi) - 1 < f(x) < f(\xi) + 1. \tag{1}$$
By reducing δ if necessary we may presume that $\xi + \delta \le b$. Moreover, since $\xi - \delta < \sup S$, $\exists x \in S$ with $\xi - \delta < x \le \xi$. Then f is bounded on $[a, x]$ so, for some constants m, M, we have
$$\forall y \in [a, x] \ m \le f(y) \le M. \tag{2}$$
Hence, by (1) and (2),
$$\forall y \in [a, \xi + \tfrac{1}{2}\delta] \ \min(m, f(\xi) - 1) \le f(y) \le \max(M, f(\xi) + 1)$$
(since $a \le y \le \xi + \tfrac{1}{2}\delta \Rightarrow a \le y \le x$ or $\xi - \delta < y < \xi + \delta$). Thus f is bounded on $[a, \xi + \tfrac{1}{2}\delta]$ and therefore $\xi + \tfrac{1}{2}\delta \in S$. But $\xi + \tfrac{1}{2}\delta > \sup S$. This contradiction shows $\xi \not< b$.

Therefore $\xi = b$ (since b is an upper bound for S, $\xi \le b$). By the continuity of f on the left at b, $\exists \delta > 0$ such that
$$\forall y \in (b - \delta, b] \ f(b) - 1 < f(y) < f(b) + 1.$$
Moreover since $b - \delta < \sup S$ $\exists x \in S$ with $b - \delta < x$. Thus $\exists m, M$ such that $\forall y \in [a, x] \ m \le f(y) \le M$. Combining these pieces of information as in the last paragraph, we see that $\forall y \in [a, b] \ \min(m, f(b) - 1) \le f(y) \le \max(M, f(b) + 1)$, that is, f is bounded on $[a, b]$. $\square$

Remarks: It is worth considering this proof carefully in the light of examples. If we allow f to be defined on an open interval (a, b) the result is false (e.g. $f(x) = 1/x$ on $(0, 1)$), or on an infinite interval (e.g. $f(x) = x$ on $[0, \infty)$).

Theorem 10.6 Let $f: [a, b] \to \mathbb{R}$ be continuous. Then f attains its supremum and infimum, that is, there are numbers ξ_1 and ξ_2 in $[a, b]$ for which $f(\xi_1) =$ $\sup\{f(x): a \le x \le b\}$ and $f(\xi_2) = \inf\{f(x): a \le x \le b\}$.

Proof: Let $M = \sup\{f(x): a \le x \le b\}$. (By Theorem 10.5 we know M exists.) Suppose that $\forall x \in [a, b]\ f(x) \ne M$ so that $\forall x \in [a, b]\ M - f(x) > 0$. Then set $g(x) = 1/(M - f(x))$ so $g: [a, b] \to \mathbb{R}$ and g is continuous. By Theorem 10.5, then, g is bounded, so $\exists M_1$ such that $\forall x \in [a, b]\ g(x) \le M_1$. Then $\forall x \in [a, b]$ $M - f(x) \ge 1/M_1$ and $f(x) \le M - 1/M_1$. But this shows that $M - 1/M_1$ is an upper bound for f which is smaller than the supremum of f, a contradiction. Thus our assumption that $\forall x \in [a, b]\ f(x) \ne M$ was wrong, so $\exists \xi_1 \in [a, b]$ such that $f(\xi_1)$ $= M$. The other case is now clear. $\square$

Example 10.7 One useful conclusion from this is that if $f: [a, b] \to \mathbb{R}$ is continuous and all its values are positive, then its infimum is positive. Before showing this we ought to say why we bother with it: a set of positive numbers can have 0 as its infimum, for example $(0, 1)$ does, so it is conceivable that f could have lots of local minima in the interval $[a, b]$, all positive, but including some arbitrarily close to 0.

Solution: By the Theorem $f: [a, b] \to \mathbb{R}$ attains its infimum, so $\exists \xi \in [a, b]$ with $f(\xi) = \inf\{f(x): a \le x \le b\}$. Since all the values of f are positive, $f(\xi) > 0$. $\square$

Notice that our results may be useful even with functions continuous on sets other than closed intervals, provided we are careful. For example, if $f: (0, \infty) \to \mathbb{R}$ is continuous, and all its values are positive then $\{f(x): a \le x \le b\}$ has a positive infimum provided $0 < a < b < \infty$, since the restriction of f to $[a, b]$ is continuous.

In everyday life it is natural to presume that in whatever process we are concerned with, a small change in the input will produce a small change in the output. If instructions call for ingredients to be mixed in the proportions 2:1 we do not interpret this as demanding that there be exactly twice as many molecules of the one as the other; we approximate this. Again, common experience tells us that with some processes we need to be more accurate in our approximation than with others. Continuity is essentially the same idea: if x is close to a then $f(x)$ will be close to $f(a)$, the idea of how close is necessary to attain a certain accuracy being measured by ε and δ .

10.4 Inverse Functions
Having said that continuity is natural, we need to pursue two distinct approaches; to try to check that continuity is present where we would expect it and to be aware of where it is absent. We have shown that the sum, product and composition of two continuous functions are continuous, and that the quotient of two continuous functions is continuous except perhaps where the denominator is zero. These results are all of the general form that performing the appropriate operations on two continuous functions yields a result which is continuous. There are, however, other

ways in which we construct functions out of other functions, or example, the square root function is obtained from the simpler functions which maps x into x^2 by "undoing" the action of the latter, or, in more formal terms, the square root function is the "inverse" of the square function.   Many other standard functions of mathematics are inverses of simpler functions, so we must relate the operation of inverting to continuity. We start by defining our terms.

Definition Suppose that A and B are two sets and $f: A \to B$. Then we say that $g: B \to A$ is the **inverse function** to f if $\forall x \in A$ $g(f(x)) = x$ and $\forall y \in B$ $f(g(y)) = y$. If such an inverse function g exists, we denote it by f^{-1} (not to be confused with $1/f$).

The function $f: A \to B$ is said to be **injective** if distinct points in A have distinct images under f. In symbols this is $\forall x_1, x_2 \in A$ $x_1 \neq x_2 \Rightarrow f(x_1) \neq f(x_2)$ (or equivalently, $f(x_1) = f(x_2) \Rightarrow x_1 = x_2$). f is said to be **bijective** if it is injective and every point in B is the image of some point in A. (In symbols, the second condition is $\forall y \in B$ $\exists x \in A$ such that $y = f(x)$.) A bijective function $f: A \to B$ establishes a correspondence between the members of the sets A and B in that every $x \in A$ is mapped by f to one member $f(x) \in B$ and every member $y \in B$ is of the form $f(x)$ for exactly one $x \in A$.

Example 10.8 Let $f_1: [0, \infty) \to [0, \infty)$ be given by $f_1(x) = x^2$ and $g_1: [0, \infty) \to [0, \infty)$ be given by $g_1(x) = \sqrt{x}$. Then $g_1 = f_1^{-1}$. Notice also that f_1 is injective, while $f_2: \mathbb{R} \to [0, \infty)$ given by $f_2(x) = x^2$ is not injective because $f_2(1) = f_2(-1)$. Mark this example: the domain of the function is as important as the rule for transforming x to $f(x)$. $\square$

We are interested in conditions which guarantee the existence of an inverse function. We first notice that if f has an inverse (g say) then f is injective, for, in the notation of the definition, $f(x_1) = f(x_2) \Rightarrow g(f(x_1)) = g(f(x_2)) \Rightarrow x_1 = x_2$. The second point is that if $f: A \to B$ is to have an inverse then every point of B must be in the image of f (since $y = f(g(y))$ in the above notation). In practice, we often pay less attention to the codomain so we may be prepared to alter it. Let $f(A)$ denote the **image** of f, that is, the set $\{y \in B: \exists x \in A \text{ s.t. } y = f(x)\}$ (or, more briefly, $\{f(x): x \in A\}$) where $f: A \to B$. Then if f is injective, we see that for all $y \in f(A)$ there is an $x \in A$ satisfying $f(x) = y$ and, since f is injective, there is *only one* such x. Define $g(y)$ to be x. Then $g: f(A) \to A$ and it is easily checked that g is the inverse of f. We have proved the following:

Lemma 10.7 Let $f: A \to B$. Then the function $f: A \to f(A)$ has an inverse function $g: f(A) \to A$ if and only if f is injective. Notice that the domain of g need not be all of B. $\square$

From the point of view of analysis, we shall be interested in showing that the inverse of a continuous function, if it exists, is continuous. The first step is to notice that the only injective continuous functions are the obvious ones.

Definitions Let $A \subset \mathbb{R}$ and $f\colon A \to \mathbb{R}$. f is said to be **increasing** if $x < y \Rightarrow f(x) \le f(y)$ and **strictly increasing** if $x < y \Rightarrow f(x) < f(y)$. We say f is **decreasing (strictly decreasing)** if $x \le y \Rightarrow f(x) \ge f(y)$ $(f(x) > f(y))$.

Lemma 10.8 Let $f\colon [a, b] \to \mathbb{R}$ be continuous and injective. Then f is either strictly increasing or strictly decreasing.

Proof: Since f is injective, $f(a) \ne f(b)$ (neglecting the trivial case where $a = b$). Thus either $f(a) < f(b)$ or $f(a) > f(b)$. We shall tackle the former case; the other is nearly identical. Suppose, therefore, that $f(a) < f(b)$.

Let $a \le x < y \le b$. Then $f(a) \le f(x) < f(b)$. For, if not, either $f(x) < f(a)$ or $f(x) \ge f(b)$. In the first case, $f(x) < f(a) < f(b)$, so $f(a)$ is intermediate between $f(x)$ and $f(b)$ and, applying the Intermediate Value Theorem to f on the interval $[x, b]$, we see that $\exists \xi \in (x, b)$ such that $f(\xi) = f(a)$. But then $\xi \ne a$, which contradicts the injectivity of f. (Note $\xi \ne x$ or b since $f(\xi)$ does not equal $f(x)$ or $f(b)$.) Thus $f(x) \not< f(a)$. Similarly if $f(x) \ge f(b)$, we have $f(a) < f(b) \le f(x)$ so, applying the Intermediate Value Theorem to $[a, x]$ we see that $\exists \xi \in [a, x]$ with $f(\xi) = f(b)$. Since $\xi \le x < b$, this contradicts the injectivity of f.

Moreover, $f(x) < f(y)$, for f is continuous and injective on $[x, b]$, so if $f(y) \le f(x)$ we have $f(y) \le f(x) < f(b)$ (since we know $f(x) < f(b)$) whence applying the Intermediate Value Theorem to $[y, b]$ shows $\exists \xi \in [y, b]$ such that $f(\xi) = f(x)$. Since $\xi \ne x$ this is impossible. Thus $f(x) < f(y)$.

Since x and y were typical points of $[a, b]$ with $x < y$ we have proved that $a \le x < y \le b \Rightarrow f(x) < f(y)$, so f is strictly increasing on $[a, b]$. $\square$

Theorem 10.9 Let $f\colon [a, b] \to \mathbb{R}$ be continuous and injective. Then $f^{-1}\colon f([a, b]) \to [a, b]$ is continuous.

Proof: Since f is continuous and injective it is either strictly increasing or strictly decreasing. We shall prove the former case. Then, assuming $a \ne b$, $f(a) < f(b)$ and $x \in [a, b] \Rightarrow f(x) \in [f(a), f(b)]$. In fact, the Intermediate Value Theorem allows us to show that every number between $f(a)$ and $f(b)$ belongs to $f([a, b])$.

Let $f(a) \le y_1 < y_2 \le f(b)$, and let $x_i = f^{-1}(y_i)$ $(i = 1, 2)$. Since f is strictly increasing, $x_1 < x_2$ (since $x_1 \ge x_2 \Rightarrow f(x_1) \ge f(x_2) \Rightarrow y_1 \ge y_2$). Thus f^{-1} is strictly increasing.

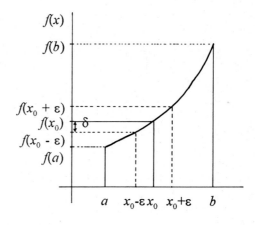

Fig. 10.4

Let $y_0 \in (f(a), f(b))$ and $f(x_0) = y_0$. Let $\varepsilon > 0$. We shall assume that ε is sufficiently small that $x_0 \pm \varepsilon \in [a, b]$; if this is not so, we choose a smaller ε for which it is true. Then, since f is strictly increasing, $f(x_0 - \varepsilon) < y_0 < f(x_0 + \varepsilon)$. Let δ be the minimum of $y_0 - f(x_0 - \varepsilon)$ and $f(x_0 + \varepsilon) - y_0$ Therefore $f(x_0 - \varepsilon) \le y_0 - \delta$ and $y_0 + \delta \le f(x_0 + \varepsilon)$. Then, since f^{-1} is strictly increasing,

$$y \in (y_0 - \delta, y_0 + \delta) \Rightarrow f(x_0 - \varepsilon) < y < f(x_0 + \varepsilon) \Rightarrow x_0 - \varepsilon < f^{-1}(y) < x_0 + \varepsilon$$

so $y \in (y_0 - \delta, y_0 + \delta) \Rightarrow |f^{-1}(y) - f^{-1}(y_0)| < \varepsilon$. Since $\varepsilon > 0$ was arbitrary we have shown f^{-1} is continuous at y_0, and since y_0 was arbitrary, f^{-1} is continuous on $(f(a), f(b))$.

Minor changes in the proof above show the continuity at the endpoints. □

Applying this result to the function $f: [0, b] \to [0, b^n]$ given by $f(x) = x^n$ (where n is a given natural number) shows that the nth root function is continuous on $[0, b^n]$. To show the root function is continuous on all of $[0, \infty)$ we need only notice that if $a \ge 0$ then it is possible to choose b so that $b^n \ge a$, hence f^{-1} is continuous on $[0, b^n]$ and, in particular, at a.

10.5 Some Discontinuous Functions

Partly as a warning, partly to see exactly how powerful the results we have proved are, we shall give some examples where continuity is absent.   These functions are pathological to our accustomed way of thinking and are not the sort of function we would expect to encounter.   However, if we produce functions by ingenious processes other than actually naming their values explicitly we need to ensure that such oddities are not present.   It is instructive to look at proofs of ordinary results to see how such peculiarities are excluded.

Example 10.9

1. *A function $f_1: \mathbb{R} \to \mathbb{R}$ which is nowhere continuous.*

Define f_1 by $f_1(x) = \begin{cases} 1 & \text{if } x \text{ is rational,} \\ 0 & \text{if } x \text{ is irrational.} \end{cases}$

To see that f_1 is discontinuous everywhere, let $c \in \mathbb{R}$ and $\varepsilon = 1$. Then for all $\delta > 0$, $(c - \delta, c + \delta)$ contains both rational and irrational numbers, and in particular an x of the opposite type to c; for this x, $|f_1(x) - f_1(c)| = 1 \ge \varepsilon$ Thus

$$\exists \varepsilon > 0 \text{ such that } \forall \delta > 0 \ \exists x \in (c - \delta, c + \delta) \text{ such that } |f_1(x) - f_1(c)| \ge \varepsilon$$

that is, f is discontinuous at c.

2. *A function $f_2: \mathbb{R} \to \mathbb{R}$ which is continuous at infinitely many points but between every two points of continuity there is a point of discontinuity and vice versa.*

Let $f_2(x) = 0$ if x is irrational

$= \dfrac{1}{q}$ if $x = \dfrac{p}{q}$ where p and q are integers with no common

factor greater than 1 and $q > 0$.

Let c be rational, where $f_2(c) = 1/q$. Let $\varepsilon = 1/q$ (>0). Then $\forall \delta > 0$ $\exists x \in (c - \delta, c + \delta)$ (e.g. any irrational x) such that $f_2(x) = 0$, hence

$|f_2(x) - f_2(c)| = 1/q \geq \varepsilon$.   This shows f_2 is discontinuous at c and hence at all rational numbers.

Now suppose c is irrational; we prove f_2 is continuous at c. Let $\varepsilon > 0$. Then if $q \in \mathbb{N}$, $1/q \geq \varepsilon \Leftrightarrow q \leq 1/\varepsilon$. The interval $(c - 1, c + 1)$ contains finitely many rational numbers with a denominator at most $1/\varepsilon$, and since none of these numbers is c, because c is irrational, the nearest of these is a positive distance δ from c. Thus if $|x - c| < \delta$, either x is irrational or x is rational and of the form p/q (where p and q have no common factor) with $q > 1/\varepsilon$, and therefore $f(x) < \varepsilon$ (since $f(x) = 0$ or $1/q$) whence $|f_2(x) - f_2(x)| < \varepsilon$. Since ε was arbitrary, f_2 is continuous at c.

3. *A function* $f_3\colon (0, 1] \to \mathbb{R}$ *which is continuous and bounded but which does not tend to a limit as* $x \to 0+$.

In this example we shall jump ahead and presume the properties of the sine function. Since none of our theory will depend on this, there is no danger of circular arguments resulting.

Let $$f_3(x) = \sin(\tfrac{1}{x}) \quad (x \in (0, 1)).$$

Then (presuming $\sin y$ to be continuous in y) we see that f_3 is the composition of two continuous functions. Also $\forall x \in (0, 1]$ $|f_3(x)| \leq 1$ so f_3 is bounded.

Suppose that $f_3(x) \to L$ as $x \to 0+$. Let $\varepsilon > 0$. Then $\exists \delta > 0$ such that $\forall x \in (0, \delta)$ $|f_3(x) - L| < \varepsilon$. Now choose $n \in \mathbb{N}$ such that $1/n < 2\pi\delta$, which is certainly possible, and choose $x = 1/(2n\pi + \pi/2)$. Then $x \in (0, \delta)$ and $f_3(x) = \sin(2n\pi + \pi/2) = 1$, so $|1 - L| < \varepsilon$. Since $\varepsilon > 0$ was arbitrary we see that

$$\forall \varepsilon > 0 \quad |1 - L| < \varepsilon, \text{ so } |1 - L| \leq 0 \text{ and hence } L = 1.$$

Again let $\varepsilon > 0$ so $\exists \delta > 0$ such that $0 < x < \delta \Rightarrow |f_3(x) - L| < \varepsilon$. Choose $n \in \mathbb{N}$ such that $1/n < 2\pi\delta$ and let $x = 1/(2n\pi + 3\pi/2)$, so $0 < x < \delta$, $f_3(x) = -1$ and hence $|-1 - L| < \varepsilon$. Since this is true for all $\varepsilon > 0$ we deduce that $L = -1$. This is a contradiction, so our assumption that $f_3(x)$ tends to a limit as $x \to 0+$ is wrong. $\square$

Problems

1. Prove directly from the definition that $\lim\limits_{x \to a} x^3 = a^3$ and $\lim\limits_{x \to 0} (1 + x^2)/(2 - x)$

    $= 1/2$.

2. Suppose that $\forall x \in (-1, 1)$ $1 - |x| \leq f(x) \leq 1 + x^2$. Prove that $f(x) \to 1$ as $x \to 0$.

3. Using the fact that $\sqrt{x} - \sqrt{y} = \dfrac{x - y}{\sqrt{x} + \sqrt{y}}$ for $x, y > 0$, or otherwise, prove from the definition of limit, that if $a > 0$ $\lim\limits_{x \to a} \sqrt{x} = \sqrt{a}$.

4. Let f and h be two continuous functions on $(a - 1, a + 1)$ and suppose that $\forall x \in (a - 1, a + 1)$, $f(x) \leq g(x) \leq h(x)$. Show that if $f(a) = h(a)$ then g is

continuous at a. (Draw a diagram!)

5.   Suppose that K is a constant and that $\forall x \in \mathbb{R}$ $|f(x)| \leq K|x|$. Prove that f is continuous at 0. (First find $f(0)$.) Deduce that the function f given by $f(0) = 0$, $f(x) = x \sin(1/x)$ $(x \neq 0)$ is continuous at 0.

6.   Let $f: \mathbb{R} \to \mathbb{R}$ be continuous at a and $f(a) > 0$. Show that there is a positive δ such that $|x - a| < \delta \Rightarrow f(x) > 0$. (Draw a diagram to see how to choose ε.)

7.   We say that $f(x) \to L$ as $x \to \infty$ if and only if $\forall \varepsilon > 0$ $\exists X$ such that $\forall x \in \mathbb{R}$ $x \geq X \Rightarrow |f(x) - L| < \varepsilon$. Prove that $1/x \to 0$ as $x \to \infty$, and, working directly from the definition, show that if $f(x) = 1 + a_2/x + a_1/x^2 + a_0/x^3$ (a_0, a_1, a_2 being constants) then $f(x) \to 1$ as $x \to \infty$. Deduce that $\exists X_1$ such that $\forall x \geq X_1$ $f(x) > 0$ and $\exists R_1$ for which $\forall x \geq R_1$ $x^3 + a_2 x^2 + a_1 x + a_0 > 0$. By considering $g(x) = 1 - a_2/x + a_1/x^2 - a_0/x^3$ as $x \to \infty$, show that $\exists R_2$ such that $\forall x \geq R_2$ $x^3 - a_2 x^2 + a_1 x - a_0 > 0$ to deduce that $\forall y \leq -R_2$ $y^3 + a_2 y^2 + a_1 y + a_0 < 0$. Finally, use the Intermediate Value Theorem to prove that the equation $x^3 + a_2 x^2 + a_1 x + a_0 = 0$ has at least one real solution.

8.   Let $f: [0, 1] \to [0, 1]$ be continuous. By considering the function h given by $h(x) = f(x) - x$, or otherwise, deduce that there is a $\xi \in [0, 1]$ for which $f(\xi) = \xi$.

9.   Suppose that $g: \mathbb{R} \to \mathbb{R}$ is continuous and that $\forall x \in \mathbb{Q}$ $g(x) = 0$. Deduce that g is identically zero.

10. Let $f: [a, b] \to \mathbb{R}$ be continuous. Prove that $\{f(x): a \leq x \leq b\}$ is a closed, bounded interval.

11*. Suppose that $f: [0, \infty) \to \mathbb{R}$ is continuous and that $f(x) \to L$ as $x \to \infty$ (definition in Question 7). Show that f is bounded. (Hint: Use the limit to establish bounds for $\{f(x): x > X\}$, for a suitable X, then deal with $\{f(x): 0 \leq x \leq X\}$.)

12*. Let $f: \mathbb{R} \to \mathbb{R}$ be continuous and bounded, and define g by $g(x) = \sup\{f(x): y \leq x\}$. Prove that g is continuous.

13*. Let $f: \mathbb{R} \to \mathbb{R}$ be continuous and neither bounded above nor below. Show that $\{f(x): x \in \mathbb{R}\} = \mathbb{R}$.

11

Differentiation

"Just as from the known Algorithm, as I call it, of this calculus, which I call differential, all other differential equations can be solved by a common method...".
G. W. Leibniz (1684).

11.1 Basic Results

Curiously, although it was the need to put calculus on a rigorous foundation which gave rise to analysis, we shall find that we have already done the substantial work, and adapting the analysis we have set up to cope with the ideas of calculus is relatively easy. We shall use the work we have done to dispose rather quickly of the standard results of calculus, although the style of our analysis will cause us to approach some topics in a way which would not have seemed natural had we tackled them from a background purely of calculus.

The first thing we have to do is define the derivative of a function. In calculus this represents the rate of change of the function, that is, the rate at which $f(x)$ is changing per unit of change in x.

Definition Let f be a real-valued function whose domain includes $(c - h, c + h)$ for some $h > 0$. Then f is said to be **differentiable** at c if $\lim_{x \to c} \dfrac{f(x) - f(c)}{x - c}$ exists, and when it does we denote its value by $f'(c)$, and call $f'(c)$ the **derivative** of f at c. A function $f \colon S \to \mathbb{R}$ is said to be **differentiable** if it is differentiable at every point of S.

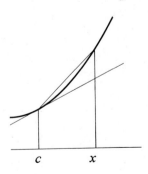

Fig. 11.1 $f'(c)$, being the limit of the gradient of the straight line from $(c, f(c))$ to $(x, f(x))$ as $x \to c$, is the gradient of the tangent to the curve $y = f(x)$ at c.

This definition does not allow us to consider the differentiability of f at a if $f: [a, b] \to \mathbb{R}$; if the need arises we demand that the obvious one-sided limit exists.

If $f: S \to \mathbb{R}$ is differentiable, then for all $c \in S$ we have defined $f'(c)$, so that $f': S \to \mathbb{R}$ is a new function also called the derivative of f. It can be important to distinguish between the function f' and its value at the point c, $f'(c)$, and this notation (Newton's) does this conveniently. The alternative notation (Leibniz's) of $\frac{df}{dx}$ is rather less well suited to making this distinction, though it has a number of other advantages, particularly that it suggests many correct results. If the function f' is differentiable at c we say f is **twice differentiable** there and denote the derivative of f' at c by f''; if f'' is differentiable we denote its derivative by f''' or $f^{(3)}$, and so on.

Comment If f were given by the formula $f(x) = Ax + B$ for constants A and B, then we would have $f'(c) = A$ for all c. In this case a change of h in the variable x (from c to $c + h$) produces a corresponding change of Ah in the value of f. That is, the rate at which $f(x)$ changes per unit change in x is A. The case of a more general function, where f' is not constant, is based on the idea that the rate at which $f(x)$ is changing relative to x is $f'(c)$ *at the point* c. As this is the rate of change at c, and to have a change we need x to differ from c, the idea has to be defined formally by a limit. From a formal point of view, we do not need to know this, but it can be useful in constructing proofs (not to mention motivating why we spend time on derivatives).

Theorem 11.1 If f is differentiable at c it is continuous at c.

Proof. Since $\qquad f(x) - f(c) = \dfrac{f(x) - f(c)}{x - c}(x - c)$, $\qquad$ by $\qquad$ Lemma 10.1

$\lim_{x \to c}(f(x) - f(c)) = f'(c).0 = 0$, i.e. $f(x) \to f(c)$ as $x \to c$. $\quad \square$

This result, though it may seem peripheral, is necessary in the standard results that follow:

Lemma 11.2 Suppose that f and g are differentiable at c and λ is a constant. Then $\lambda f, f + g, fg$ and (provided $g(c) \neq 0$), f/g are all differentiable at c, their derivatives being, respectively, $\lambda f'(c)$, $f'(c) + g'(c)$, $f'(c)g(c) + f(c)g'(c)$, and $\dfrac{f'(c)g(c) - f(c)g'(c)}{(g(c))^2}$.

Proof. These are routine uses of Lemma 10.1, but notice the need for Theorem 11.1 in establishing it:

$$\frac{f(x)g(x) - f(c)g(c)}{x - c} = \frac{f(x) - f(c)}{x - c}g(x) + \frac{g(x) - g(c)}{x - c}f(c) \to f'(c)g(c) + g'(c)f(c)$$

where the continuity of g is used in the first part. $\quad \square$

The consideration of differentiating $f{\circ}g$ leads to a problem whose solution stimulates an idea that turns out to have a wider usefulness. The natural impulse is to write $\{f(g(x)) - f(g(c))\}/\{x - c\}$ as the product of $\{f(g(x)) - f(g(c))\}/\{g(x) - g(c)\}$ and $\{g(x) - g(c)\}/\{x - c\}$, but this gives a meaningless expression if $g(x) = g(c)$ which may, of course, happen for values of x other than c. The way to avoid this difficulty is to notice that if f is differentiable at c then

$$f(x) - f(c) = (x - c)f'(c) + r(x)$$

where $r(x)/(x - c) \to 0$ as $x \to c$, and that, conversely, if there is a constant α such that

$$f(x) = f(c) + \alpha(x - c) + r(x) \text{ where } r(x)/(x - c) \to 0 \text{ as } x \to c$$

then f is differentiable at c and $f'(c) = \alpha$. In this sense, f is differentiable if it is of the form of a linear function, i.e. one of the form $Ax + B$ for constants A and B , plus a 'small' correction. That is, f is differentiable if its graph differs from a straight line graph through $(c, f(c))$ by a 'small' correction.

Lemma 11.3 Chain Rule. Let g be differentiable at a, and f be differentiable at $b = g(a)$. Then (assuming f and g are compatible so that $f{\circ}g$ exists) $f{\circ}g$ is differentiable at a and $(f{\circ}g)'(a) = f'(g(a)) g'(a)$.

Proof: Let $r_1(x) = g(x) - g(a) - (x - a) g'(a)$ and $r_2(y) = f(y) - f(b) - (y - b) f'(b)$. Then $r_1(x)/(x - a) \to 0$ as $x \to a$ and $r_2(y)/(y - b) \to 0$ as $y \to b$. Also

$$(f{\circ}g)(x) - (f{\circ}g)(a) = (g(x) - g(a))f'(b) + r_2(g(x))$$

$$= (x - a)g'(a)f'(b) + f'(b)r_1(x) + r_2(g(x))$$

so if we set $r_3(x) = f'(b)r_1(x) + r_2(g(x))$ it only remains to show that $r_3(x)/(x - a) \to 0$ as $x \to a$. Since $r_1(x)/(x - a) \to 0$ as $x \to a$ we need only show that $r_2(g(x))/(x - a) \to 0$.

Let $\varepsilon > 0$. Since $r_2(y)/(y - b) \to 0$ as $y \to b$, $\exists \delta_1 > 0$ such that

$$0 < |y - b| < \delta_1 \;\Rightarrow\; |r_2(y)|/|y - b| < \varepsilon/(|g'(a)| + 1)$$

$$\Rightarrow\; |r_2(y)| < |y - b|\varepsilon/(|g'(a)| + 1) .$$

Since $r_2(b) = 0$, we deduce that

$$|y - b| < \delta_1 \Rightarrow |r_2(y)| \le |y - b|\varepsilon/(|g'(a)| + 1) . \tag{1}$$

Now since g is differentiable, $\exists \delta_2 > 0$ such that

$$0 < |x - a| < \delta_2 \;\Rightarrow\; \left|\frac{g(x) - g(a)}{x - a} - g'(a)\right| < 1$$

$$\Rightarrow\; |g(x) - g(a)| < (|g'(a)| + 1)|x - a| .$$

Thus, if $\delta = \min(\delta_2, \delta_1/(|g'(a)| + 1))$,

$$0 < |x - a| < \delta \;\Rightarrow |g(x) - g(a)| < (|g'(a)|+1)\delta < \delta_1 \tag{2}$$

$$\Rightarrow |r_2(g(x))| \le |g(x) - g(a)|\varepsilon/(|g'(a)|+1) < \varepsilon|x - a| ,$$

because (2) shows us that we may substitute $g(x)$ for y in (1). Since $\varepsilon > 0$ was arbitrary we have shown $r_2(g(x))/(x - a) \to 0$ as $x \to a$. $\square$

It should be clear to the reader by now that the proof above was not obtained without a preliminary attempt, to see what the appropriate constants should be. Notice that inequality (1) allows $y = b$, so that when we substitute $y = g(x)$ we do not need $g(x)$ and $g(a)$ to be different.

The Lemma just proved is called the **chain rule**, and is more expressively put by writing in Leibniz's notation: $\frac{d}{dx}(f(g(x))) = \frac{df}{dg}\frac{dg}{dx}$. The Newtonian notation makes it a little more explicit where the various functions are to be evaluated. For the sake of efficiency we shall clear one more piece of general theory out of the way before proceeding.

Lemma 11.4 Let $f: [a, b] \to \mathbb{R}$ be differentiable and possess an inverse function $g: [c, d] \to [a, b]$. Then, if $f'(x_0) \neq 0$, g is differentiable at $y_0 = f(x_0)$ and $g'(y_0) = 1/f'(x_0)$.

Proof. We know that g is continuous by Theorem 10.8. Also if $y \neq y_0$

$$\frac{g(y)-g(y_0)}{y-y_0} = \frac{g(y)-g(y_0)}{f(g(y))-f(g(y_0))} .$$

Since f is differentiable and $f'(x_0) \neq 0$, $(x - x_0)/(f(x) - f(x_0)) \to 1/f'(x_0)$ as $x \to x_0$, and, noticing that f is bijective, so that if $x \neq x_0$, $f(x) \neq f(x_0)$,

$$\forall \varepsilon > 0 \; \exists \delta > 0 \text{ such that } 0 < |x - x_0| < \delta \Rightarrow \left| \frac{x - x_0}{f(x) - f(x_0)} - \frac{1}{f'(x_0)} \right| < \varepsilon .$$

Then, by the continuity and injectivity of g, $\exists \delta_1 > 0$ such that

$$0 < |y - y_0| < \delta_1 \Rightarrow 0 < |g(y) - g(y_0)| < \delta \Rightarrow \left| \frac{g(y) - g(y_0)}{y - y_0} - \frac{1}{f'(y_0)} \right| < \varepsilon$$

(substituting $g(y)$ for x above). Thus g is differentiable at y_0 and $g'(y_0) = 1/f'(x_0)$. $\square$

The main result above is not the value of $g'(x_0)$, which we could calculate by the chain rule, but the knowledge that it exists. We may use this to show that the function $g: [0, \infty) \to [0, \infty)$ given by $g(x) = x^{1/n}$ is differentiable and $g'(x) = \frac{1}{n}x^{\frac{1}{n}-1}$, since g is the inverse of f, where $f(x) = x^n$. (Fill in the details!)

As is familiar from calculus, the derivative yields information about the function, although care is occasionally needed to distinguish those situations where we require to know the derivative at a range of points from those where the knowledge is only required at a single point.

Example 11.1 Suppose that $f: [a, b] \to \mathbb{R}$ is continuous, and is differentiable at all points of (a, b). If $f(x_0)$ is the maximum value of $f(x)$ for $a \leq x \leq b$ then either x_0 is one of the endpoints or $f'(x_0) = 0$. (The corresponding result for the minimum is also true of course.)

Solution: Theorem 10.6 guarantees that there is an $x \in [a, b]$ for which $f(x_0)$ is the maximum of $\{f(x): a \leq x \leq b\}$. Suppose x_0 is *not* an endpoint so $a < x_0 < b$. Then if $x < x_0$, $(f(x) - f(x_0))/(x - x_0) \geq 0$ (numerator non-positive, since $f(x_0)$ is the maximum value). Thus $\lim_{x \to x_0-} (f(x) - f(x_0))/(x - x_0) \geq 0$ so that $f'(x_0) \geq 0$.

For $x > x_0$, we have $\lim_{x \to x_0+} (f(x) - f(x_0))/(x - x_0) \leq 0$ so that $f'(x_0) \geq 0$.

Combining these inequalities, we see that $f'(x_0) = 0$. $\square$

It is worth noticing that the maximum of $f(x)$ need not occur where $f'(x_0) = 0$, as is shown by considering $f(x) = x$ on $[0, 1]$. It is easy to overlook the possibility that the maximum may occur at an endpoint! In the solution to Example 11.1 we used the fact that x_0 is an interior point in evaluating both the left-hand and right-hand limits at x_0. If x_0 is an endpoint only one of the two inequalities can be deduced.

11.2 The Mean Value Theorem and its Friends

In deducing more information about a function from a knowledge of its derivatives we need some simple results which connect f with f' without involving limits explicitly. The principal theorem of this sort is the Mean Value Theorem, for which Rolle's Theorem is a preamble.

Theorem 11.5 Rolle's Theorem. Suppose that $a < b$, that $f : [a, b] \to \mathbb{R}$ is continuous, that f is differentiable at all points of $[a, b]$ and that $f(a) = f(b)$. Then there is a point $\xi \in (a, b)$ for which $f'(\xi) = 0$.

Proof: The idea is suggested by Fig. 11.2; ξ is the point at which f is maximum or minimum.

By Theorem 10.5 there are points x_0, x_1 at which f attains its minimum and maximum, so $\forall x \in [a, b]$ $f(x_0) \leq f(x) \leq f(x_1)$. This is the key to the proof. If either x_0 or x_1 is interior to $[a, b]$ then f' is zero there by Example 11.1 above, so we choose ξ to be x_0 or x_1, and $f'(\xi) = 0$. The only remaining possibility is that x_0 and x_1 are both endpoints of $[a, b]$. In this

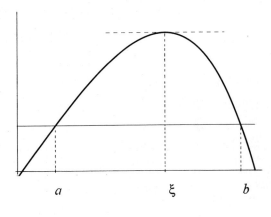

Fig. 11.2

case, $f(x_0) = f(x_1)$ (since $f(a) = f(b)$) and f is constant on $[a, b]$ whence $f'(x) = 0$ for $a < x < b$ and we may choose $\xi = (a + b)/2$. $\square$

Theorem 11.6 The Mean Value Theorem. Suppose that $a < b$, that $f : [a, b] \to \mathbb{R}$ is continuous and that f is differentiable at all points of (a, b). Then there is a point $\xi \in (a, b)$ for which $f'(\xi) = \dfrac{f(b) - f(a)}{b - a}$.

Remark: The number $(f(b) - f(a))/(b - a)$ is the average gradient of f between a and b, so the theorem states that at some point, f' takes its average (mean) value. (See Fig. 11.3.)

Proof: We reduce this to Rolle's Theorem (which is a special case). f satisfies all the hypotheses of Rolle's Theorem except that $f(a)$ and $f(b)$ may differ, so let $g(x) = f(x) + kx$ (for $a \le x \le b$) where k is a constant. Then if we choose $k = -(f(a) - f(b))/(b - a)$ we have $f(a) = f(b)$, and the resulting function g is continuous and differentiable as required, so Rolle's Theorem guarantees the existence of a point $\xi \in (a, b)$ for which $g'(\xi) = 0$. Since $g'(\xi) = f'(\xi) + k$ we see from this that $f'(\xi) = (f(a) - f(b))/(b - a)$. $\square$

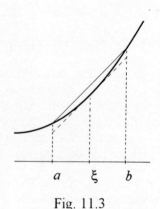

Fig. 11.3

Example 11.2 The Mean Value Theorem is the simplest method of relating properties of f to those of f' .   For example, if we know that for all $x \in (a, b) \, f'(x) > 0$ then it follows that f is strictly increasing on $[a, b]$.   To see this, choose $x, y \in [a, b]$ with $x < y$.   Then applying the Mean Value Theorem to f on the interval $[x, y]$ shows that, for some $\xi \in (x, y)$, $f(y) - f(x) = (y - x)f'(\xi) > 0$.   Thus $f(y) > f(x)$. This is a typical use of the Mean Value Theorem; some property of f' which is known to be true at all points of some interval is deployed to deduce information about f.   We need to know about $f'(x)$ for all x in some interval since we do not usually have more precise details of the location of the 'ξ' of the theorem.   Notice also that the 'a' and 'b' of the theorem may be chosen to be any two points in the interval on which the function is defined. $\square$

Less obviously, we can use the Mean Value Theorem's ideas to produce a very useful result about limits.   The limit, as $x \to a$, of $f(x)/g(x)$ is $f(a)/g(a)$ provided both functions are continuous and $g(a) \ne 0$.   It may happen, however, that $f(a)$ and $g(a)$ are both zero and the limit may exist even though our simple results on continuity do not show this; consider $\lim_{x \to 0} x/x$.   "L'Hôpital's Rule" (which was proved first by Bernoulli, not L'Hôpital) is a useful tool in this situation.   We first need to modify the Mean Value Theorem.

Lemma 11.7 Cauchy's Mean Value Theorem.   Let $a < b$, $f, g: [a, b] \to \mathbb{R}$ be two continuous functions, both differentiable at all points of (a, b) , and suppose that $\forall x \in (a, b) \, g'(x) = 0$.   Then there is a point $\xi \in (a, b)$ for which

$$\frac{f(b) - f(a)}{g(b) - g(a)} = \frac{f'(\xi)}{g'(\xi)} .$$

Proof: Let $\phi(x) = (g(b) - g(a))f(x) - (f(b) - f(a))g(x)$.   Then ϕ is continuous on $[a, b]$, differentiable on (a, b) and $\phi(a) = \phi(b)$ so by Rolle's Theorem there is a

$\xi \in (a, b)$ for which $\phi'(\xi) = 0$. Therefore

$$\phi'(\xi) = (g(b) - g(a))f'(\xi) - (f(b) - f(a))g'(\xi) = 0 . \tag{*}$$

Then, by assumption, $g'(\xi) \neq 0$ and $g(b) - g(a) \neq 0$ (since otherwise Rolle's Theorem applied to g would show $g'(x) = 0$ for some $x \in (a, b)$) so the result is obtained by dividing and rearranging the equation (*). □

Theorem 11.8 L'Hôpital's Rule. Suppose that $f, g: [a, b] \to \mathbb{R}$ are continuous on $[a, b]$ and differentiable on (a, b) and that $f(a) = g(a) = 0$. Then if $f'(x)/g'(x)$ tends to a limit as $x \to a+$ so does $f(x)/g(x)$ and

$$\lim_{x \to a+} \frac{f(x)}{g(x)} = \lim_{x \to a+} \frac{f'(x)}{g'(x)} .$$

The corresponding results for left-hand and two-sided limits also hold.

Proof. Suppose that $f'(x)/g'(x) \to L$ as $x \to a+$. Let $\varepsilon > 0$. Then

$$\exists \delta > 0 \text{ such that } 0 < x - a < \delta \Rightarrow |f'(x)/g'(x) - L| < \varepsilon . \tag{1}$$

Now choose $x \in (a, a + \delta)$ and apply the Cauchy Mean Value Theorem to $[a, x]$. There is a number ξ with $a < \xi < x$ such that

$$\frac{f'(\xi)}{g'(\xi)} = \frac{f(x) - f(a)}{g(x) - g(a)} = \frac{f(x)}{g(x)} .$$

Thus $\left| \dfrac{f(x)}{g(x)} - L \right| = \left| \dfrac{f'(\xi)}{g'(\xi)} - L \right| < \varepsilon$ by (1), since $0 < \xi - a < x - a \leq \delta$. Since x was a typical point of $(a, a + \delta)$ we have proved that $\forall x \in (a, a + \delta)$ $|f(x)/g(x) - L| < \varepsilon$ and, noticing now that $\varepsilon > 0$ was arbitrary, we have shown that $\lim_{x \to a+} f(x)/g(x) = L$. □

Example 11.3 Find $\lim_{x \to 0} \dfrac{\sqrt{(1 + x)} - x/2 - 1}{x^2}$.

Both numerator and denominator are differentiable and have the value 0 at $x = 0$, so by L'Hôpital's Rule, our limit exists if

$$\lim_{x \to 0} \frac{1/(2\sqrt{(1 + x)}) - 1/2}{2x} = \lim_{x \to 0} \frac{1/\sqrt{(1 + x)} - 1}{4x}$$

exists. This limit is equally indeterminate as it stands, so we apply L'Hôpital again and the second limit will exist if $\lim_{x \to 0} \dfrac{(-\frac{1}{2})(1 + x)^{-3/2}}{4}$ exists. This limit clearly does exist and has the value $-1/8$, so the two previous limits are shown to exist in turn and have the value $-1/8$. □

In the statements of the Mean Value Theorem and its near relatives, we assumed that $a < b$. This is not an issue if we allocate the labels a and b to two numbers of our choice since a may be chosen to be the smaller, but it may be a nuisance if a and b have predetermined meanings. Notice, however, that the result of the Mean Value Theorem does not rely on the order of a and b in that $(f(b) - f(a))/(b - a)$ is unaltered if we interchange the roles of a and b. For this reason we may relax our requirement that $a < b$ to one that $a \neq b$ provided we alter the statement about ξ to be that ξ lie strictly between a and b; the word 'strict' denotes that ξ is not allowed to be equal to either a or b. The same comment applies to Taylor's Theorem below, the last elaboration of the Mean Value Theorem.

Theorem 11.9 Taylor's Theorem. Suppose $a \neq b$, that f is defined on the closed interval defined by a and b and that f possesses derivatives of order up to n at all points of the interval. Then

$$f(b) = f(a) + (b-a)f'(a) + \frac{(b-a)^2}{2!}f''(a) + \cdots + \frac{(b-a)^{n-1}}{(n-1)!}f^{(n-1)}(a) + R_n$$

where, for some point ξ between a and b,

$$R_n = \frac{(b-a)^n}{n!}f^{(n)}(\xi) \ .$$

Proof: We shall apply the Mean Value Theorem to a suitable function, some care being needed to choose one which does not give the intermediate derivatives in the remainder term. Define $\phi: [a, b] \to \mathbb{R}$ by

$$\phi(x) = f(x) + (b-x)f'(x) + \cdots + \frac{(b-x)^{n-1}}{(n-1)!}f^{(n-1)}(x) \ .$$

ϕ is continuous and differentiable on $[a, b]$, since each part of it is, and $\phi'(x) = (b - x)^{n-1}f^{(n)}(x)/(n - 1)!$, so, to dispose of the $(b - \xi)^{n-1}$ term which would otherwise occur, we let $\psi(x) = (b - x)^n$ and apply Cauchy's Mean Value Theorem, yielding the existence of a point ξ for which

$$\frac{R_n}{-(b-a)^n} = \frac{\phi(b)-\phi(a)}{\psi(b)-\psi(a)} = \frac{\phi'(\xi)}{\psi'(\xi)} = \frac{(b-\xi)^{n-1}f^{(n)}(\xi)}{-(n-1)!n(b-\xi)^{n-1}} \ .$$

Hence $R_n = \dfrac{(b-a)^n}{n!}f^{(n)}(\xi)$, as promised. $\square$

The whole value of Taylor's Theorem is in the expression for the 'remainder' R_n. If we suppose f satisfies the hypotheses of the theorem, then if $a < x \leq b$ the restriction of f to $[a, x]$ is n times differentiable, and applying the theorem shows that $f(x) = f(a) + (x-a)f'(a) + \cdots + \dfrac{(x-a)^{n-1}}{(n-1)!}f^{(n-1)}(a) + R_n$ where, for some ξ_x between a and x, $R_n = (x - a)^n f^{(n)}(\xi_x)/n!$. Thus f may be approximated by the polynomial expression on the right, the error in this approximation being R_n. Whether this is useful or not depends on what we can say about R_n, of course. For 'ordinary' functions we have much choice as to the value of n which we select, and it is usual to choose n as small as possible for the purposes in hand; we shall see this in examples.

Had we used the ordinary Mean Value Theorem instead of Cauchy's the expression obtained for R_n would have been $(b - a)(b - \xi)^{n-1} f^{(n)}(\xi_x)/(n - 1)!$, which is not as useful as the form we obtained above.

Example 11.4 Suppose that f is n times differentiable in $[a - h, a + h]$, $h > 0$ and that $f'(a) = f''(a) = ... = f^{(n-1)}(a) = 0 \neq f^{(n)}(a)$. Then if $f^{(n)}(a)$ is continuous,

$$f \text{ has a local maximum at } a \text{ if } n \text{ is even and } f^{(n)}(a) < 0 ,$$

$$f \text{ has a local minimum at } a \text{ if } n \text{ is even and } f^{(n)}(a) > 0 ,$$

$$f \text{ has a point of inflexion at } a \text{ if } n \text{ is odd .}$$

Solution: Here we use the remainder of order n since $f^{(n)}(a)$ is the lowest order derivative which is non-zero at a.   Let $x \in [a - h, a + h]$, so by Taylor,

$$f(x) = f(a) + \frac{(x-a)^n}{n!} f^{(n)}(\xi) \text{ where } \xi \text{ is some point, depending on } x, \text{ between}$$

a and x . Since $f^{(n)}$ is continuous, there is an interval $(a - \delta, a + \delta)$ such that if ξ belongs to this interval, $f^{(n)}(\xi)$ has the same sign as $f^{(n)}(a)$. (Example 10.5). Therefore, for all $x \in [a - \delta, a + \delta]$, $f(x) - f(a) = (x - a)^n f^{(n)}(\xi)/n!$ has the same sign as $(x - a)^n f^{(n)}(a)$. If n is even, so that $(x - a)^n \geq 0$, $f(x) - f(a)$ has constant sign for $x \in (a - \delta, a + \delta)$ and if $f^{(n)}(a) > 0$ this shows $f(x) \geq f(a)$ so there is a local minimum at a, while $f^{(n)}(a) < 0$ indicates a local maximum. If n is odd, $(x - a)^n$ is positive for $a < x \leq a + \delta$ and negative for $-\delta \leq x < a$ so $f(x) - f(a)$ changes sign in the interval $(a - \delta, a + \delta)$, no matter how small δ is, and f has a point of inflexion at a . □

This example is typical of many uses of Taylor's Theorem in that we need to use the remainder of the correct order; the hypotheses tell us nothing directly about R_{n-1} or, even if f is $(n + 1)$-times differentiable, R_{n+1}.

We can use our knowledge of the behaviour of functions to simplify the treatment of sequences.   Earlier we considered sequences of the form $a_1 = \alpha$, $a_{n+1} = f(a_n)$ where f is a given function.   Some of our work there can be simplified. We first need a lemma:

Lemma 11.10 Suppose that f is continuous at a and that for each $n \in \mathbb{N}$ a_n belongs to the domain of f. Then if $a_n \to a$ as $n \to \infty$, $f(a_n) \to f(a)$ as $n \to \infty$.

Proof: Let $\varepsilon > 0$. Since f is continuous at a, $\exists \delta > 0$ such that $|x - a| < \delta \Rightarrow$ $|f(x) - f(a)| < \varepsilon$. Since $a_n \to a$ as $n \to \infty$, $\exists N$ such that $\forall n \geq N$ $|a_n - a| < \delta$. Therefore $\forall n \geq N$ $|f(a_n) - f(a)| < \varepsilon$.

Since $\varepsilon > 0$ was arbitrary, we have shown that $\forall \varepsilon > 0$ $\exists N$ such that $\forall n \geq N$ $|f(a_n) - f(a)| < \varepsilon$, as required. □

From this it follows that the limit, a, of the sequence defined by $a_{n+1} = f(a_n)$, if there is one, will satisfy $a = f(a)$.   Also, we may investigate the increasing or decreasing nature of (a_n) by noticing that $a_{n+1} - a_n = f(a_n) - f(a_{n-1}) = (a_n - a_{n-1}) f'(\xi_n)$ for some ξ_n between a_n and a_{n-1} , by the Mean Value Theorem.

Example 11.5 Let $a_1 = 1$, $a_1 = 1$, $a_{n+1} = \frac{1}{2}(a_n + \frac{2}{a_n})$. Discuss the behaviour of (a_n) as $n \to \infty$.

Solution: If we set $f(x) = \frac{1}{2}(x + \frac{2}{x})$ we see that f is differentiable on $(0, \infty)$ and $f'(x) = \frac{1}{2}(1 - \frac{2}{x^2})$.

It is obvious that $x > 0 \Rightarrow f(x) > 0$, so by induction, $a_n > 0$ for all n since $a_n > 0 \Rightarrow a_{n+1} = f(a_n) > 0$ and $a_1 > 0$. Sketching the graph of f shows that it has a minimum at $a = \sqrt{2}$ and, omitting the details, we notice that $x > 0 \Rightarrow f(x) \geq \sqrt{2}$. From this we see that $\forall n \geq 1 \ a_{n+1} = f(a_n) \geq \sqrt{2}$, thus $\forall n \geq 2 \ a_n \geq \sqrt{2}$. (See Fig. 11.4.)

Also $a_{n+1} - a_n = f'(\xi_n)(a_n - a_{n-1})$ so, since ξ_n lies between a_{n-1} and a_n, we see that $\forall n \geq 3 \ \xi_n \geq \sqrt{2}$ so $f'(\xi_n) \geq 0$ and $a_{n+1} - a_n$ has the same sign as $a_n - a_{n-1}$, hence the same sign as $a_3 - a_2$. Since $a_2 = 3/2$ and $a_3 = 17/12$, we see that after the second term, (a_n) decreases and is bounded below by 0, so there is an a such that $a_n \to a$ as $n \to \infty$. Since a satisfies $a = f(a)$, $a^2 = 2$ so $a_n \to \sqrt{2}$ as $n \to \infty$ (because $a_n \geq 0$ for all n hence $a \geq 0$).

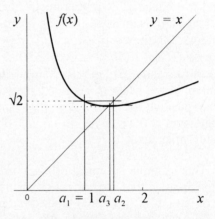

Fig. 11.4

This may also be seen by noticing from the graph that if $x \geq \sqrt{2}$, $f(x) \leq x$ hence $a_n > \sqrt{2} \Rightarrow a_{n+1} \leq a_n$. . Thus, plotting a_n on the x-axis, if we mark the intercept of the line $x = a_n$ with the graph of f and draw the horizontal line from this point $(a_n, f(a_n))$ to the line $y = x$ we obtain the point $(f(a_n), f(a_n))$, i.e. (a_{n+1}, a_{n+1}) . Dropping the vertical line to the x-axis shows the position of a_{n+1} . This method is very helpful but care must be taken to draw the graph sufficiently accurately; it is important to notice whether, say, a minimum lies to the left or the right of the intersection of the graph of f and the line $y = x$. This is easier to illustrate in the next example. □

Example 11.6 Let $a_1 = \alpha$, $a_{n+1} = \mu a_n(1 - a_n)$ where $0 \leq \alpha \leq 1$ and $0 < \mu \leq 2$. Then, letting $f(x) = \mu x(1 - x)$, we see that $f(x) = x$ if and only if $x = 0$ or $x = 1 - 1/\mu$, so the only possible limits are 0 and $1 - 1/\mu$. f has a maximum at $x = 1/2$, with $f(1/2) = \mu/4$. It is easy to show by induction that $\forall n \in \mathbb{N}$ $0 \leq a_n \leq 1$.
Case 1: $0 < \mu \leq 1$. Here the only solution of $x = f(x)$ which lies in $[0, 1]$ is 0, so this is the only potential limit; we prove $a_n \to 0$ as $n \to \infty$. Since $a_{n+1} - a_n = (\mu - 1)a_n - \mu a_n^2$, if $\mu \leq 1$ (a_n) is decreasing; it is also bounded below by 0 . Filling in the details, we see that $a_n \to 0$ (See Fig. 11.5.)

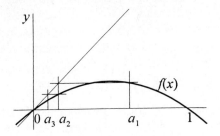

Fig. 11.5. $f(x) = \mu x(1-x)$ with $\mu = 0.7$; here $a_n \to 0$ as $n \to \infty$.

Case 2: $1 < \mu \le 2$. Since the maximum value of f occurs at $x = 1/2 > 1 - 1/\mu$, we see that $0 \le x \le 1 - 1/\mu \Rightarrow f'(x) \ge 0$ so f is increasing on $[0, 1 - 1/\mu]$ by Example 11.2. Therefore, $0 \le x \le 1 - 1/\mu \Rightarrow 0 = f(0) \le f(x) \le f(1 - 1/\mu) = 1 - 1/\mu$, so $a_n \in [0, 1 - 1/\mu] \Rightarrow a_{n+1} \in [0, 1 - 1/\mu]$. Thus if $0 \le \alpha \le 1 - 1/\mu$ then for all n $0 \le a_n \le 1 - 1/\mu$. Since $a_{n+1} - a_n = \mu a_n(1 - \frac{1}{\mu} - a_n)$ we deduce that for all

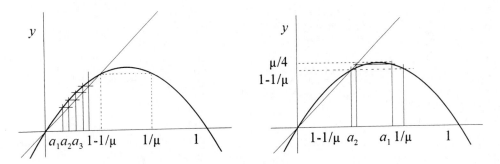

Fig. 11.6. $\mu = 1.5$. With $0 < a_1 < 1 - 1/\mu$, a_n increases and tends to $1 - 1/\mu$. If $1 - 1/\mu \le a_1 \le 1/\mu$, a_n decreases and tends to $1 - 1/\mu$.

$n \in \mathbb{N}$ $a_{n+1} - a_n \ge 0$, that is, (a_n) is increasing and bounded above; if $\alpha = 0$ then $a_n \to 0$ otherwise the limit cannot be 0 (it must be at least a_1 because the sequence is increasing) whence it is $1 - 1/\mu$, for 0 and $1 - 1/\mu$ are the only possible limits. If $1/\mu \le \alpha \le 1$ then $0 \le a_2 \le 1 - 1/\mu$, because f is decreasing on $[\frac{1}{2}, 1]$ so, by the above analysis, the sequence tends to 0 if $a_2 = 0$ (i.e. $\alpha = 1$) and to $1 - 1/\mu$ otherwise. This leaves the case where $1 - 1/\mu < \alpha < 1/\mu$, and Fig. 11.6 shows that in this event, $\forall n \in \mathbb{N}$, $1 - 1/\mu < a_n < 1/\mu$ and (a_n) decreases, because $f(x) \le x$ for $x \in (1 - 1/\mu, 1/\mu)$ and is bounded below. Here $a_n \to 1 - 1/\mu$ (since $a_n \to 0$ is impossible, and 0 and $1 - 1/\mu$ are the only two potential limits). The implication $1 - 1/\mu < x < 1/\mu \Rightarrow 1 - 1/\mu < f(x) < \mu/4 \le 1/\mu$ supplies the main part of the last argument. Readers should fill in all the details to convince themselves that everything above is valid. $\square$

Had we chosen $\mu > 2$ in the above example, we would have the maximum of f occurring at $x = 1/2$, between the points where $f(x) = x$. This means that $x \in (0, 1 - 1/\mu) \not\Rightarrow f(x) \in (0, 1 - 1/\mu)$ and (a_n) need not be increasing. Not all is lost, however, as we may use the Mean Value Theorem to show that the distance of a_n from its limit tends to zero. Let $a = 1 - 1/\mu$ so that $f(a) = a$. Then

$$|a_{n+1} - a| = |f(a_n) - f(a)| = |f'(\xi_n)||a_n - a| \qquad (*)$$

where ξ_n is between a_n and a . Provided we can show that for all points ξ in an interval containing all the a_n , $|f'(\xi)| \leq \gamma$ where γ is a constant less than 1, we will be able to obtain a useful result.

From Fig. 11.7, it is clear that $\quad 0 < a_n \leq 1 - 1/\mu \Rightarrow a_{n+1} \geq a_n$, whence we see that if $0 < \alpha \leq 1 - 1/\mu$, the first few terms of the sequence will increase in value. Since $f(x)$ has a maximum at $x = 1/2$ and this maximum value is $\mu/4$, it is also obvious that $\forall n \geq 2$ $a_n = f(a_n) \leq \mu/4$.

For definiteness, let us choose $\mu = 5/2$. Then let $a = 3/5$ so that $f(a) = a$. Since $f(1/2) = 5/8$ we see that $\forall n \geq 2$ $a_n \leq 5/8$. Moreover, $f'(x) \leq 0$ for $x \in [1/2, 5/8]$ so, by Problem 7, f is decreasing on the interval $[1/2, 5/8]$, and $f(5/8) = 75/128 > 1/2$ so we see that $1/2 \leq x \leq 5/8 \Rightarrow f(5/8) \leq f(x) \leq f(1/2) \Rightarrow 1/2 \leq f(x) \leq 5/8$. Thus $a_n \in [1/2, 5/8] \Rightarrow a_{n+1} \in [1/2, 5/8]$.

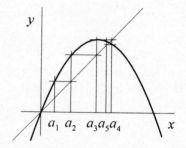

Fig. 11.7   The function $f(x) = \mu x(1 - x)$ for $\mu = 5/2$; in this case $1 - 1/\mu - 3/5$.

If $\alpha = 0$ or 1 then $\forall n \geq 2$ $a_n = 0$ and $a_n \to 0$ as $n \to \infty$. If $0 < a_n \leq 3/5$ then $a_{n+1} \geq a_n$, so if $0 < \alpha \leq 3/5$ it follows that either (a_n) is increasing or there is a value of N such that $a_N > 3/5$. Since $\forall n \geq 2$ $a_n \leq 5/8$, the first case implies that (a_n) is increasing and bounded above, hence (a_n) tends to some limit. This limit will satisfy $f(x) = x$, hence is either 0 or $3/5$, and since the limit cannot be less than α (since $\forall n \in \mathbb{N}$ $a_n \geq a_1 = \alpha$) , $a_n \to 3/5$ as $n \to \infty$. In the second case we see that $\forall n \geq N$ $a_n \in [1/2, 5/8]$ by the last paragraph. Then

$$|a_{n+1} - a| = |f(a_n) - f(a)| = |f'(\xi_n)||a_n - a| \qquad (*)$$

for some ξ_n between a and a_n , hence in $[1/2, 5/8]$. Since $f'(x) = 5/2 - 5x$ we see that $|f'(x)| \leq |f'(5/8)| = 5/8$ for all $x \in [1/2, 5/8]$. From (*) then, $\forall n \geq N$ $|a_{n+1} - a| \leq (5/8)|a_n - a|$, so $\forall n \geq N$ $|a_n - a| \leq (5/8)^{n-N}|a_N - a|$, whence $a_n \to a$. The cases for $\alpha > 3/5$ can be seen immediately on noticing that $0 \leq a_2 < 3/5$.

The method just used will tackle values of μ in the range $2 < \mu < 1 + \sqrt{3}$. Considerably more cunning is needed to deal with $1 + \sqrt{3} \leq \mu \leq 3$. For values of μ between 3 and 4, the sequence does not usually tend to a limit; consider the case $\mu = 7/2$ and $\alpha = 3/7$. The behaviour for such μ is very complicated and difficult to determine and great care is needed in inferring results from diagrams. (Some idea of this can be gained by testing the values on a computer.)

In fact, the behaviour of this sequence, given by $a_{n+1} = \mu a_n(1 - a_n)$, and others like it, has been the subject of much research in recent years. For $1 \le \mu \le 3$ we saw that, apart from the cases where $\alpha = 0$ or 1 , $a_n \to a = 1 - 1/\mu$, this being the non-zero solution of the equation $x = f(x)$ which a limit of (a_n) must satisfy. For $\mu > 3$, $|f'(a)| > 1$ and there is an interval $(a - \delta, a + \delta)$ on which $|f'|$ is greater than 1. By the Mean Value Theorem, in the form $|a_{n+1} - a| = |f'(\xi_n)||a_n - a|$, we see that if a_n and a_{n+1} both lie in $(a - \delta, a + \delta)$ then $|f'(\xi_n)| > 1$ and, unless $a_n = a$, a_{n+1} is *further* from a than a_n is. The sequence is "repelled" from a and, apart from exceptional cases where there is a value of N with $a_N = a$, the sequence cannot tend to a . For μ slightly greater than 3 it turns out that alternate terms tend to two distinct limits, the sequence of even-numbered terms tending to p, say, while the odd-numbered terms tend to q. In this case $f(p) = q$ and $f(q) = p$, both numbers satisfying the equation $x = f(f(x))$. The limit a could be said to have "split" into two, alternate terms of the sequence tending to these two. For larger μ the two points p and q split into four, then eight, sixteen and so on, in these cases the sequence (a_n) for large n, being nearly periodic, the values being close to a number of points in turn. For still larger values of μ the behaviour becomes chaotic and the sequence need have little discernible pattern, while the behaviour appears to vary unpredictably with changes in the value of α . For a readable account of this, see May (1976).

Example 11.7 Another example occurs with Newton's method for finding the root of an equation $f(x) = 0$. Suppose that $a < b$ and that f is twice differentiable on $[a, b]$, that $f(a)$ and $f(b)$ have opposite signs and that both f' and f'' are non-zero and have constant sign on $[a, b]$. Let (a_n) be obtained by choosing a_1 to be whichever of a or b has the property that f and f'' have the same sign, and letting $a_{n+1} = a_n - f(a_n)/f'(a_n)$. Then (a_n) tends to the solution of $f(x) = 0$ lying between a and b.

(Some relaxation of the conditions may be allowed, but at the expense of some complication, and perhaps doubt in determining which solution of $f(x) = 0$ is the limit.)

Solution: We shall tackle the case illustrated in Fig. 11.8, where $f(a) < 0 < f(b)$ and $f'(x) > 0$, $f''(x) > 0$ on $[a, b]$. By the Intermediate Value Theorem there is at least one solution of the equation $f(x) = 0$ between a and b, and the positivity of f' ensures that there is no more than one: call it x_0 . The diagram suggests that $\forall n \in \mathbb{N}$ $x_0 \le a_n \le b$ and that (a_n) decreases.

Let $\phi(x) = x - f(x)/f'(x)$ so that $a_{n+1} = \phi(a_n)$, and

$$x > x_0 \Rightarrow \phi'(x) = f(x)f''(x)/(f'(x))^2 > 0 .$$

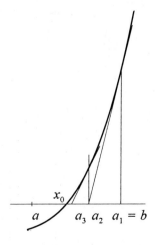

Fig. 11.8

Then

$$a_n > x_0 \Rightarrow a_{n+1} - x_0 = \phi(a_n) - \phi(x_0) = (a_n - x_0)\phi'(\xi_n) > 0 \Rightarrow a_{n+1} > x_0$$

(where ξ_n is some point between x_0 and a_n). Also if $x_0 \le a_n \le b$ then $a_{n+1} = a_n - f(a_n)/f'(a_n) \le a_n$. Therefore $x_0 \le a_n \le b \Rightarrow x_0 \le a_{n+1} \le b$ and $a_{n+1} \le a_n$. Induction now shows that (a_n) decreases and we see that $a_n \to x_0$, since (a_n) is decreasing and bounded below, and the limit, say x, to which it must tend, satisfies $\phi(x) = x$. $\square$

The convergence in Example 11.7 is eventually 'rapid' in that

$$a_{n+1} - x_0 = \phi(a_n) - \phi(x_0) = \frac{(a_n - x_0)^2}{2}\phi''(\xi_n)$$

for some ξ_n between x_0 and a_n (by Taylor's Theorem applied to ϕ using the fact that $\phi'(x_0) = 0$) so that if $|\phi''(x)| \le M$ for $x \in [a, b]$, $|a_{n+1} - x_0| \le \frac{1}{2}M|a_n - x_0|^2$ and if a_n is very close to x_0, a_{n+1} will be considerably closer. A precise evaluation of the consequences of this is best left to a course on numerical mathematics where the topic is important.

As a final example of the use of Taylor's Theorem we shall consider a few finite series. In common with most examples of Taylor's Theorem, we use the number of terms in the theorem appropriate to our problem. The motivation is the set of induction results giving

$$\sum_{k=1}^{n} k = \frac{1}{2}n(n+1), \quad \sum_{k=1}^{n} k^2 = \frac{1}{6}n(n+1)(2n+1), \quad \sum_{k=1}^{n} k^3 = \frac{1}{4}n^2(n+1)^2$$

and so on. In all these, the coefficient of the highest power of n in $\sum_{k=1}^{n} k^\alpha$ is $1/(\alpha+1)$; we shall show this is general.

Let $f(x) = x^{\alpha+1}/(\alpha+1)$ where $\alpha > 0$ so that f has derivatives of all orders. By Taylor's Theorem on $[k, k+1]$ with remainder of the form $f''(\xi)/2!$, for $k \in \mathbb{N}$ and $\alpha \ge 1$,

$$((k+1)^{\alpha+1} - k^{\alpha+1})/(\alpha+1) = k^\alpha + \frac{1}{2}\alpha\xi_k^{\alpha-1}$$

for some $\xi_k \in (k, k+1)$, so, summing from $k = 1$ to $n-1$,

$$(n^{\alpha+1} - 1)/(\alpha+1) = \sum_{k=1}^{n-1} k^\alpha + R_n$$

where $R_n = \frac{1}{2}\alpha \sum_{k=1}^{n-1} \xi_k^{\alpha-1} \le \frac{1}{2}\alpha \sum_{k=1}^{n-1}(k+1)^{\alpha-1} \le \frac{1}{2}\alpha \cdot n \cdot n^{\alpha-1}$.

Therefore $\sum_{k=1}^{n} k^\alpha = n^\alpha + (n^{\alpha+1} - 1)/(\alpha+1) - R_n = n^{\alpha+1}/(\alpha+1) + R_n'$ where

$R_n' = n^\alpha - 1/(\alpha+1) - R_n$ so that $|R_n'| \le (1 + \frac{1}{2}\alpha)n^\alpha \ \forall n \in \mathbb{N}$. Therefore the leading term is $n^{\alpha+1}/(\alpha+1)$ (for $R_n'/n^{\alpha+1} \to 0$ as $n \to \infty$).

This rather neatly illustrates that one gets out of Taylor's Theorem what one puts in. If we take an extra term, we obtain

$$((k+1)^{\alpha+1} - k^{\alpha+1})/(\alpha+1) = k^\alpha + \tfrac{1}{2}\alpha k^{\alpha-1} + \alpha(\alpha-1)\eta_k^{\alpha-2}/6$$

where $\eta_k \in (k, k+1)$. Then summing from $k=1$ to $n-1$ gives

$$(n^{\alpha+1} - 1)/(\alpha+1) \;=\; \sum_{k=1}^{n-1} k^\alpha + \tfrac{1}{2}\alpha \sum_{k=1}^{n-1} k^{\alpha-1} + R_n$$

where here $|R_n| \le \alpha(\alpha-1)(n^{\alpha-1})/6$ if $\alpha \ge 2$. Adjusting, and noticing that

$$\sum_{k=1}^{n-1} k^{\alpha-1} \;=\; -n^{\alpha-1} + \sum_{k=1}^{n} k^{\alpha-1} \;=\; -n^{\alpha-1} + n^\alpha/\alpha + S_n$$

where we have used the result just proved with $\alpha - 1$ in place of α, we see that for some constant M $|S_n| \le Mn^{\alpha-1}$, and we obtain

$$\sum_{k=1}^{n} k^\alpha \;=\; n^{\alpha+1}/(\alpha+1) + \tfrac{1}{2}n^\alpha + T_n$$

where, for some constant M', $\forall n \in \mathbb{N}$ $|T_n| \le M'n^{\alpha-1}$ (for $\alpha \ge 2$). Notice the curiosity that the coefficient of n^α does not depend on α. This method can, of course, be taken further.

Problems

1. Let $f(x) = |x|$ $\forall x \in \mathbb{R}$. Show that f is differentiable at all points of $\mathbb{R}$ except 0 and find its derivative.

2. Suppose that $f_1, f_2, \ldots, f_n$ are differentiable and that, for $i = 1, 2, \ldots, n$, $f_i(c) \ne 0$. Let $h(x) = f_1(x)f_2(x)$ and show that
$$h'(c)/h(c) = f_1'(c)/f_1(c) + f_2'(c)/f_2(c).$$
Show also that if $k(x) = f_1(x)f_2(x)\ldots f_n(x)$ then
$$k'(c)/k(c) = \sum_{i=1}^{n} f_i'(c)/f_i(c).$$

3. Let f, g and h be differentiable. Calculate the derivatives of the functions whose value at x is $f(2x)$, $f(x^2)$, $f(g(x)h(x))$, $f(g(h(x)))$.

4. Let $f: [a, b] \to \mathbb{R}$ be continuous, and be differentiable on (a, b). Use the Mean Value Theorem to show that if $\forall x \in (a, b)$ $f'(x) = 0$, then f is constant. Deduce that if $\forall x \in (a, b)$ $f'(x) = g'(x)$ and $f(a) = g(a)$, then $f = g$. (Hint: Consider $f - g$.)

5. Suppose that $f, g: \mathbb{R} \to \mathbb{R}$ are differentiable and that $\forall x \in \mathbb{R}$ $|f'(x)| \le M$. Prove that $\forall x, y \in \mathbb{R}$ $|f(x) - f(y)| \le M|x - y|$.

6. Suppose that $g: \mathbb{R} \to \mathbb{R}$ and that $\forall x, y \in \mathbb{R}$ $|g(x) - g(y)| \le M|x - y|^2$, where

M is a constant.   Find the derivative of g and prove that g is constant.

7.   Suppose that $f: [a, b] \to \mathbb{R}$ and that f is differentiable at all points of (a, b).   Show that if $\forall x \in (a, b)$ $f'(x) < 0$ then f is strictly decreasing, that is, $y < z \Rightarrow f(y) > f(z)$ for all $y, z \in [a, b]$) .   (See Example 11.2.)
      If in the above situation there is a point $c \in (a, b)$ at which we do not know the value of f', that is, if we know that $\forall x \in (a, c) \cup (c, b)$ $f'(x) < 0$, then f is still strictly decreasing; show this.

8.   Suppose that $f: [a, b] \to \mathbb{R}$, that f is differentiable on (a, b) and that $a < c < b$.   Given that for some $\delta > 0$ $f'(x) > 0$ for all $x \in (c - \delta, c)$ and $f'(x) < 0$ for all $x \in (c, c + \delta)$, show that if $0 < |y - c| < \delta$ then $f(y) < f(c)$. (In other words, show that f has a local maximum at c and that $f(c)$ is greater than all other points in $(c - \delta, c + \delta)$.)

9.   In each case below, the function described has a stationary value at $x = 0$; decide whether it is a maximum, minimum or point of inflexion: x^7, $\cos x - 1 + x^2/2$, $x^2/(1 + x)^2$, $x^2 - \sin^2 x$, $\sin x - \tan x$.

10.   Let f be twice differentiable and suppose that $f(a) = f(b) = f(c)$, where $a < b < c$.   Prove that there is a point ξ between a and c for which $f''(\xi) = 0$.

11.   Show that if f is strictly increasing and differentiable on (a, b) then $\forall x \in (a, b)$ $f'(x) \geq 0$.   Give an example to show that $f'(x)$ may be zero for some points x.

12.   (i) Let $a < b < c$ and suppose that $\phi: (a, c) \to \mathbb{R}$ is differentiable.   Let $\phi'(b) > 0$.   Deduce that $\exists \delta > 0$ such that $0 < |x - b| < \delta \Rightarrow \phi(x) - \phi(b)$ and $x - b$ have the same sign; Example 10.5 should help. Hence show that $b < x < b + \delta \Rightarrow \phi(x) > \phi(b)$ and $b - \delta < x < b \Rightarrow \phi(x) < \phi(b)$.
      (ii) Let $a < b < c$, $f: (a, c) \to \mathbb{R}$ be twice differentiable, $f'(b) = 0$ and $f''(b) > 0$.   Show that $\exists \delta > 0$ such that $b < x < b + \delta \Rightarrow f'(x) > 0$ and hence that for the same δ , $b < x < b + \delta \Rightarrow f(x) > f(b)$). Deduce that if f has a local maximum at b , $f''(b) \leq 0$.   (Unlike the result proved in the text, we do not assume f'' is continuous here.)

13.   The function $y: [a, b] \to \mathbb{R}$ is continuous, and twice differentiable at all points of (a, b).   It is also known that $y(a) = y(b) = 0$.   By considering the point at which y attains its maximum or minimum, prove that if y is not identically zero, either there is a $\xi \in (a, b)$ for which $y''(\xi) \leq 0$ and $y(\xi) > 0$ or there is a $\xi \in (a, b)$ with $y''(\xi) \geq 0$ and $y(\xi) < 0$.   (Use Question 12.)
      If, in addition, y satisfies the differential equation
$$y''(x) + f(x)y'(x) - g(x)y(x) = 0 \quad (\forall x \in (a, b))$$

where $\forall x \in (a, b)$ $g(x) > 0$, show that y is identically zero.

14.   Evaluate the limits as $x \to 0$ of:

$$\frac{\sqrt{(1+x)}-1}{\sqrt{(1-x)}-1}, \quad \frac{x^2-x}{\sqrt{(1+x)}-1}, \quad \frac{\sin x}{x}, \quad \frac{\sqrt{(1+x^2)}-1}{x\sin x}.$$

15.   Let $a_1 = \alpha$ and $\forall n \in \mathbb{N}$ $a_{n+1} = 1 + 1/(1 + a_n)$. Show that if $\alpha \geq 0$ then $\forall n \geq 2$ $a_n \geq 1$, and $|a_{n+1} - \sqrt{2}| \leq (1/4)|a_n - \sqrt{2}|$. Deduce that $a_n \to \sqrt{2}$ as $n \to \infty$.

16.   Define (a_n) by $a_1 = 1$, $a_{n+1} = 1 + 1/a_n$. By considering the function f given by $f(x) = 1 + 1/x$ and showing that $3/2 \leq x \leq 2 \Rightarrow 3/2 \leq f(x) \leq 2$, or otherwise, show that $a_n \to (1 + \sqrt{5})/2$ as $n \to \infty$. Investigate the limit when different choices are made for $a_1 > 0$.

17.   Let $f(x) = 2\sqrt{x}$ (for $x \geq 0$). By applying Taylor's Theorem to f, show that for $n \in \mathbb{N}$ $2\sqrt{(n+1)} - 2\sqrt{n} = \dfrac{1}{\sqrt{n}} - \dfrac{1}{4\xi_n^{3/2}}$ for some $\xi_n \in (n, n+1)$. By showing that $\sum \xi_n^{-3/2}$ converges, deduce that if $S_n = 1 + 1/\sqrt{2} + \ldots + 1/\sqrt{n}$, then $S_n - 2\sqrt{(n+1)}$ tends to a limit as $n \to \infty$.

18.   Assuming the required properties of the sine function, define f by $f(0) = 0$, $f(x) = x^2\sin(1/x)$ (for $x \neq 0$), and show that $f'(0) = 0$. Observe that f' is not continuous at 0.

19.* Suppose that $f: (a, b) \to \mathbb{R}$ is differentiable. Show that f' has the intermediate value property even if the derivative is discontinuous, as follows: First show that if $f: [c, d] \to \mathbb{R}$ is not injective then $\exists \xi \in [c, d]$ such that $f'(\xi) = 0$. Deduce that if $\forall x \in (c, d)$ $f'(x) \neq 0$, then f is strictly increasing or strictly decreasing on $[c, d]$. Now suppose that c and d are two typical points of (a, b) and that $f'(c) < 0 < f'(d)$ and deduce the existence of $\xi \in (c, d)$ with $f'(\xi) = 0$.

12

Functions Defined by Power Series

12.1 Introduction

We are led naturally to the study of those functions which can be represented as the sums of power series for two reasons. Taylor's Theorem allows us to express a function f as a sum of finitely many terms of the form $(x - a)^n f^{(n)}(a)/n!$ plus a remainder, and we may observe that in some cases the remainder tends to zero as the number of terms increases; this may be thought of as expressing f as a sum of the simpler functions $(x - a)^n$. More significantly, we arrive at power series by attempting to find functions with desirable properties, usually arising from differential equations. If we wish to find a function equal to its own derivative, then if this function were expressible in the form $\sum_{n=0}^{\infty} a_n x^n$, and if the derivative turned out to equal the expression obtained by differentiating term by term, $\sum_{n=1}^{\infty} n a_n x^{n-1}$, then the equality of the function and its derivative would be guaranteed if we were to choose the coefficients a_n so that these two series were identical, that is, $a_n = 1/n!$. Assuming this outline programme is correct and, in particular, that the interchange of the limits involved in taking derivatives of infinite sums can be justified, this yields a simple technique for producing various special functions required in everyday mathematics.

In what follows we shall consider functions of the form $\sum_{n=0}^{\infty} a_n x^n$ where a_n does not depend on n and the term $a_0 x^0$ is understood to mean a_0. By a simple change of variable this allows us to consider functions of the form $\sum a_n (x - a)^n$.

The first thing to recall is the result of Lemma 8.11, that if the power series $\sum a_n w^n$ converges, then $\sum a_n x^n$ converges if $|x| < |w|$. We make this systematic:

Definition The power series $\sum a_n x^n$ is said to have **radius of convergence** R if it converges for all real x with $|x| < R$ and diverges for all real x with $|x| > R$. (Notice that nothing is said about $x = \pm R$.)

Theorem 12.1 A power series $\sum a_n x^n$ either converges for all $x \in \mathbb{R}$ or has a radius of convergence.

Proof. We introduce the set $A = \{x \geq 0 : \sum a_n x^n \text{ converges}\}$. The two cases arise according as A is bounded above or not.

Suppose A is bounded above. Since A is non-empty $(0 \in A)$, it has a supremum; call it R. We show that R is the radius of convergence.

Let $-R < x < R$. Since $|x| < \sup A \; \exists w \in A$ with $|x| < w$. Then since $\sum a_n w^n$ converges, Lemma 8.11 shows that $\sum a_n x^n$ converges absolutely. Thus, for all $x \in (-R, R) \; \sum a_n x^n$ converges.

Now let $x \in \mathbb{R}$ with $|x| > R$. Then $\sum a_n x^n$ must diverge, for if not, let $y = (|x| + R)/2$ so that $|y| < |x|$ and by Lemma 8.11 $\sum a_n y^n$ would converge, giving $y \in A$ and $y > \sup A = R$, a contradiction.

For the remaining case, suppose A is not bounded above. Let $x \in \mathbb{R}$. Then $|x|$ is not an upper bound for A, so $\exists w \in A$ such that $w > |x|$, whence Lemma 8.11 proves that $\sum a_n x^n$ converges. Since x was arbitrary, we have shown that the series converges for all $x \in \mathbb{R}$. $\square$

Note: The proof above actually shows more than we stated: the series is *absolutely* convergent for $|x| < R$ in the first case and for all $x \in \mathbb{R}$ in the second.

Example 12.1 The ratio test shows that $\sum x^n$ has radius of convergence 1 and that $\sum x^n /n!$ converges (absolutely) for all real x. $\square$

12.2 Functions Defined by Power Series
Theorem 12.2 Suppose that the power series $\sum a_n x^n$ converges absolutely for all $x \in (-R, R)$ and define f by

$$f(x) = \sum_{n=0}^{\infty} a_n x^n \quad (-R < x < R) .$$

f is continuous.

Proof: Caution is needed here. The individual functions $a_n x^n$ summed in the series are all continuous, but we take a limit informing the sum to infinity, so we need to prove f continuous and avoid interchanging the order of limits in this proof. The crux of the proof is the reduction to a finite sum.

Let $x_0 \in (-R, R)$; we prove that f is continuous at x_0.

Let $\varepsilon > 0$. Choose r with $|x_0| < r < R$, which is certainly possible. Since $\sum |a_n r^n|$ converges, there is an N such that

$$\sum_{n=N+1}^{\infty} |a_n r^n| = \left| \sum_{n=0}^{\infty} |a_n r^n| - \sum_{n=0}^{N} |a_n r^n| \right| < \varepsilon/3 . \tag{1}$$

Now the function $x \mapsto \sum_{n=0}^{N} a_n x^n$ is the sum of $N + 1$ continuous functions and is therefore continuous, so there is a $\delta_1 > 0$ for which

$$|x - x_0| < \delta_1 \Rightarrow \left| \sum_{n=0}^{N} a_n x^n - \sum_{n=0}^{N} a_n x_0^n \right| < \varepsilon/3 . \tag{2}$$

We now need to consider only x satisfying $|x| \le r$ and $|x - x_0| < \delta_1$, so let $\delta = \min(\delta_1, r - |x_0|)$. Then $|x - x_0| < \delta \Rightarrow |x| < r$ and $|x - x_0| < \delta_1$ so that, if $|x - x_0| < \delta$,

$$|f(x) - f(x_0)| = \left| \sum_{n=0}^{\infty} a_n x^n - \sum_{n=0}^{\infty} a_n x_0^n \right|$$

$$\leq |\sum_{n=N+1}^{\infty} a_n x^n| + |\sum_{n=0}^{N} a_n x^n - \sum_{n=0}^{N} a_n x_0^n| + |\sum_{n=N+1}^{\infty} a_n x_0^n|$$

$$< \sum_{n=N+1}^{\infty} |a_n||x^n| + \varepsilon/3 + \sum_{n=N+1}^{\infty} |a_n||x_0^n| \qquad \text{by (2)}$$

$$\leq 2 \sum_{n=N+1}^{\infty} |a_n|r^n + \varepsilon/3 \quad (\text{since } |x| \leq r, \ |x_0| \leq r)$$

$$< \varepsilon \qquad \qquad \text{by (1)}.$$

Thus $\forall \varepsilon > 0 \ \exists \delta > 0$ such that $|x - x_0| < \delta \Rightarrow |f(x) - f(x_0)| < \varepsilon$, and f is continuous at x_0. $\square$

Note: The key step is the observation that for all x with $|x| \leq r$ $\sum_{n=N+1}^{\infty} |a_n x^n| \leq \sum_{n=N+1}^{\infty} |a_n r^n| < \varepsilon/3$, that is, we have a single value of N which ensures this inequality for all of these values of x. The result of the theorem is that, in these specific circumstances, we may interchange the limits involved in

$$\lim_{x \to x_0} (\sum_{n=0}^{\infty} a_n x^n) = \sum_{n=0}^{\infty} \lim_{x \to x_0} (a_n x^n) .$$

The fact that the sum to infinity of a power series behaves as we might expect tends to obscure the significance of this result. Notice that in Problem 2 we construct a series of continuous functions whose sum to infinity is discontinuous.

The continuity of sums of power series is the key to the following 'identity theorem' allowing us to equate the terms of two power series. It is the continuity which allows us to deduce the step we need for $x = 0$ when the preliminary information was obtained subject to the restriction $x \neq 0$.

Theorem 12.3 Let $\sum a_n x^n$ and $\sum b_n x^n$ be two power series and let $f(x) = \sum_{n=0}^{\infty} a_n x^n$ and $g(x) = \sum_{n=0}^{\infty} b_n x^n$, each valid when the appropriate series converges. If there is an $R > 0$ such that $\forall x \in (-R, R) \ f(x) = g(x)$, the series are identical, i.e. $\forall n \geq 0 \ a_n = b_n$.

Proof: Suppose that $\forall x \in (-R, R) \ f(x) = g(x)$, where $R > 0$. Then setting $x = 0$ shows that $a_0 = b_0$.

Now suppose that $a_n = b_n$ for $n = 0, 1, 2, ..., N$. Then, for $x \in (-R, R)$,

$$0 = f(x) - g(x) = \sum_{n=0}^{\infty} (a_n - b_n)x^n = \sum_{n=N+1}^{\infty} (a_n - b_n)x^n = x^{N+1}h(x)$$

where $h(x) = \sum_{n=0}^{\infty} (a_{n+N+1} - b_{n+N+1})x^n$. Clearly $h(x) = 0$ if $x \in (-R, R)\backslash\{0\}$. Since h is the sum of a power series, convergent for $|x| < R$ (and thus absolutely convergent), h is continuous, so $h(0) = \lim_{x \to 0} h(x) = 0$. Substituting shows

$a_{N+1} = b_{N+1}$.

It follows by induction that $\forall n \geq 0$ $a_n = b_n$. □

The next step in our development of a calculus for power series is to consider differentiation, in particular, to consider the series obtained by differentiating each term of the power series and that which yields the original series when we differentiate the terms of the new series.

Theorem 12.4 The series $\Sigma a_n x^n$, $\Sigma n a_n x^{n-1}$, $\Sigma (a_n/(n+1))x^{n+1}$ either all have the same radius of convergence or all converge for all $x \in \mathbb{R}$.

Proof: Since all the series converge for $x = 0$, we need only consider $x \neq 0$.
Suppose that $\Sigma n a_n x^{n-1}$ converges absolutely for all $x \in (-R_2, R_2)$. Let $b_n = n a_n/x$ so that $n a_n x^{n-1} = b_n x^n$. Since $|a_n| \leq |n a_n| = |b_n x|$ for $n \geq 1$, $\Sigma |a_n x^n|$ is convergent for $x \in (-R_2, R_2)$ by comparison with $\Sigma |x| |b_n x^n|$.
Now suppose, instead, that $\Sigma |a_n x^n|$ is absolutely convergent for all $x \in (-R_1, R_1)$. Let $0 < |x| < R_1$ and choose w such that $|x| < |w| < R_1$. Then $\Sigma |a_n w^n|$ is convergent. Since $w/x > 1$, Bernoulli's Inequality shows that $\forall n \in \mathbb{N}$ $|w/x|^n > n(|w/x| - 1)$ whence $|w|^n > n|x|^{n-1}(|w| - |x|)$ so it follows that $|n a_n x^{n-1}| < |a_n w^n|/(|w| - |x|)$. By comparison with $\Sigma |a_n w^n|$, we see that $\Sigma |n a_n x^{n-1}|$ converges. Since x was typical, we deduce that $\Sigma |n a_n x^{n-1}|$ converges for all $x \in (-R_1, R_1)$.
The two paragraphs above show that if either of the series $\Sigma |a_n x^n|$ and $\Sigma |n a_n x^{n-1}|$ converges for all $x \in \mathbb{R}$ then so does the other. (Suppose the first series converges for all $x \in \mathbb{R}$; then R_1 may be chosen as large as we wish, and, given $x \in \mathbb{R}$, we may choose $R_1 > |x_0|$ to deduce that the second series converges at $x = x_0$.) The remaining case is where both series have a radius of convergence, say R_1 and R_2 respectively. Then by the first paragraph, $R_1 \geq R_2$ (since $\Sigma a_n x^n$ converges for $|x| < R_2$) while the second paragraph shows $R_2 \geq R_1$.
The result for $\Sigma (a_n/(n+1))x^{n+1}$ is obtained from the above by setting $c_0 = 0$, $c_n = a_{n-1}/n$ $(n \geq 1)$ and considering $\Sigma c_n x^n$ and $\Sigma n c_n x^{n-1}$. □

Note: By applying the Theorem to $\Sigma n a_n x^{n-1}$ we can deduce that $\Sigma n(n-1)a_n x^{n-2}$ has the same radius of convergence as $\Sigma a_n x^n$, and then, in turn, so has $\Sigma n(n-1)(n-2)a_n x^{n-3}$.

Theorem 12.5 Suppose that for all $x \in (-R, R)$ the series $\Sigma a_n x^n$ converges absolutely and that $f(x) = \sum_{n=0}^{\infty} a_n x^n$. Then f possesses derivatives of all orders, $f'(x) = \sum_{n=1}^{\infty} n a_n x^{n-1}$ and the kth derivative is given by $f^{(k)}(x) = \sum_{n=k}^{\infty} n(n-1)...(n-k+1)a_n x^{n-k}$.

Proof: We shall prove the result for f' . The general case then follows by observing that f' is given by the sum of a power series so the theorem applies to f', and so on.
Choose $x \in (-R, R)$ and r satisfying $|x| < r < R$. These we shall keep fixed.

Considering the function $y \mapsto y^n$ we see that there is a ξ_n between x and y for which $y^n = x^n + (y - x)nx^{n-1} + \frac{1}{2}(y - x)^2 n(n - 1)\xi_n^{n-2}$. The ξ_n here depends on n and y, but for all $y \in (-r, r)$, $|\xi_n| < r$ so, since all the series converge by comparison with $\sum|a_n r^n|$, $\sum|na_n r^{n-1}|$ and $\sum|n(n - 1)a_n r^{n-2}|$, which converge by Theorem 12.4,

$$f(y) - f(x) \;\; = \;\; \sum_{n=0}^{\infty} a_n (y^n - x^n) = \sum_{n=1}^{\infty} a_n (y^n - x^n)$$

$$= \;\; \sum_{n=1}^{\infty} na_n (y - x)x^{n-1} + \sum_{n=1}^{\infty} \tfrac{1}{2} n(n - 1)a_n (y - x)^2 \xi_n^{n-2} .$$

Thus $\left| \dfrac{f(y) - f(x)}{y - x} - \displaystyle\sum_{n=1}^{\infty} na_n x^{n-1} \right| \leq |y - x| \sum_{n=1}^{\infty} \tfrac{1}{2} n(n - 1)|a_n \xi_n^{n-2}|$

$$\leq M|y - x| , \tag{1}$$

where $M = \sum_{n=1}^{\infty} \tfrac{1}{2} n(n - 1)|a_n r^{n-2}|$, a constant independent of y. It is now clear that $\lim_{y \to x} (f(y) - f(x))/(y - x) = \sum_{n=1}^{\infty} na_n x^{n-1}$. $\quad\square$

Again, notice that the introduction of r allowed us to obtain the estimate with the number M independent of y, and thus (1) is valid for all $y \in (x - \delta, x + \delta)$ where $\delta = r - |x|$. The result is another in which the interchange of two limiting operations (summing to infinity and differentiating) is justified.

12.3 Some Standard Functions of Mathematics

Suppose we seek a function f which is equal to its own derivative. If we presume that the function can be expressed in the form $f(x) = \sum_{n=0}^{\infty} a_n x^n$, where the series converges for $x \in (-R, R)$, then $f'(x) = \sum_{n=1}^{\infty} na_n x^{n-1}$ so $f' = f$ implies that $\forall x \in (-R, R)$ $\sum_{n=0}^{\infty} a_n x^n = \sum_{n=1}^{\infty} na_n x^{n-1} = \sum_{n=0}^{\infty} b_n x^n$ where $b_n = (n + 1)a_{n+1}$. By Theorem 12.3, then, $\forall n \geq 0$ $a_n = b_n = (n + 1)a_{n+1}$. Thus $a_n = a_0/n!$ and $f(x) = a_0 \sum_{n=0}^{\infty} x^n/n!$. At this stage we know that if there is a function satisfying the equation $f' = f$ and given by the power series, then it is of the form given; we still have to check that there is such a function. This is now easy, for we can check that the series converges for all $x \in \mathbb{R}$ and so the calculations are all justified in retrospect.

Definition We define the **exponential function** $\exp: \mathbb{R} \to \mathbb{R}$ by

$$\exp(x) \;\; = \;\; \sum_{n=0}^{\infty} \frac{x^n}{n!}$$

We now know that the exponential function is continuous, differentiable as often as we wish, and that it equals its own derivative. To go further we shall use a general theorem whose proof requires a good deal of stamina and organisation. The aim is to consider the result of multiplying two power series $(\sum_{n=0}^{\infty} a_n x^n)(\sum_{n=0}^{\infty} b_n x^n)$. Ignoring, for the moment, all qualms about justification, we may hope this product can be expressed as a power series in x, and that the coefficient of x^n in it is obtained by collecting the terms from the two brackets, $a_r x^r$ being matched with $b_{n-r} x^{n-r}$; the expected coefficient is then $a_0 b_n + a_1 b_{n-1} + \dots + a_n b_0$. This result is true, though we shall prove it only where the series are absolutely convergent. It may be false if neither series converges absolutely.

Theorem 12.6   The Cauchy Product Theorem. Let x be a number for which the two power series $\sum a_n x^n$ and $\sum b_n x^n$ are both absolutely convergent. Define $c_n = \sum_{r=0}^{n} a_r b_{n-r}$. Then the series $\sum c_n x^n$ is absolutely convergent and

$$\sum_{n=0}^{\infty} c_n x^n = (\sum_{n=0}^{\infty} a_n x^n)(\sum_{n=0}^{\infty} b_n x^n) \ .$$

Proof. We first need to identify the nature of our task. Since

$$c_k = a_0 b_k + a_1 b_{k-1} + \dots + a_k b_0 \ ,$$

$$c_k x^k = \sum_{i=0}^{k} (a_i x^i)(b_{k-i} x^{k-i})$$

$$= \sum \{(a_i x^i)(b_j x^j) : \quad i, j \geq 0, \quad i+j=k\} \ .$$

(In what follows, i, j, k and n will always denote integers.) Therefore, $c_k x^k$ is the sum of all those terms of the form $(a_i x^i)(b_j x^j)$ where the pair of integers (i, j), plotted on a graph, lies on that portion of the line $i+j = k$ in the first quadrant ($i \geq 0, j \geq 0$ ensuring this). We now see that $\sum_{n=0}^{N} c_n x^n$ equals the sum of all terms $(a_i x^i)(b_j x^j)$ where $i+j$ takes in turn the values $0, 1, \dots, N$, that is, the sum of all the terms whose indices (i, j) lie in the triangular region defined by $i \geq 0, j \geq 0$, $i+j \leq N$ (See Fig. 12.1). $(\sum_{i=0}^{\infty} a_i x^i)(\sum_{j=0}^{\infty} b_j x^j)$ is

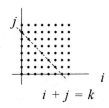

$i + j = k$

Fig. 12.1.

the limit as $N \to \infty$ of $(\sum_{i=0}^{N} a_i x^i)(\sum_{j=0}^{N} b_j x^j)$. Now the latter expression is the sum of all the terms of the form $(a_i x^i)(b_j x^j)$ where the indices lie in the square region $0 \leq i, j \leq N$.

Let $s_n = \sum_{i=0}^{n} a_i x^i$, $t_n = \sum_{j=0}^{n} b_j x^j$ and $u_n = \sum_{k=0}^{n} c_k x^k$ for $n = 0, 1, \dots$. From the discussion above, $s_N t_N$ will not generally be equal to u_M for any M. The technique is to use the absolute convergence of $\sum a_i x^i$ and $\sum b_j x^j$ to prove first that $\sum c_k x^k$ is absolutely convergent, so that by choosing N and n large enough, $s_N t_N$ and u_n will be close to their respective limits. If we at the same time ensure that $n \geq 2N$ the terms $(a_i x^i)(b_j x^j)$ summed in $s_N t_N$ will all contribute to the sum forming u_n, so $u_n - s_N t_N$ will consist of the sum of those terms $(a_i x^i)(b_j x^j)$ where (i, j) lies in the region outside the small square but inside the triangle in Fig. 12.2. It is here that we use the absolute convergence: because the sum of all terms $|(a_i x^i)(b_j x^j)|$ which have one index greater than N can be shown to be small, the same is true of our chosen selection of these terms since the terms omitted, being moduli, are non-negative. Let us start.

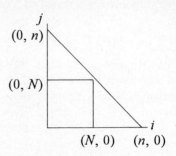

Fig. 12.2

Let

$$s'_n = \sum_{i=0}^{n} |a_i x^i|, \quad t'_n = \sum_{j=0}^{n} |b_j x^j| \text{ and } u'_n = \sum_{k=0}^{n} |c_k x^k|. \text{ Since } \sum |a_i x^i| \text{ and } \sum |b_j$$

$x^j|$ converge there are numbers s, s', t, t' such that $s_n \to s$, $s_n' \to s'$, $t_n \to t$, and $t_n' \to t'$ as $n \to \infty$. Moreover, since (s_n') and (t_n') are increasing, $\forall n \in \mathbb{N}$ $s_n' \leq s'$ and $t_n' \leq t'$.

For each $k \in \mathbb{N}$, the triangle inequality shows that

$$|c_k x^k| = \left| \sum \{(a_i x^i)(b_j x^j) : i \geq 0, \ j \geq 0, \ i + j = k\} \right|$$

$$\leq \sum \{|a_i x^i||b_j x^j| : i \geq 0, \ j \geq 0, \ i + j = k\} .$$

Hence, for all n,

$$u'_n = \sum_{k=0}^{n} |c_k x^k| \leq \sum_{k=0}^{n} \left(\sum \{|a_i x^i||b_j x^j| : i \geq 0, \ j \geq 0, \ i + j = k\} \right)$$

$$= \sum \{|a_i x^i||b_j x^j| : i \geq 0, \ j \geq 0, \ i + j \leq n\}$$

$$\leq \sum \{|a_i x^i||b_j x^j| : 0 \leq i \leq n, \ 0 \leq j \leq n\}$$

$$= s'_n t'_n \leq s' t' .$$

Since this is true for all $n \in \mathbb{N}$, (u_n') is bounded above and hence $\sum |c_k x^k|$ converges, since (u_n') is increasing.

To find the sum of $\sum c_k x^k$ we need more effort. Notice first that we may presume s' and t' are both positive since if, say, $s' = 0$ then $\forall n \geq 0$, $a_n = 0$ and the whole result is trivial.

Let $\varepsilon > 0$. Then there exist N_1, N_2 and N_3 such that

$$\left. \begin{array}{l} \forall n \geq N_1 \quad |s'_n - s'| < \varepsilon/(4t') \ , \\[4pt] \forall n \geq N_2 \quad |t'_n - t'| < \varepsilon/(4s') \ , \\[4pt] \forall n \geq N_3 \quad |s_n t_n - st| < \varepsilon/2 \ . \end{array} \right\} \tag{1}$$

and

Let $N = \max(N_1, N_2, N_3)$ so that all three inequalities in (1) hold if $n \geq N$. Recalling that if $n \geq 2N$, all the terms $(a_i x^i)(b_j x^j)$ which occur in $s_N t_N$ occur in u_n (see Fig. 12.2), we have

$$\forall n \geq 2N \ u_n = \sum \{(a_i x^i)(b_j x^j): \ i, j \geq 0, \ i+j \leq n\}$$

$$= \sum \{(a_i x^i)(b_j x^j): \ 0 \leq i, j \leq N\}$$

$$+ \sum \{(a_i x^i)(b_j x^j): \ i, j \geq 0, \ i+j \leq n, \ \max(i, j) > N\} \ .$$

Since the first sum is equal to $s_N t_N$, we have $\forall n \geq 2N$

$$|u_n - s_N t_N| \ \leq \ \sum \{|a_i x^i||b_j x^j|: \ i, j \geq 0, \ i+j \leq n, \ \max(i, j) > N\} \qquad (W)$$

$$\leq \ \sum \{|a_i x^i||b_j x^j|: \ 0 \leq i, j \leq n, \ \max(i, j) > N\} \qquad (X)$$

$$= \ \sum \{|a_i x^i||b_j x^j|: \ 0 \leq i \leq n, \ N < j \leq n\} \qquad (Y)$$

$$+ \ \sum \{|a_i x^i||b_j x^j|: \ N < i \leq n, \ 0 \leq j \leq N\} \qquad (Z)$$

$$= \ s'_n (t'_n - t'_N) + (s'_n - s'_N)t'_N \ . \qquad (2)$$

To see this, the sets of indices (i, j) occurring in the sums (W), (X), (Y), and (Z) are indicated in Fig. 12.3.

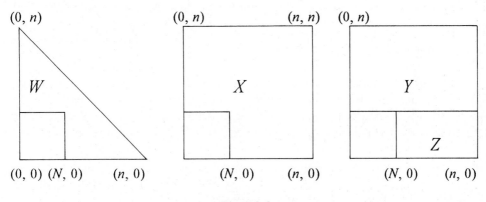

Fig. 12.3.

Finally,

$$\forall n \geq 2N \ |u_n - st| \ \leq \ |u_n - s_N t_N| + |s_N t_N - st|$$

$$< \ s'_n (t'_n - t'_N) + (s'_n - s'_N)t'_N + \tfrac{1}{2}\varepsilon \qquad \text{(by (1) and (2))}$$

$$\leq \ s'(t' - t'_N) + (s' - s'_N)t' + \tfrac{1}{2}\varepsilon$$

$$< \ s' \varepsilon/(4s') + t' \varepsilon/(4t') + \tfrac{1}{2}\varepsilon = \varepsilon \qquad \text{(by (1))} \ .$$

Since $\varepsilon > 0$ was arbitrary we have shown that

$$\forall \varepsilon > 0 \ \exists N \text{ such that } \forall n \geq 2N \ |u_n - st| < \varepsilon,$$

so $u_n \to st$ as $n \to \infty$, as required. $\square$

Remarks: The result need not be about power series, just the corresponding product result for absolutely convergent series. This is obtained by setting $x = 1$ above.

This allows us to produce the required properties of the exponential function:

Theorem 12.7 (i) The exponential function is continuous, possesses derivatives of
　　all orders and $\frac{d}{dx}(\exp x) = \exp x$.

(ii)　$\forall x, y \in \mathbb{R} \;\; \exp(x + y) = (\exp x)(\exp y)$.

(iii)　$\exp(0) = 1$　and　$\forall x \in \mathbb{R} \;\; \exp x > 0$.

Proof: (i) is immediate from Theorems 12.2 and 12.5.
(ii) Since the series for $\exp x$ and $\exp y$ converge absolutely we can apply the
Cauchy Product Theorem. Let $a_n = x^n/n!$ and $b_n = y^n/n!$ so that $\sum a_n$ and $\sum b_n$
are absolutely convergent. It follows that

$$(\exp x)(\exp y) = (\sum_{n=0}^{\infty} a_n)(\sum_{n=0}^{\infty} a_n) = \sum_{n=0}^{\infty} c_n$$

where　$c_n = \sum_{i=0}^{n} a_i b_{n-i} = \sum_{i=0}^{n} \frac{x^i y^{n-i}}{i!(n-i)!} = \sum_{i=0}^{n} \frac{1}{n!}\binom{n}{i} x^i y^{n-i} = \frac{(x+y)^n}{n!}$.

(iii) The value of $\exp(0)$ is obvious. Also it is clear that
$x \geq 0 \Rightarrow \exp x = 1 + \sum_{n=1}^{\infty} x^n/n! \geq 1$. By parts (i) and (ii), $(\exp x)(\exp(-x)) = 1$,
whence $x < 0 \Rightarrow \exp x = 1/\exp(-x) > 0$. □

　　　We define the number e by $\exp(1)$. Then since $1 + 1 = 2$, it follows that
$\exp(2) = e^2$ and it is easily checked that if n is an integer $\exp n = e^n$, while a little
more work will check that $\exp(x) = e^x$ for rational values of x . We therefore
define e^x for irrational powers x by the equation $e^x = \exp x$, noticing that this
coincides with our old definition for rational x. The number e is about
2.71828...; the series converges so rapidly that it is easy to evaluate it to a high
accuracy. (See Problems.) The letter e is used for this number in honour of the
great eighteenth century Swiss mathematician Leonhard Euler.

　　　The results of Theorem 12.7 are well known, but the following growth
properties are equally useful - and their proof illustrates some simple but effective
techniques. We first need some definitions, extending those we know.

Definitions Let $f: (a, \infty) \to \mathbb{R}$. We say that $f(x) \to \infty$ as $x \to \infty$ if and only if
$\forall R \in \mathbb{R} \;\exists X$ such that $\forall x \geq X \; f(x) > R$. Similarly $f(x) \to \infty$ as $x \to c$ means
that $\forall R \in \mathbb{R} \;\exists \delta > 0$ such that $\forall x \in (c - \delta, c + \delta) \; f(x) > R$, with the analogous
statements for one-sided limits. $f(x) \to -\infty$ if $-f(x) \to \infty$. Also if $g: (-\infty, b) \to \mathbb{R}$
we say that $g(x) \to L$ as $x \to -\infty$ if and only if $\forall \varepsilon > 0 \;\exists X$ such that $\forall x \leq X$
$|g(x) - L| < \varepsilon$.

Theorem 12.8 (i) The exponential function is strictly increasing, $e^x \to \infty$ as
　　$x \to \infty$, and $e^x \to 0$ as $x \to -\infty$,

(ii)　$\{e^x : x \in \mathbb{R}\} = (0, \infty)$ and

(iii)　for all integers k , $e^x/x^k \to \infty$ as $x \to \infty$ and $x^k e^{-x} \to 0$ as $x \to \infty$.

Proof. (i) That exp is strictly increasing is immediate on noticing that its derivative is positive at all points. Now, for $x > 0$, $e^x = 1 + x + \sum_{n=2}^{\infty} x^n/n! > x$, so $\forall R \in \mathbb{R} \; \exists X$ (e.g. $X = |R|$) such that $x \geq X \Rightarrow e^x > R$. It follows that $e^x \to \infty$ as $x \to \infty$. The remaining part is clear on noticing that

$$\forall \varepsilon > 0 \quad x < -1/\varepsilon \Rightarrow e^{-x} > e^{1/\varepsilon} > 1/\varepsilon \Rightarrow e^x = 1/e^{-x} < \varepsilon .$$

(ii) We know that $x \in \mathbb{R} \Rightarrow e^x \in (0, \infty)$. Let $y \in (0, \infty)$. Then by the properties of part (i), $\exists x_0$ and $x_1 \in \mathbb{R}$ such that $e^{x_0} < y$ and $e^{x_1} > y$. The Intermediate Value Theorem now shows $\exists x \in \mathbb{R}$ such that $e^x = y$. This proves (ii).

(iii) Let k be an integer. If $k < 0$ it is clear that $e^x/x^k \to \infty$ (since both factors tend to ∞), so we may suppose $k \geq 0$. For $x > 0$,

$$e^x = \sum_{n=0}^{\infty} \frac{x^n}{n!} > \frac{x^{k+1}}{(k+1)!} ,$$

so $e^x/x^k > x/(k+1)!$. Therefore $\forall R > 0 \; \exists X$ (e.g. $X = (k+1)!R$) such that $x \geq X \Rightarrow e^x/x^k > R$ showing that $e^x/x^k \to \infty$ as $x \to \infty$. The final part is now easily deduced from the observation that $x^k e^{-x}$ is the reciprocal of e^x/x^k. $\square$

The intuitive result from Theorem 12.8 is that in a product of an exponential and a power, the exponential dominates the behaviour.

Since the exponential function is strictly increasing and has image $(0, \infty)$, it has an inverse function.

Definition The **logarithm function** log: $(0, \infty) \to \mathbb{R}$ is defined to be the inverse function of exp.

Theorem 12.9 The logarithm function is continuous, strictly increasing and differentiable and:

(i)　$\frac{d}{dx} \log x = \frac{1}{x}$,

(ii)　$\forall x, y > 0 \;\; \log(xy) = \log x + \log y$,

(iii)　$\log 1 = 0$ and $\log x > 0 \Leftrightarrow x > 1$,

(iv)　$\log x \to \infty$ as $x \to \infty$ but for all $k \in \mathbb{N}$, $x^{-k} \log x \to 0$ as $x \to \infty$,

(v)　$\log x \to -\infty$ as $x \to 0+$ but for all $k \in \mathbb{N}$, $x^k \log x \to 0$ as $x \to 0+$.

Proof. (i) The differentiability arises from Lemma 11.4 on inverse functions, which also shows that log is strictly increasing. This result, or the chain rule, now allows the value of the derivative to be verified, since $\forall x \in \mathbb{R} \;\; x = e^{\log x}$.

(ii) follows on noticing that $\forall x, y \in \mathbb{R} \;\; e^{\log xy} = xy = (e^{\log x})(e^{\log y}) = e^{\log x + \log y}$.

(iii) is immediate since $\exp 0 = 1$ and log is strictly increasing.

(iv) Since exp is strictly increasing, $\log x > R \Leftrightarrow x = e^{\log x} > e^R$ which immediately shows that $\log x \to \infty$ as $x \to \infty$.

As $x^{-k} \leq x^{-1}$ for $k \geq 1$ and $x \geq 1$, to show that $x^{-k} \log x \to 0$ as $x \to \infty$, it is enough to prove that $x^{-1} \log x \to 0$ as $x \to \infty$. Let $\varepsilon > 0$. We know already that $ye^{-y} \to 0$ as $y \to \infty$, so $\exists Y$ such that $y \geq Y \Rightarrow |ye^{-y}| < \varepsilon$. Then $x \geq e^Y \Rightarrow$

$\log x \geq Y \Rightarrow |x^{-1} \log x| = |\log x \; e^{-\log x}| < \varepsilon$. We have shown $x^{-1} \log x \to 0$ as $x \to \infty$. (This process is the formal equivalent of substituting $x = e^y$ in $x^{-1} \log x$ and noticing that we know the behaviour of the resulting expression $y e^{-y}$.)

(v) $\log \frac{1}{x} = -\log x$ so that $\log x = -\log \frac{1}{x}$ and as $x \to 0+$, $\frac{1}{x} \to \infty$. This shows us, using (iv), that we can make $\log \frac{1}{x}$ as large as we like by choosing $\frac{1}{x}$ sufficiently large, which we can ensure by choosing x in turn to be sufficiently small. This is the motivation for the proof.

Let $R \in \mathbb{R}$. Then by (iv) $\exists Y$ such that $\forall y \geq Y \; \log y > -R$.

$$0 < x < \frac{1}{Y} \Rightarrow \frac{1}{x} > Y \Rightarrow \log \frac{1}{x} > -R \Rightarrow \log x < R .$$

This shows that $\log x \to -\infty$ as $x \to 0+$.

The case of $x^k \log x$ is similar, for we notice that $x^k \log x = -\left(\frac{1}{x}\right)^{-k} \log\left(\frac{1}{x}\right)$

and by (iv) $\left(\frac{1}{x}\right)^{-k} \log\left(\frac{1}{x}\right)$ can be made as small as we wish by ensuring $\frac{1}{x}$ is large enough. Let $\varepsilon > 0$. Then by (iv) $\exists Y$ such that $\forall y \geq Y \; |y^{-k} \log y| < \varepsilon$.
Then

$$0 < x < \frac{1}{Y} \Rightarrow \frac{1}{x} > Y \Rightarrow \left|\left(\frac{1}{x}\right)^{-k} \log\left(\frac{1}{x}\right)\right| < \varepsilon$$

$$\Rightarrow \left|x^k \log x\right| < \varepsilon .$$

Therefore, since ε was arbitrary, $x^k \log x \to 0$ as $x \to 0+$. □

The restriction of k to being an integer in Theorem 12.9 is not essential – the result remains true if k is a positive real number. This is quite easy to deduce from the result above (see Problem 3), but a direct proof would have been less easy to present since the comparison with $x^{-1} \log x$ in part (iv) would not always be helpful.

We use the log function to define irrational powers of real numbers. Let $a > 0$ and $q \in \mathbb{Q}$. Then, since $a = e^{\log a}$, it is easily checked that $a^q = e^{q \log a}$. We extend this:

Definition If $a > 0$ and $x \in \mathbb{R}$ we **define** $a^x = e^{x \log a}$.

Notice that this coincides with the old definition of a^x for rational x, but that it is now clear that the function $x \mapsto a^x$ is differentiable, and its derivative can be evaluated.

Before we show what else can be done with our techniques, let us prove the General Binomial Theorem. The proof introduces an idea at least as useful as the result. Rather than prove directly that $(1 + x)^\alpha$ equals the sum of the series (which we could attempt by calculating the remainder in Taylor's Theorem) we define a new function and use our theorems to find its properties. This oblique approach is useful elsewhere.

Theorem 12.10 The General Binomial Theorem. For $\alpha \in \mathbb{R}$ and $n \in \mathbb{N}$ let $\binom{\alpha}{0} = 1$ and $\binom{\alpha}{n} = \frac{\alpha(\alpha-1)...(\alpha-n+1)}{n!}$. Then, provided $|x| < 1$,

$$(1+x)^{\alpha} = \sum_{n=0}^{\infty} \binom{\alpha}{n} x^n .$$

Proof. The formula for $\binom{\alpha}{n}$ shows that if α is a non-negative integer then for $n \geq \alpha + 1$, $\binom{\alpha}{n} = 0$ and the sum is a finite sum (that is, there are finitely many non-zero terms). Then for $0 \leq n \leq \alpha$ $\binom{\alpha}{n} = \frac{\alpha!}{n!(\alpha-n)!}$ the binomial coefficient we have already seen. There is nothing new in this case.

If, however, α is not a non-negative integer, then for all $n \in \mathbb{N}$ $\binom{\alpha}{n} \neq 0$ and the ratio test shows that the series has radius of convergence 1, so define $f: (-1, 1] \to \mathbb{R}$ by $f(x) = \sum_{n=0}^{\infty} \binom{\alpha}{n} x^n$. Since f is given by a power series, $f'(x) = \sum_{n=1}^{\infty} \binom{\alpha}{n} n x^{n-1}$. By observing that

$$(n+1)\binom{\alpha}{n+1} + n\binom{\alpha}{n} = \alpha\binom{\alpha}{n}$$

it can be verified that $(1+x)f'(x) = \alpha f(x)$. In solving this differential equation we see that

$$\frac{d}{dx}\left((1+x)^{-\alpha} f(x)\right) = -\alpha(1+x)^{-\alpha-1} f(x) + (1+x)^{-\alpha} f'(x) = 0$$

so that $(1+x)^{-\alpha}f(x)$ is a constant. Substituting $x = 0$ shows that the constant is 1, yielding the result. We leave the details to the reader. □

The technique just used, of defining a function by means of a power series and then finding its properties by observing that it satisfies a simple differential equation, is quite useful. It helps, of course, when we have a good idea what properties we are looking for! In other cases, different features of the series give rise to the properties involved; in the case of the sine and cosine functions we cannot readily use differential equations since the solution of these equations requires a knowledge of the functions we seek. We can, however, use power series to obtain solutions of differential equations where no other technique is available.

Example 12.2 We can now define more of the standard functions of mathematics, in this case the principal trigonometric functions sine and cosine. We shall define them as functions rather than trigonometric ratios and consider at the end how the functions we have chosen are related to angles. This is a simpler approach than beginning with angles.

We **define** the two functions sin and cos by

$$\sin x = \sum_{n=0}^{\infty} \frac{(-1)^n x^{2n+1}}{(2n+1)!} , \qquad \cos x = \sum_{n=0}^{\infty} \frac{(-1)^n x^{2n}}{(2n)!} .$$

By using the ratio test we see that both series converge for all $x \in \mathbb{R}$ and

Theorem 12.5 shows that

$$\text{(i)} \quad \frac{d}{dx}(\sin x) = \cos x \quad \text{and} \quad \frac{d}{dx}(\cos x) = -\sin x .$$

(Notice that a little care is needed in that the coefficient of x^0 in $\sin x$ is zero, so

$$\frac{d}{dx}\sin x = \sum_{n=0}^{\infty} \frac{(-1)^n (2n+1)x^{2n}}{(2n+1)!} .)$$

(ii) By using the Cauchy Product Theorem, we show that if $x, y \in \mathbb{R}$

$\sin x \cos y = \sum_{n=0}^{\infty} c_n$ where

$$c_n = \sum_{j=0}^{n} \frac{(-1)^j x^{2j+1}}{(2j+1)!} \cdot \frac{(-1)^{n-j} y^{2(n-j)}}{(2n-2j)!} = \sum_{j=0}^{n} \frac{(-1)^n}{(2n+1)!} \binom{2n+1}{2j+1} x^{2j+1} y^{2n-2j} .$$

Similarly $\cos x \sin y = \sum_{n=0}^{n} c_n$ where $c_n = \dfrac{(-1)^n}{(2n+1)!} \sum_{j=0}^{n} \binom{2n+1}{2j} x^{2j} y^{2n+1-2j}$,

whence

$$\sin x \cos y + \cos x \sin y \quad = \quad \sum_{n=0}^{\infty} \left(\sum_{k=0}^{2n+1} \frac{(-1)^n}{(2n+1)!} \binom{2n+1}{k} x^k y^{2n+1-k} \right)$$

$$= \quad \sum_{n=0}^{\infty} \frac{(-1)^n (x+y)^{2n+1}}{(2n+1)!}$$

$$= \quad \sin(x+y) .$$

Therefore we have proved that, for all $x, y \in \mathbb{R}$,

$$\sin(x+y) = \sin x \cos y + \cos x \sin y .$$

(iii) Fixing y and differentiating both sides of (ii) with respect to x gives

$$\cos(x+y) = \cos x \cos y - \sin x \sin y .$$

(iv) By substituting directly into the series we see that $\cos 0 = 1$, $\sin 0 = 0$, $\cos(-x) = \cos x$, and $\sin(-x) = -\sin x$, the last two for all $x \in \mathbb{R}$. Using these and substituting $y = -x$ in (iii) now yields

$$(\cos x)^2 + (\sin x)^2 = 1$$

The tricky step is to show that sin and cos are periodic. This can be deduced from the addition formulae (ii) and (iii) once we have established some points at which these functions are 0. Since $\cos 0 = 1$ and cos is continuous, $\cos x > \frac{1}{2}$ in some interval $(-\delta, \delta)$, and so in the same interval $\frac{d}{dx}\sin x > \frac{1}{2}$.

Therefore $\sin x > \frac{1}{2}x$ for all $x \in (0, \delta)$. By the continuity of the derivative of $\sin$ x (that is, $\cos x$) the value of $\sin x$ will increase with x until x reaches a value (if there is one) where $\cos x = 0$. The existence of such a point can be deduced most simply by noticing that $\cos 2 < 0$; once we know this the Intermediate Value Theorem guarantees us a point ξ where $\cos \xi = 0$. Now by the proof of the Alternating Series Theorem (Theorem 8.10) if (a_n) is a decreasing sequence of positive numbers and $s_n = a_1 - a_2 + \ldots + (-1)^{n-1}a_n$, then

$$s_{2n} \le \sum_{j=1}^{\infty} (-1)^{j-1} a_j \le s_{2n+1} \ .$$

Now let $a_n = 2^{2n}/(2n)!$, so that $\sum_{j=1}^{\infty} (-1)^{j-1} a_j = 1 - \cos 2$ and, noticing that (a_n) is decreasing, we deduce that $s_2 \le 1 - \cos 2 \le s_3$. Since $s_2 = 4/3$ we deduce that $\cos 2 \le -1/3 < 0$, as required.

By looking at the power series for $\frac{\sin x}{x}$ and doing a similar calculation we deduce that for $0 < x \le 2$ (to ensure that the estimate in the alternating series theorem applies)

$$\tfrac{2}{3} \le 1 - \frac{x^2}{6} \le \frac{\sin x}{x} = \sum_{j=1}^{\infty} \frac{(-1)^{j-1} x^{2j-2}}{(2j-1)!}$$

so that $0 < x \le 2 \Rightarrow \sin x > 0$. It follows that $\cos x$ decreases strictly on $[0, 2]$ so we deduce that:

(v) there is a unique $\xi \in (0, 2]$ for which $\cos \xi = 0$.

We now **define** the number π to be 2ξ , where ξ is as just defined. It is easy to check using the addition formulae (ii) and (iii) and property (iv) that $\sin \xi = 1$ (for $\sin^2 \xi = 1 - \cos^2 \xi = 1$ and $\sin x > 0$ for $x \in (0, 2]$), and in turn, $\cos \pi = -1$, $\sin \pi = 0$, $\cos 2\pi = 1$, $\sin 2\pi = 0$. Then we have

(vi) For all $x \in \mathbb{R}$ $\sin(x + 2\pi) = \sin x$ and $\cos(x + 2\pi) = \cos x$.

Using (v) allows us to show that 2π is the smallest positive number a with the property that $\forall x \in \mathbb{R}$ $\sin(x + a) = \sin x$ and $\cos(x + a) = \cos x$.

To relate all of the above to angles requires the idea of the length of a curve, so we shall postpone this until we have considered integration. $\square$

12.4 Further Examples
Example 12.3 Suppose we wish to solve the differential equation:

$$x^2 \frac{d^2 y}{dx^2} + x \frac{dy}{dx} + (x^2 - 1) y(x) = 0 \ .$$

(The equation is a particular case of Bessel's equation, and it arises frequently in certain applied mathematics. For details of this, see Dunning-Davies (1982), Chapter 13.)

If we assume, temporarily, that $y = \sum_{n=0}^{\infty} a_n x^n$ and that this series converges for all $x \in (-R, R)$, where $R > 0$, then by applying our theorems on power series we obtain

$$\sum_{n=2}^{\infty} n(n-1) a_n x^n + \sum_{n=1}^{\infty} n a_n x^n + \sum_{n=0}^{\infty} a_n (x^{n+2} - x^n) = 0 \ .$$

A little care is needed for low values of n here, since the lowest power of x in the first series is 2, in the second 1, and so on. By writing the x^0 and x^1 terms separately and collecting powers of x^n we deduce that

$$-a_0 x^0 + \sum_{m=2}^{\infty} ((m^2 - 1)a_m + a_{m-2})x^m = 0 \ .$$

Since this is true for all $(-R, R)$, we may use Theorem 12.3 to deduce that the coefficients are all zero, that is, $a_0 = 0$ and $\forall m \geq 2$ $(m^2 - 1)a_m + a_{m-2} = 0$. These two equations show us immediately that $a_m = 0$ for all even m, while, for odd m, $a_m = -a_{m-2}/(m^2 - 1)$. Thus $a_3 = -a_1/(4.2)$, $a_5 = a_1/(6.4.4.2)$ and we may check by induction that

$$\forall n \in \mathbb{N} \quad a_{2n+1} = (-1)^n a_1/(2^{2n} n!(n + 1)!) \ .$$

Therefore

$$y(x) = a_1 \sum_{n=0}^{\infty} (-\tfrac{1}{4})^n \frac{x^{2n+1}}{n!(n + 1)!} \ .$$

The ratio test confirms that this series converges for all $x \in \mathbb{R}$, which gives us the information we need to check the validity of the whole process. We define a function $J_1(x)$ by

$$J_1(x) = \sum_{n=0}^{\infty} (-1)^n \frac{(x/2)^{2n+1}}{n!(n + 1)!} \qquad (x \in \mathbb{R})$$

and check, via the arithmetic above, that it satisfies the differential equation. The determination of the properties of J_1 , and a whole sequence of similar functions, is quite involved so we shall not pursue the matter. (For more information, see Dunning-Davies (1982).) Notice, however, that we have only produced one solution of the differential equation, where experience of differential equations leads us to expect two independent solutions. Close examination of the logic shows that if there is a second independent solution, it cannot be of the form of a function given by a power series. ☐

Example 12.4 Consider the function $f: \mathbb{R} \to \mathbb{R}$ defined by $f(0) = 0$, $f(x) = \exp(-1/x^2)$ $(x \neq 0)$. Since $e^{-y} \to 0$ as $y \to \infty$, we see that if $\varepsilon > 0$

$$\exists Y \text{ such that } y \geq Y \Rightarrow |e^{-y}| < \varepsilon$$

whence $0 < |x| < 1/\sqrt{Y} \Rightarrow 1/x^2 \geq Y \Rightarrow |f(x)| < \varepsilon$. Therefore f is continuous at 0.

Since $\sqrt{y}e^{-y} \to 0$ as $y \to \infty$ (Problem 3), we see that if $\varepsilon > 0$ $\exists Y$ such that $y \geq Y \Rightarrow |\sqrt{y}e^{-y}| < \varepsilon$, whence $0 < |x| < 1/\sqrt{Y} \Rightarrow |f(x)/x| < \varepsilon$. This shows f is differentiable at 0 and $f'(0) = 0$; at all non-zero x , $f'(x) = (2/x^3)\exp(-1/x^2)$ by the chain rule.

Pursuing these ideas shows that f possesses derivatives of all orders at 0 and that $\forall n \in \mathbb{N}$ $f^{(n)}(0) = 0$. If f were expressible as a power series $\sum a_n x^n$ for $x \in (-R, R)$ then we see by Theorem 9.5 that $a_n = f^{(n)}(0)/n! = 0$. In this case the series $\sum f^{(n)}(0) x^n /n!$ converges for all $x \in \mathbb{R}$ but for $x \neq 0$ its sum is not $f(x)$. The mere existence of all the derivatives of f is no guarantee that $\sum f^{(n)}(0) x^n /n!$ converges to $f(x)$. (This may also be seen by calculating the remainder R_n in Taylor's Theorem; in this case $R_n(x) = f(x)$ for all n, so $R_n(x) \not\to 0$.) This also shows that even if two functions, g and h, and all their derivatives are equal at one point a, the two functions need not be equal at any other point; consider $g(x) = h(x) + \exp(-1/x^2)$. ☐

As a final theme, we observe that we can very easily obtain an estimate of the difference between the sum of the first N terms and the sum to infinity of a power series.

Theorem 12.11 Suppose that the power series $\sum a_n x^n$ converges absolutely for all $x \in (-R, R)$, and that $0 < r < R$. Then, for each $N \in \mathbb{N}$, there is a constant K

such that
$$\forall x \in [-r, r] \quad \left| \sum_{n=0}^{\infty} a_n x^n - \sum_{n=0}^{N-1} a_n x^n \right| \le K|x|^N .$$

Proof: $\left| \sum_{n=0}^{\infty} a_n x^n - \sum_{n=0}^{N-1} a_n x^n \right| = \left| \sum_{n=N}^{\infty} a_n x^n \right| \le |x|^N \sum_{n=N}^{\infty} |a_n| |x|^{n-N} \le K|x|^N$ where

$K = \sum_{n=N}^{\infty} |a_n| r^{n-N}$. □

The point here is that K, though dependent on N and r, does not depend on x. This gives us scope for applications, the value of N being chosen for the purpose in hand.

Example 12.5 If $x \in \mathbb{R}$, $n\log(1 + x/n) \to x$ as $n \to \infty$, and $(1+\frac{x}{n})^n \to e^x$ as $n \to \infty$.

Solution: For $|y| < 1$, $\log(1+y) = \sum_{n=1}^{\infty} (-1)^{n-1} y^n / n$ (Problem 6). Therefore, by Theorem 12.11, there is a constant K such that
$$|y| \le \tfrac{1}{2} \Rightarrow |\log(1 + y) - y| \le K|y|^2 .$$

Let $x \in \mathbb{R}$. Then if $n \ge 2|x|$, $|x/n| \le \tfrac{1}{2}$ and $|\log(1 + x/n) - x/n| \le K|x/n|^2$ whence $|n\log(1 + x/n) - x| \le K|x|^2/n$. It is now clear that
$$\forall n > \max(2|x|, K|x|^2/\varepsilon) \quad |n\log(1 + x/n) - x| < \varepsilon .$$
Therefore $n\log(1 + x/n) \to x$ as $n \to \infty$.

Since $\exp$ is continuous and $(1 + x/n)^n = \exp(n\log(1 + x/n))$ we see that $(1 + x/n)^n \to e^x$ as $n \to \infty$. □

Problems
1. Calculate the radius of convergence of the following series, where there is one, or show the series converges for all $x \in \mathbb{R}$: $\sum x^n/n$, $\sum x^n /2^n$, $\sum ((2n)!/(n!)^2)x^n$, $\sum n^n x^n$, $\sum ((2n)!/(3n)!)x^n$, $\sum n x^{3n+1}$, $\sum 2^n x^{2n}$.

2. Show that the series $\sum x/(1 + |x|)^n$ is convergent for all real x, and find $\sum_{n=0}^{\infty} x/(1+|x|)^n$. Notice that the sum is discontinuous at $x = 0$.

3. Let $\alpha > 0$. By using the result $x^{-1}\log x \to 0$ as $x \to \infty$, deduce that, for all

sufficiently large x, $(\alpha \log x)/x < 1/2$ and hence $\alpha \log x - x < -x/2$. Hence show that $x^\alpha e^{-x} \to 0$ as $x \to \infty$. (α here need not be an integer.) Show also that $x^{-\alpha} e^x \to \infty$ as $x \to \infty$.

4. Calculate $\lim_{x \to 0+} x^x$ and $\lim_{x \to \infty} x^{1/x}$.

5. Define $f : \mathbb{R} \to \mathbb{R}$ by $f(x) = \sum_{n=0}^{\infty} a_n x^n$ (where the series is presumed to converge). Show that $a_n = f^{(n)}(0)/n!$.

6. Let $f(x) = \log(1 + x)$ (for $x > -1$) and show that $f^{(n)}(0) = (-1)^{n-1}(n - 1)!$ (for $n \geq 1$), and $f(0) = 0$. This suggests we define $g : (-1, 1) \to \mathbb{R}$ by $g(x) = \sum_{n=1}^{\infty} (-1)^{n-1} x^n / n$. Check this series has radius of convergence 1 and that $g'(x) = 1/(1 + x)$. By considering $(1 + x)e^{-g(x)}$, or otherwise, show that $g(x) = \log(1 + x)$.

7. Let $\sin x$ be defined by the equation $\sin x = \sum_{n=0}^{\infty} \dfrac{(-1)^n x^{2n+1}}{(2n + 1)!}$, and show that

$$\lim_{x \to 0} \frac{\sin x}{x} = 1 .$$

8. Show that there is a constant K such that $|y| \leq \frac{1}{2} \Rightarrow$ $|\sqrt{(1 + y)} - (1 + \frac{1}{2}y)| \leq K|y|^2$, and use this to show that

$$\lim_{x \to 0} \frac{\sqrt{(1+x)} - \sqrt{(1-x)}}{x} = 1 .$$

9. Show that if $x > 0$, $0 \leq x - \log(1 + x) \leq x^2/2$; use the Mean Value Theorem rather than the series for $\log(1 + x)$. Deduce that if we set $a_n = (1 + \frac{1}{2} + \ldots + \frac{1}{n}) - \log(n + 1)$ then (a_n) is increasing and bounded above. Deduce that $1 + \frac{1}{2} + \ldots + \frac{1}{n} - \log n$ tends to a limit as $n \to \infty$.

10. Suppose that $\forall x \in (-R, R)$ $y(x) = \sum_{n=0}^{\infty} a_n x^n$ and that y satisfies the differential equation $x^2(dy/dx) = y - x$. Find the coefficients a_n and the radius of convergence of this series.

11.* Define $f(x) = e^{-1/x}$ (for $x > 0$) , $f(x) = 0$ (for $x \leq 0$). Show that f is differentiable at 0 and $f'(0) = 0$. Show also that for $x > 0$, $f'(x) = (1/x^2)e^{-1/x}$ and that f' is continuous at 0, and for each $n \in \mathbb{N}$ show that $f^{(n)}(0) = p_n(1/x)e^{-1/x}$ for some polynomial p_n of degree at most $2n$. Deduce that for all natural numbers n, f is n times differentiable at 0, and $f^{(n)}(0) = 0$.

13

Integration

"But just as much as it is easy to find the differential of any given quantity, so it is difficult to find the integral of a given differential. Moreover, sometimes we cannot even say with certainty whether the integral of a given quantity can be found or not ...."
Johann Bernoulli (1691)

13.1 The Integral

The idea of the integral as the area under a curve is an old one, and it was known before the rise of calculus. The great feature which established the calculus was the observation that integration and differentiation are related so that if it is possible to find a function F whose derivative is f then

$$\int_a^b f(x)\,dx = F(b) - F(a) ,$$

where $\int_a^b f(x)\,dx$ is the area under the curve $y = f(x)$ between $x = a$ and $x = b$, solving the problem of integrating f. This method of integration, by finding a function whose derivative is the function we wish to integrate, is sometimes called 'anti-differentiation'.

While this provides an easy method for dealing with many integrals, it is not effective in all cases. That word "if" in the statement "if it is possible to find a function F ..." is important! Since the process of finding a function whose derivative is f is fairly unsystematic, relying on noticing suitable substitutions and so on, it is of no help at all in those cases where we are unable to find the desired function. Indeed, it can be shown that certain functions, such as e^{-x^2}, do not have an indefinite integral which can be expressed in terms of algebraic functions, exponentials and logarithms. We need an idea of integral which will allow us to deal with all the functions likely to occur in mathematics, and one where, even if in specific cases we cannot evaluate the integral exactly, we can still calculate it approximately to within some prescribed accuracy. The method is to return to the idea of the area under a curve and work with that.

Let $f: [a, b] \to \mathbb{R}$ be a bounded function, and choose a finite sequence of numbers $x_0, x_1, ..., x_n$ such that $a = x_0 < x_1 < ... < x_n = b$. Then choose numbers M_i for which $x_{i-1} \le x \le x_i \Rightarrow f(x) \le M_i$. The intuitive idea of area now tells us that the area under the curve $y = f(x)$ between $x = x_{i-1}$ and $x = x_i$ will not exceed $M_i(x_i - x_{i-1})$, the area of the rectangle of height M_i and base $[x_{i-1}, x_i]$. Therefore the area under f between $x = a$ and $x = b$ will be no more than $\sum_{i=1}^n M_i(x_i - x_{i-1})$. In this procedure we require of M_i only that it should satisfy $\forall x \in [x_{i-1}, x_i]\ M_i \ge f(x)$, which leaves us the option of choosing M_i very much larger than the values of $f(x)$. Obviously the most useful estimate of the area under

f is obtained by choosing M_i as small as possible, that is, $M_i = \sup\{f(x): x_{i-1} \leq x \leq x_i\}$. (Notice, in passing, that since f is given to be bounded, there is no doubt about the existence of the supremum.) The value of the sum $\sum_{i=1}^{n} M_i(x_i - x_{i-1})$ we obtain by this process will depend on the choice of the points $x_0, ..., x_n$ which we use to subdivide the interval $[a, b]$, and we would expect that, by taking more points of dissection, we can make the sum smaller, and thus a closer approximation to the area under the curve f. The independent variable here is the dissection of the interval, a complicated

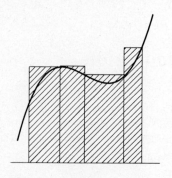

Fig. 13.1.

object, so we will not look for a limit but seek the 'smallest' value of these sums or, in case the set of all these sums has no smallest member, the infimum over all the sums obtained.

Before we rush on with this, notice that we could just as well approximate the integral from below. In this case, if $a = x_0 < x_1 < ... < x_n = b$ and $m_i = \inf\{f(x): x_{i-1} \leq x \leq x_i\}$, then $\sum_{i=1}^{n} m_i(x_i - x_{i-1})$ should not exceed the area under the curve $y = f(x)$ between $x = a$ and $x = b$, and taking more points of dissection ought to make the sum closer to the area under the curve. This leads us to consider the supremum of the set of all sums

$$\sum_{i=1}^{n} m_i(x_i - x_{i-1}) .$$

These two ways of evaluating what we think of as the area under a curve are equally valid and either could be used as the definition of the integral, *provided they give*

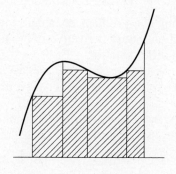

Fig. 13.2.

the same answer. Intuitively, it is hard to imagine the two processes giving different answers, at least for 'ordinary' functions, but we must not presume that there are no peculiar functions which would yield different answers. The definition below acknowledges this issue.

Definition Let $a < b$ and let $f: [a, b] \to \mathbb{R}$ be a bounded function. A **dissection** of the interval $[a, b]$ is a finite sequence of points $x_0, ..., x_n$ such that $a = x_0 < x_1 < ... < x_n = b$. Suppose that $D = \{x_0, x_1, ..., x_n\}$ is a dissection. Then by the **upper sum**, $S(D)$, we mean

$$S(D) = \sum_{i=1}^{n} M_i(x_i - x_{i-1}) ,$$

where $M_i = \sup\{f(x): x_{i-1} \leq x \leq x_i\}$, and by the **lower sum**, $s(D)$, we mean

$$s(D) \;=\; \sum_{i=1}^{n} m_i(x_i - x_{i-1}) \;,$$

where $m_i = \inf\{f(x): x_{i-1} \le x \le x_i\}$. Where it is necessary to distinguish the relevant function we shall denote these sums by $S(f, D)$ and $s(f, D)$ respectively.

We say that the function f is **integrable** (on $[a, b]$) if and only if $\inf\{S(D): D \text{ is a dissection}\} = \sup\{s(D): D \text{ is a dissection}\}$, and when f is integrable we denote the common value of these two quantities by $\int_a^b f$ or $\int_a^b f(x)\mathrm{d}x$. $\square$

Lemma 13.1 Let $f: [a, b] \to \mathbb{R}$ be a bounded function and let D_1 and D_2 be two dissections of $[a, b]$. If D_2 contains every point of dissection that belongs to D_1 then $S(D_2) \le S(D_1)$ and $s(D_2) \ge s(D_1)$. Also

$$\sup\{s(D): D \text{ is a dissection}\} \le \inf\{S(D): D \text{ is a dissection}\}.$$

Proof. We shall prove that $S(D_2) \le S(D_1)$ in the case where D_2 contains one more point of dissection that D_1. The general case is obtained by applying this result a finite number of times, since D_2 can, in general, have only finitely many more points than D_1.

Let $D_1 = \{x_0, x_1, \ldots, x_n\}$ and $D_2 = \{x_0, \ldots, x_{j-1}, y, x_j, \ldots, x_n\}$ where $a = x_0 < x_1 < \ldots < x_{j-1} < y < x_j < \ldots < x_n = b$. Then $S(D_1) = \sum_{i=1}^{n} M_i(x_i - x_{i-1})$ where $M_i = \sup\{f(x): x_{i-1} \le x \le x_i\}$, $S(D_2)$ being given by a similar expression with the term $M_j(x_j - x_{j-1})$ replaced by $\overline{M}_j(y - x_{j-1}) + M'_j(x_j - y)$ where $\overline{M}_j = \sup\{f(x): x_{j-1} \le x \le y\}$ and $M'_j = \sup\{f(x): y \le x \le x_j\}$. Therefore

$$
\begin{aligned}
S(D_1) - S(D_2) &= M_j(x_j - x_{j-1}) - \overline{M}_j(y - x_{j-1}) - M'_j(x_j - y) \\
&= (M_j - \overline{M}_j)(y - x_{j-1}) + (M_j - M'_j)(x_j - y) \ge 0
\end{aligned}
$$

since $M_j \ge \overline{M}_j$ and $M_j \ge M'_j$ from their definitions.

The proof that $s(D_2) \ge s(D_1)$ is virtually identical.

Now let D_3 and D_4 be two typical dissections and form the dissection D_5 containing those points which are either in D_3 or in D_4. By what we have just proved, $S(D_5) \le S(D_3)$ and $s(D_5) \ge s(D_4)$. Moreover, from the definition of upper and lower sums, $s(D_5) \le S(D_5)$ (since $m_i \le M_i$ in the notation of the definition). Therefore $s(D_4) \le S(D_3)$.

Since D_3 and D_4 were typical in the above paragraph, we see that if we choose D_3, then for all dissections D_4, $s(D_4) \le S(D_3)$ hence $\sup\{s(D_4): D_4 \text{ is a dissection}\} \le S(D_3)$. Since this result is true for all D_3, $\sup\{s(D_4): D_4 \text{ is a dissection}\} \le \inf\{S(D_3): D_3 \text{ is a dissection}\}$. $\square$

Note: The inequality between two upper sums or between two lower sums requires that the two dissections be comparable, that is, one contains all the points of the other. The inequality between an upper sum and a lower sum does not require that the dissections be comparable; every upper sum is at least as large as every lower sum.

This allows us a simple criterion for integrability:

Lemma 13.2 Let $f: [a, b] \to \mathbb{R}$ be a bounded function. Then f is integrable if and only if $\forall \varepsilon > 0 \; \exists$ a dissection D such that $S(D) - s(D) < \varepsilon$.

Proof: Let f be integrable and suppose that $\varepsilon > 0$. From the definitions of sup and inf, there exist dissections D_1 and D_2 satisfying

$$S(D_1) < \inf\{S(D): D \text{ is a dissection}\} + \varepsilon/2$$

and
$$s(D_2) > \sup\{s(D): D \text{ is a dissection}\} - \varepsilon/2 \ .$$

Let I denote the common value of $\inf\{S(D)\}$ and $\sup\{s(D)\}$ (f is integrable) and let D_3 be the dissection consisting of points which occur in D_1 or in D_2 . Then $S(D_3) \le S(D_1) < I + \varepsilon/2 < s(D_2) + \varepsilon \le s(D_3) + \varepsilon$, yielding $S(D_3) - s(D_3) < \varepsilon$. Since $\varepsilon > 0$ was arbitrary we have proved one implication.

Conversely, suppose that $\forall \varepsilon > 0 \; \exists D$ such that $S(D) - s(D) < \varepsilon$. Let $\varepsilon > 0$ and choose D_0 such that $S(D_0) - s(D_0) < \varepsilon$. Then

$$\inf\{S(D): D \text{ is a dissection}\} \le S(D_0) \quad < s(D_0) + \varepsilon$$

$$\le \sup\{s(D): D \text{ is a dissection}\} + \varepsilon$$

Since ε was arbitrary we deduce that $\forall \varepsilon > 0 \; \inf\{S(D)\} < \sup\{s(D)\} + \varepsilon$, whence $\inf\{S(D)\} \le \sup\{s(D)\}$. The opposite inequality follows from Lemma 13.1, whence f is integrable. $\square$

Example 13.1 Let $f: [0, 1] \to \mathbb{R}$ be defined by $f(x) = x$; f is integrable.

Solution: Let $D_n = \{0, 1/n, 2/n, \ldots, 1\}$, so, letting $x_j = j/n$, we have $M_j = \sup\{f(x): x_{j-1} \le x \le x_j\} = j/n$ and $m_j = \inf\{f(x): x_{j-1} \le x \le x_j\} = (j - 1)/n$. Then

$$S(D_n) - s(D_n) = \sum_{j=1}^{n} (M_j - m_j)(x_j - x_{j-1}) = \sum_{j=1}^{n} 1/n^2 = 1/n . \qquad \text{Thus, given}$$

$\varepsilon > 0$, if we choose $n > 1/\varepsilon$ we obtain D_n for which $S(D_n) - s(D_n) < \varepsilon$.

In fact, $\inf\{S(D): D \text{ all dissections } D\} \le \inf\{S(D_n): n \in \mathbb{N}\} = 1/2$ (since

$$S(D_n) = \sum_{j=1}^{n} j/n^2 = \tfrac{1}{2}(1 + 1/n)) . \qquad \text{Also, since} \qquad s(D_n) = \tfrac{1}{2}(1 - 1/n) ,$$

$\sup\{s(D): \text{ all dissections } D\} \ge \sup\{s(D_n): n \in \mathbb{N}\} = 1/2$, so

$$1/2 \le \sup\{s(D): \text{ all } D\} \le \inf\{S(D): \text{ all } D\} \le 1/2$$

and there is equality throughout, showing that $\int_0^1 f = 1/2$. $\square$

The criterion of Lemma 13.2 in fact tells us more about approximations to the integral than we have so far stated.

Lemma 13.3 Let $f: [a, b] \to \mathbb{R}$ be integrable. Then if $\varepsilon > 0$ there is a dissection D_0 such that $S(D_0) - s(D_0) < \varepsilon$ and if D is a dissection whose points of subdivision include those of D_0, we have

$$\left| \int_a^b f - \sum_{j=1}^n f(\xi_j)(x_j - x_{j-1}) \right| < \varepsilon$$

where $D = \{x_0, \ldots, x_n\}$ and the ξ_j are numbers satisfying $x_{j-1} \leq \xi_j \leq x_j$ ($j = 1, 2, \ldots, n$).

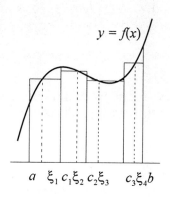

Fig. 13.3.

Proof: Let $\varepsilon > 0$. By Lemma 13.2 there is dissection D_0 with $S(D_0) - s(D_0) < \varepsilon$. Choose D to be a dissection containing all the points of D_0, and it follows that

$$S(D) \leq S(D_0) < s(D_0) + \varepsilon \leq \int_a^b f + \varepsilon \qquad \text{and,}$$

similarly, $s(D) > \int_a^b f - \varepsilon$. Now choosing ξ_j with $x_{j-1} \leq \xi_j \leq x_j$ we have, for each j

$$m_j = \inf\{f(x): x_{j-1} \leq x \leq x_j\} \leq f(\xi_j) < \sup\{f(x): x_{j-1} \leq x \leq x_j\} = M_j$$

so $s(D) \leq \sum_{j=1}^n f(\xi_j)(x_j - x_{j-1}) \leq S(D)$, whence

$$\int_a^b f - \varepsilon < \sum_{j=1}^n f(\xi_j)(x_j - x_{j-1}) < \int_a^b f + \varepsilon \ ,$$

which is what we require. □

The definition of integral that we have made has the virtue that it is very general but as a technique for calculating the value of the integral exactly it is far from good. Our programme will be to show that the class of functions which have integrals is usefully wide, then to establish the properties of the integral, with the hope that the properties will allow us a reasonable method of calculating most integrals that arise.

Showing that a wide class of functions is integrable.
Lemma 13.4 If $f: [a, b] \to \mathbb{R}$ is either increasing or decreasing, it is integrable.

Proof: We shall prove the case where f is increasing; the other is virtually identical.

Suppose that $D = \{x_0, \ldots x_n\}$ is a dissection $[a, b]$. Then, because f is increasing, $\sup\{f(x): x_{j-1} \leq x \leq x_j\} = f(x_j)$ and $\inf\{f(x): x_{j-1} \leq x \leq x_j\} = f(x_{j-1})$ so

$$S(D) - s(D) = \sum_{j=1}^n (f(x_j) - f(x_{j-1}))(x_j - x_{j-1}) \ .$$

The issue is to show that, given $\varepsilon > 0$, a dissection can be chosen for which this quantity is less than ε.

Let $\varepsilon > 0$. Assuming $f(b) \neq f(a)$, so that $f(b) > f(a)$ since f is increasing, choose the dissection $D = \{x_0, \ldots x_n\}$ such that, for each j, $x_j - x_{j-1} < \varepsilon/(f(b) - f(a))$, which is obviously possible if $[a, b]$ is divided into sufficiently many equal parts, for example. Then

$$S(D) - s(D) = \sum_{j=1}^{n} (f(x_j) - f(x_{j-1}))(x_j - x_{j-1})$$

$$< \{\varepsilon / (f(b) - f(a))\} \sum_{j=1}^{n} (f(x_j) - f(x_{j-1})) = \varepsilon .$$

Since $\varepsilon > 0$ was arbitrary, Lemma 13.2 now shows that f is integrable. If $f(b) = f(a)$, then f is constant, because increasing, so f is integrable. $\square$

In the proof just given, $S(D) - s(D)$ was expressed in the form $\sum_{j=1}^{n}(M_j - m_j)(x_j - x_{j-1})$ and we proved that this was small by making all of the $(x_j - x_{j-1})$ small and ensuring that we knew $\sum(M_j - m_j)$. The alternative tactic of ensuring that, for all j, $M_j - m_j$ is small seems useful since we know $\sum(x_j - x_{j-1})$. This requires a little more work. Now

$$M_j - m_j \quad = \sup\{f(x)\colon x \in [x_{j-1}, x_j]\} - \inf\{f(x)\colon x \in [x_{j-1}, x_j]\}$$

$$= \sup\{f(x)\colon x \in [x_{j-1}, x_j]\} + \sup\{-f(y)\colon y \in [x_{j-1}, x_j]\}$$

$$= \sup\{f(x) - f(y)\colon x, y \in [x_{j-1}, x_j]\}$$

$$= \sup\{|f(x) - f(y)|\colon x, y \in [x_{j-1}, x_j]\}$$

(*Caution.* Notice that although $\sup\{-f(x)\colon x \in [x_{j-1}, x_j]\}$ is equal to $\sup\{-f(y)\colon y \in [x_{j-1}, x_j]\}$ since the symbols x and y are "dummy" variables, that is, the value of the supremum does not depend on x or y, the quantity we wish here is $\sup\{f(x) - f(y)\colon x, y \in [x_{j-1}, x_j]\}$ where x and y are chosen to vary independently. This is generally not the same thing as $\sup\{f(x) - f(x)\colon x \in [x_{j-1}, x_j]\}$, which equals 0.)

We therefore define the **oscillation** of f on the set A, written $\underset{A}{\operatorname{osc}} f$, by

$$\underset{A}{\operatorname{osc}} f = \sup\{|f(x) - f(y)|\colon x, y \in A\} = \sup\{f(x)\colon x \in A\} - \inf\{f(y)\colon y \in A\}$$

If f is continuous at c and $\varepsilon > 0$ then we know that there is a $\delta > 0$ for which $x \in (c - \delta, c + \delta) \Rightarrow |f(x) - f(c)| < \varepsilon/2$ whence $x, y \in (c - \delta, c + \delta) \Rightarrow |f(x) - f(y)| < \varepsilon$ so that $\underset{(c-\delta, c+\delta)}{\operatorname{osc}} f \le \varepsilon$. More importantly, we may extend this result:

Theorem 13.5 Let $a < b$ and $f\colon [a, b] \to \mathbb{R}$ be continuous. Then if $\varepsilon > 0$ there is a dissection $D = \{x_0, \dots x_n\}$ of $[a, b]$ such that, for each j, $\underset{[x_{j-1}, x_j]}{\operatorname{osc}} f \le \varepsilon$.

Proof. Let $\varepsilon > 0$. By our remarks above, $\exists \delta > 0$ such that $\underset{(a, a+\delta)}{\operatorname{osc}} f \le \varepsilon$ so if we choose $a < x < a + \delta$, the trivial dissection of $[a, x]$ (that is, the dissection $\{a, x\}$) has the property that the oscillation on its one subinterval does not exceed ε.

Let $A = \{x \in [a, b]\colon [a, x]$ has a dissection D such that on every subinterval of D, $\operatorname{osc} f \le \varepsilon\}$. The previous paragraph shows that for some $\delta > 0$,

$(a, a + \delta) \subset A$ so that A is non-empty and $\sup A > a$. ($\sup A$ exists since A is also bounded above by b.) Let $\xi = \sup A$.

Firstly we show that $\xi \le b$. Suppose that $\xi < b$. Then since f is continuous at ξ, $\exists \delta > 0$ such that $\underset{(\xi - \delta, \xi + \delta)}{\mathrm{osc}} f \le \varepsilon$. Since $\xi - \delta < \xi$ there is an element, say x, of A with $x > \xi - \delta$. Then there is a dissection, say $\{x_0, ..., x_n\}$, of $[a, x]$ with $\underset{[x_{j-1}, x_j]}{\mathrm{osc}} f \le \varepsilon$ for each j. Let $x_{n+1} = \xi + \delta/2$, so that

$\underset{[x_n, x_{n+1}]}{\mathrm{osc}} f \le \varepsilon$ since $[x_n, x_{n+1}] \subset (\xi - \delta, \xi + \delta)$ and every subinterval of the dissection $\{x_0, ..., x_{n+1}\}$ has $\mathrm{osc}\, f \le \varepsilon$, so $x_{n+1} = \xi + \delta/2 \in A$. But this is a contradiction since $\xi + \delta/2 > \sup A$. Therefore $\xi \ge b$.

We know that $\xi \ge b$, hence $\xi = b$ since b is an upper bound for A. We need to show $b \in A$. Because f is continuous at b, $\exists \delta > 0$ such that $\underset{(b - \delta, b]}{\mathrm{osc}} f \le \varepsilon$. Since $b - \delta < b = \sup A$, there is a member of A, say x, with $x > b - \delta$. Then there is a dissection $\{x_0, ... x_n\}$ of $[a, x]$ for which, for $j = 1, 2,$ $..., n$, $\underset{[x_{j-1}, x_j]}{\mathrm{osc}} f \le \varepsilon$ whence $\mathrm{osc}\, f \le \varepsilon$ on each subinterval of $\{x_0, ..., x_n, b\}$, and $b \in A$ as required.

Since $\varepsilon > 0$ was arbitrary we have proved the result for all $\varepsilon > 0$. $\square$

Corollary If $f: [a, b] \to \mathbb{R}$ is continuous, it is integrable.

Proof. Let $\varepsilon > 0$. By Theorem 13.5 there is a dissection $D = \{x_0, ... x_n\}$ of $[a, b]$ with $\underset{[x_{j-1}, x_j]}{\mathrm{osc}} f \le \varepsilon/(2(b - a))$ for all j. Then

$$S(D) - s(D) \;=\; \sum_{j=1}^{n} (M_j - m_j)(x_j - x_{j-1}) \;=\; \sum_{j=1}^{n} (\underset{[x_{j-1}, x_j]}{\mathrm{osc}} f)(x_j - x_{j-1})$$

$$\le\; \varepsilon/(2(b-a)) \sum_{j=1}^{n} (x_j - x_{j-1}) \;=\; \varepsilon/2 < \varepsilon.$$

By Lemma 13.2, f is integrable. $\square$

Although it is not about integration, it is useful to state another important corollary of Theorem 13.5 here.

Theorem 13.6 The Uniform Continuity Theorem. Let $f: [a, b] \to \mathbb{R}$ be continuous. Then for all $\varepsilon > 0$ there is a $\delta > 0$ such that

$$\forall x, y \in [a, b] \; |x - y| < \delta \Rightarrow |f(x) - f(y)| < \varepsilon.$$

Proof. Suppose that $\varepsilon > 0$. Then, since $\varepsilon/3 > 0$, Theorem 13.5 ensures that there is a dissection $D = \{x_0, ..., x_n\}$ such that, for $j = 1, 2, ..., n$, $\underset{[x_{j-1}, x_j]}{\mathrm{osc}} f \le \varepsilon/3$. Let $\delta = \min(x_1 - x_0, ..., x_n - x_{n-1})$, so that $\delta > 0$, and if $|x - y| < \delta$ then x and y either lie in the same subinterval of D or lie in adjacent subintervals. Let c be the common endpoint of the two adjacent intervals to which x and y belong, or an endpoint if x and y belong to the same interval; then

$$|f(x) - f(y)| \le |f(x) - f(c)| + |f(c) - f(y)| \le \varepsilon/3 + \varepsilon/3 < \varepsilon$$

as x and c belong to the same subinterval and so do c and y. We have shown that $\forall \varepsilon > 0 \ \exists \delta > 0$ such that $\forall x, y \in [a, b] \ |x - y| < \delta \Rightarrow |f(x) - f(y)| < \varepsilon$. $\square$

Remarks: Notice what the Uniform Continuity Theorem says: if f is defined and continuous on a closed bounded interval then, given $\varepsilon > 0$, there is a $\delta > 0$ (and the *same* δ at all points) such that for any two points of $[a, b]$ less than δ apart, $|f(x) - f(y)| < \varepsilon$. This is a stronger statement than continuity, since if f is continuous we know that f is continuous at each point y, that is, given $\varepsilon > 0$ and $y \in [a, b]$, there is a $\delta > 0$ such that if $|x - y| < \delta$ then $|f(x) - f(y)| < \varepsilon$; in this case the existence of δ is guaranteed once we fix ε and y, so δ may depend on y.

Theorems 13.5 and 13.6 depend on the fact that $[a, b]$ is closed and they are false if this is relaxed. Consider $f(x) = 1/x$ where $f: (0, 1] \to \mathbb{R}$. Let $\delta > 0$ and choose $n \in \mathbb{N}$ with $1/n < \delta$. Then if we set $x = 1/(2n)$ and $y = 1/n$ we have $|x - y| = 1/(2n) < \delta$ and $|f(x) - f(y)| = n \ge 1$. Thus $\forall \delta > 0 \ \exists x, y \in (0,1)$ with $|x - y| < \delta$ and $|f(x) - f(y)| \ge 1$, so f does not have the 'uniform' continuity property above (since it does not hold for $\varepsilon = 1$).

Theorem 13.7 Let f and g be integrable functions on $[a, b]$ and λ be a real number. Then (i) $f + g$, λf, fg and $|f|$ are all integrable, (ii) if there is a constant $m > 0$ for which $\forall x \in [a, b] \ |g(x)| \ge m$ is integrable, and

(iii) $\int_a^b (f + g) = \int_a^b f + \int_a^b g$ and $\int_a^b \lambda f = \lambda \int_a^b f$.

Proof: To prove the various functions integrable, we shall use Lemma 13.2. To do this we shall need to show, for a given function h and $\varepsilon > 0$, that for some D,

$$S(h; D) - s(h; D) = \sum_{j=1}^{n} (\underset{[x_{j-1}, x_j]}{\operatorname{osc} h}) (x_j - x_{j-1}) < \varepsilon. \quad \text{We shall do this by relating}$$

osc h to the oscillation of the functions defining it.

$$\underset{[x_{j-1}, x_j]}{\operatorname{osc}} (f + g) = \sup\{|f(x) + g(x) - (f(y) + g(y))| : x, y \in [x_{j-1}, x_j]\}$$

$$\le \sup\{|f(x) - f(y)| + |g(x) - g(y)| : x, y \in [x_{j-1}, x_j]\}$$

$$\le \sup\{|f(x) - f(y)| : x, y \in [x_{j-1}, x_j]\}$$

$$+ \sup\{|g(x) - g(y)| : x, y \in [x_{j-1}, x_j]\}$$

$$= \underset{[x_{j-1}, x_j]}{\operatorname{osc} f} + \underset{[x_{j-1}, x_j]}{\operatorname{osc} g}.$$

Let $\varepsilon > 0$. Since f and g are integrable on $[a, b]$ there are dissections D_1 and D_2 such that $S(f; D_1) - s(f; D_1) < \varepsilon/2$ and $S(g; D_2) - s(g; D_2) < \varepsilon/2$. Letting D_3 be the dissection whose points of subdivision are those occurring in either D_1

or D_2 , we have $S(f; D_3) - s(f; D_3) \le S(f; D_1) - s(f; D_1) < \varepsilon/2$ and, similarly, $S(g; D_3) - s(g; D_3) < \varepsilon/2$. Thus if $D_3 = \{x_0, ..., x_n\}$,

$$S(f + g; D_3) - s(f + g; D_3) = \sum_{j=1}^{n} \operatorname*{osc}_{[x_{j-1}, x_j]} (f + g)(x_j - x_{j-1})$$

$$\le \sum_{j=1}^{n} (\operatorname*{osc}_{[x_{j-1}, x_j]} f + \operatorname*{osc}_{[x_{j-1}, x_j]} g)(x_j - x_{j-1})$$

$$= S(f; D_3) - s(f; D_3) + S(g; D_3) - s(g; D_3) < \varepsilon .$$

Since $\varepsilon > 0$ was arbitrary we have shown that $f + g$ is integrable.

For λf we notice that $\operatorname{osc} \lambda f = |\lambda| \operatorname{osc} f$. Then, for $\varepsilon > 0$, choose D_1 such that $S(f; D_1) - s(f; D_1) < \varepsilon/|\lambda|$, whence $S(\lambda f; D_1) - s(\lambda f; D_1) < \varepsilon$. Thus λf is integrable, since the above holds for all $\varepsilon > 0$. (The case $\lambda = 0$ is trivial.)

Let M_f and M_g be upper bounds for $|f|$ and $|g|$ respectively. Then for all $x, y \in [a, b]$,

$$|f(x)g(x) - f(y)g(y)| \le |f(x)||g(x) - g(y)| + |f(x) - f(y)||g(y)|$$
$$\le M_f |g(x) - g(y)| + M_g |f(x) - f(y)| .$$

Therefore, on all subsets A of $[a, b]$,

$$\operatorname*{osc}_{A} fg \le M_f \operatorname*{osc}_{A} g + M_g \operatorname*{osc}_{A} f ,$$

giving

$$S(fg; D) - s(fg; D) \le M_f (S(g; D) - s(g; D)) + M_g (S(f; D) - s(f; D)) .$$

Given $\varepsilon > 0$ it is only necessary to choose D_1 and D_2 such that $S(f; D_1) - s(f; D_1) < \varepsilon/(M_f + M_g)$ and $S(g; D_2) - s(g; D_2) < \varepsilon/(M_f + M_g)$, giving $S(fg; D_3) - s(fg; D_3) < \varepsilon$ where D_3 contains all the points of D_1 and D_2.

Since $\||f(x)| - |f(y)\|| \le |f(x) - f(y)|$, $\operatorname{osc} |f| \le \operatorname{osc} f$, and the usual routine shows that $|f|$ is integrable.

The integrability of $1/g$ follows from noticing that, with the given condition, $|1/g(x) - 1/g(y)| \le |g(x) - g(y)|/m^2$, so $\operatorname{osc} (1/g) \le (\operatorname{osc} g)/m^2$.

For part (iii), we need to estimate the values of the integrals. Let $\varepsilon > 0$. Choose D_1 and D_2 so that $S(f; D_1) - s(f; D_1) < \varepsilon/2$ and $S(g; D_2) - s(g; D_2) < \varepsilon/2$. Let D_3 be the dissection whose points of subdivision are those occurring in either D_1 or D_2 . Then

$$\int_a^b (f + g) \le S(f + g; D_3) \le S(f; D_3) + S(g; D_3) < s(f; D_3) + s(g; D_3) + \varepsilon$$

$$\le \int_a^b f + \int_a^b g + \varepsilon .$$

Therefore

$$\forall \varepsilon > 0 \ \int_a^b (f + g) \le \int_a^b f + \int_a^b g + \varepsilon , \quad \text{whence} \quad \int_a^b (f + g) \le \int_a^b f + \int_a^b g .$$

Arguing similarly in terms of $s(f + g; D_3)$ we may show the reverse inequality also holds and so $\int_a^b (f + g) = \int_a^b f + \int_a^b g$.

To evaluate $\int_a^b \lambda f$, notice that $S(\lambda f; D) = \lambda S(f; D)$ if $\lambda \ge 0$ and

$S(\lambda f; D) = \lambda s(f; D)$ if $\lambda \le 0$, from which the equality of $\int_a^b \lambda f$ and $\lambda \int_a^b f$ may be seen from the definition of integral. $\square$

This gives us a wide class of integrable functions: those which are increasing or continuous or which can be expressed in terms of sums, products, etc. of such functions. To find a function which is not integrable we have to resort to something fairly exotic.

Example 13.2 Define $f: [0, 1] \to \mathbb{R}$ by $f(x) = 1$ if x is rational and $f(x) = 0$ if x is irrational. f is not integrable.

To see this, let $D = \{x_0, \ldots x_n\}$. Then since each interval $[x_{j-1}, x_j]$ contains both rational and irrational numbers (as $x_{j-1} < x_j$), $\sup\{f(x): x \in [x_{j-1}, x_j]\} = 1$ and $\inf\{f(x): x \in [x_{j-1}, x_j]\} = 0$. This gives $S(f; D) = \sum_{j=1}^n (x_j - x_{j-1}) = 1$ and $s(f; D) = 0$. Thus $\inf S(f; D) = 1 \ne 0 = \sup s(f; D)$. $\square$

Despite the existence of non-integrable functions, we shall not find this presents a problem. The issue becomes troublesome only when one wishes to consider limits of sequences of functions, which we shall not do in this book.

The point we have reached is that all continuous functions, together with those which are increasing or decreasing, and sums and products of finitely many such functions are all integrable so that our results will apply to them automatically. This, with a little help from the results about manipulating integrals, essentially takes care of the question of existence of the integral for our purposes.

Manipulation of Integrals

Lemma 13.8 Let $a < b < c$. A function $f: [a, c] \to \mathbb{R}$ is integrable on $[a, c]$ if and only if it is integrable on both $[a, b]$ and $[b, c]$. When f is integrable on $[a, c]$, $\int_a^c f = \int_a^b f + \int_b^c f$.

Proof: Suppose that f is integrable on $[a, b]$ and $[b, c]$. Let $\varepsilon > 0$. Choose dissections $D_1 = \{x_0, \ldots, x_n\}$ of $[a, b]$ and $D_2 = \{y_0, \ldots, y_n\}$ of $[b, c]$ satisfying $S(D_1) - s(D_1) < \varepsilon/2$ and $S(D_2) - sD_2) < \varepsilon/2$. Then $D_3 = \{x_0, \ldots, x_n, y_1, \ldots, y_n\}$ is a dissection of $[a, c]$ and $S(D_3) = S(D_1) + S(D_2)$, $s(D_3) = s(D_1) + s(D_2)$, whence $S(D_3) - s(D_3) < \varepsilon$. Thus

$$\int_a^c f \le S(D_3) = S(D_1) + S(D_2) < s(D_1) + s(D_2) + \varepsilon \le \int_a^b f + \int_b^c f + \varepsilon$$

and

$$\int_a^c f \ge s(D_3) = s(D_1) + s(D_2) > S(D_1) + S(D_2) - \varepsilon > \int_a^b f + \int_b^c f - \varepsilon$$

Since ε was arbitrary, we see that f is integrable on $[a, b]$ and $\forall \varepsilon > 0$ $\left| \int_a^c f - (\int_a^b f + \int_b^c f) \right| < \varepsilon$ and hence $\int_a^c f = \int_a^b f + \int_b^c f$.

Conversely, suppose that f is integrable on $[a, c]$ and let $\varepsilon > 0$. Then there is a dissection D for which $S(D) - s(D) < \varepsilon$. Let D_3 be the dissection obtained from D by inserting b as a point of dissection, so $S(D_3) - s(D_3) < \varepsilon$. Now let D_1 consist of those points of dissection of D_1 lying in $[a, b]$, and D_2 consist of those in $[b, c]$. Since $(S(D_1) - s(D_1)) + (S(D_2) - s(D_2)) = S(D_3) - s(D_3) < \varepsilon$ and both brackets on the left are non-negative, $S(D_1) - s(D_1) < \varepsilon$ and $S(D_2) - s(D_2) < \varepsilon$. Since ε was arbitrary, f is integrable on $[a, b]$ and $[b, c]$. The equality of the integrals follows from the first part. □

Lemma 13.9 Suppose that $f(x) = g(x)$ for all but finitely many points of $[a, b]$. Then f is integrable if and only if g is, and $\int_a^b f = \int_a^b g$ when either function is integrable. In particular, if $h(x) = 0$ at all but finitely many points of $[a, b]$, then $\int_a^b h = 0$.

Proof. We shall prove a special case and deduce the general one from it. Suppose $h(x) = 0$ for all $x \in [c, d]$ except one of the endpoints; for definiteness, let $h(c) = \alpha > 0$. Then if $D_\delta = \{c, c + \delta, d\}$ $S(D_\delta) = \alpha\delta$ and $s(D_\delta) = 0$, so

$$\inf\{S(D): \text{all dissections } D\} \leq \inf\{S(D_\delta): \delta > 0\} = 0 = \sup s(D) .$$

Thus h is integrable (because we already know that $\sup s(D) \leq \inf S(D)$) and $\int_c^d h = 0$. The result for a function h which is non-zero at finitely many points may now be deduced by splitting the interval $[a, b]$ at the points at which h is non-zero, observing that the integral of h on each subinterval is zero and using Lemma 13.8.

If $f(x) = g(x)$ for all but finitely many points x, then $f - g$ is zero at all but finitely many points, so the preceding paragraph shows that $f - g$ is integrable and $\int_a^b (f - g) = 0$. The rest is easy. □

Example 13.3 Let $f: [a, b] \rightarrow \mathbb{R}$. f is said to be **piecewise continuous** if it is continuous at all but finitely many points of $[a, b]$ and if at all of the exceptional points the left- and right-hand limits of $f(x)$ exist (or one of them does if the exceptional point is a or b).

Notice the meaning of this. There are points $c_1, ..., c_n$ with $c_1 < c_2 < ... < c_n$ such that on each of the intervals $[a, c_1], [c_1, c_2] ..., [c_{n-1}, c_n] , [c_n, b] , f$ is continuous on the interior of the interval (that is, on $(a, c_1), (c_1, c_2) , ...$ etc.) and $f(x)$ tends to a (one-sided) limit as x tends to $a, c_1, ..., c_n, b$ from the left or from the right. In particular if we define g_1 by

$$g_1(x) = \begin{cases} \lim_{y \to a+} f(y) & (x = a) \\ f(x) & (a < x < c_1) \\ \lim_{y \to c_1-} f(y) & (x = c_1) \end{cases}$$

then g_1 is continuous on $[a, c_1]$. Therefore g_1 is integrable on $[a, c_1]$ and since f is equal to g_1 at all but at most two points of $[a, c_1]$, f is integrable on $[a, c_1]$. Similarly f is integrable on $[c_1, c_2] , ... , [c_n, b]$, and hence on $[a, b]$.

We have shown that every piecewise continuous function on $[a, b]$ is integrable. For many purposes piecewise continuous functions are as general as one would wish to consider. $\square$

Lemma 13.10 Suppose f and g are integrable on $[a, b]$ and that $\forall x \in [a, b]$ $f(x) \geq g(x)$; then $\int_a^b f \geq \int_a^b g$. If h is integrable on $[a, b]$ then $\left|\int_a^b h\right| \leq \int_a^b |h|$.

Proof. If f and g are integrable and $\forall x \in [a, b]$ $f(x) \geq g(x)$, then $f - g$ is integrable and $\forall x \in [a, b]$ $f(x) - g(x) \geq 0$. Thus, on each subinterval of a dissection $\inf(f - g) \geq 0$, so, for all dissections D, $s(f - g; D) \geq 0$. Therefore

$$\int_a^b (f - g) \geq s(f - g; D) \geq 0,$$

and $\int_a^b f = \int_a^b (f - g) + \int_a^b g \geq \int_a^b g$.

For the last part, if h is integrable, so is $|h|$ and $\forall x \in [a, b]$ $-|h(x)| \leq h(x) \leq |h(x)|$ whence $-\int_a^b |h| \leq \int_a^b h \leq \int_a^b |h|$ from which we see that

$$\left|\int_a^b h\right| \leq \int_a^b |h|. \quad \square$$

This result may be thought of as the analogue for integrals of the triangle inequality.

Evaluating Integrals
Having finished the technicalities, we may proceed to the substantial results. To simplify matters, we shall **define** $\int_a^b f = -\int_b^a f$ where $b < a$. This does not affect the truth of the results already stated except in the case of Lemma 13.10 where the inequalities presume $a < b$. Lemma 13.10 could be restated, for example, as $\left|\int_a^b h\right| \leq \int_\alpha^\beta |h|$. where $\alpha = \min(a, b)$ and $\beta = \max(a, b)$. We also define $\int_a^a f = 0$ for all a.

Theorem 13.11 The Mean Value Theorem for Integrals. Suppose that $f: [a, b] \to \mathbb{R}$ is continuous. There is a point $\xi \in [a, b]$ for which

$$\int_a^b f = (b - a) f(\xi).$$

Proof. Let $m = \inf\{f(x): a \leq x \leq b\}$ and $M = \sup\{f(x): a \leq x \leq b\}$, so, by the inequality result in Lemma 13.10, $m(b - a) \leq \int_a^b f \leq M(b - a)$. Therefore $(b - a)^{-1} \int_a^b f$ is intermediate between m and M. Since f is continuous, m and M are values attained by f in $[a, b]$, hence by the Intermediate Value Theorem $\exists \xi \in [a, b]$ with $f(\xi) = (b - a)^{-1} \int_a^b f$. (This assumes $a < b$; if not, consider $\int_b^a f$, show that $(a - b)^{-1} \int_b^a f = f(\xi)$ and use $-\int_b^a f = \int_a^b f$.) $\square$

Theorem 13.12 The Fundamental Theorem of Calculus. Let $f: [a, b] \to \mathbb{R}$ be continuous and define F by $F(x) = \int_a^x f(t)\,dt$. Then F is differentiable and

$F' = f$. Conversely, suppose that g is continuous on $[a, b]$ and differentiable on (a, b). Then if g' is integrable, $\int_a^b g'(t)dt = g(b) - g(a)$.

Proof: Choose $x \in [a, b]$ and fix it. Then for all h for which $x + h \in [a, b]$

$$F(x+h) - F(x) = \int_x^{x+h} f(t)dt .$$

By the Mean Value Theorem for Integrals there is a ξ between x and $x + h$ for which $\int_x^{x+h} f(t)dt = hf(\xi)$.

Choose $\varepsilon > 0$. By the continuity of f at x, $\exists \delta > 0$ such that $|y - x| < \delta \Rightarrow$ $|f(y) - f(x)| < \varepsilon$. For this δ, if $0 < |h| < \delta$ then the ξ above satisfies $|\xi - x| < |h| < \delta$, so

$$\left| \frac{F(x+h) - F(x)}{h} - f(x) \right| = \left| f(\xi) - f(x) \right| < \varepsilon .$$

Since ε was arbitrary, we have shown $F'(x) = f(x)$ (the derivative being the one-sided limit if $x = a$ or b). Since x was typical we have finished the first part.

For the second part, let g be as stated and $\varepsilon > 0$. By the integrability of g', there is a dissection $D = \{c_0, c_1, ..., c_n\}$ for which $S(g'; D) - s(g'; D) < \varepsilon$. By the Mean Value Theorem for differentiation (Theorem 11.6), for each j there is a $\xi_j \in [c_{j-1}, c_j]$ with $g(c_j) - g(c_{j-1}) = g'(\xi_j)(c_j - c_{j-1})$, so

$$g(b) - g(a) = \sum_{j=1}^{n} g'(\xi_j)(c_j - c_{j-1}) .$$

This last expression lies between $S(g'; D)$ and $s(g'; D)$ so

$$\int_a^b g' - \varepsilon \leq S(g'; D) - \varepsilon < s(g'; D) \leq g(b) - g(a)$$

$$\leq S(g'; D) < s(g'; D) + \varepsilon \leq \int_a^b g' + \varepsilon$$

whence $g(b) - g(a)$ and $\int_a^b g'$ differ by less than ε. Since ε was arbitrary, $\int_a^b g' = g(b) - g(a)$. $\square$

Notice that the second part above has the (essential) requirement that the derivative be integrable, and that it relates integration to the 'calculus' method of finding a function whose derivative is the function to be integrated. Thus if the derivative is sufficiently well behaved, the integration 'undoes' the action of differentiating. The first part is slightly different; if f is continuous, it guarantees that F is differentiable and differentiation 'undoes' the action of integrating.

Example 13.4 The Fundamental Theorem of Calculus provides us with an easy way of evaluating some integrals. For example, since xe^x is the derivative of $(x - 1)e^x$, we see that, putting $g(x) = (x - 1)e^x$ in the second part of the theorem,

$$\int_a^b xe^x dx = (b-1)e^b - (a-1)e^a .$$

We need a little caution in considering piecewise continuous functions. Let

$$f(x) = \begin{cases} x & (0 \le x \le 1) \\ x+1 & (1 < x \le 2) \end{cases}.$$

Since f is piecewise continuous it is integrable by Example 13.3.   Now the Fundamental Theorem, applied to the restriction of f to $[0, 1]$ shows that $\int_0^1 f = \frac{1}{2}1^2 - \frac{1}{2}0^2 = \frac{1}{2}$.

If we now set $g(x) = x + 1$ for $1 \le x \le 2$ then $g(x) = f(x)$ at all but one point of $[1, 2]$ so that $\int_1^2 g = \int_1^2 f$ and by the Fundamental Theorem

$$\int_1^2 g = \frac{1}{2}(2+1)^2 - \frac{1}{2}(1+1)^2 = \frac{5}{2}$$

(because g is the derivative of $\frac{1}{2}(x+1)^2$).   Therefore

$$\int_0^2 f = \int_0^1 f + \int_1^2 f = \frac{1}{2} + \frac{5}{2} = 3 .$$

Notice that $f'(x) = 1$ at all $x \in [0, 2]\backslash\{1\}$, but that $f(2) - f(0) \ne \int_0^2 f'$. Examples with discontinuities such as this should be treated cautiously!   □

Corollary 1 Differentiation under the Integral Sign.   Let $\phi(x, t) = \sum_{j=1}^n f_j(x)g_j(t)$ where, for each j, f_j is differentiable and g_j is continuous.

Then $\dfrac{d}{dx}(\int_a^x \phi(x, t)dt) = \phi(x, x) + \int_a^x \dfrac{\partial\phi}{\partial x}(x, t)dt$.

Proof: By the Theorem and the rules for differentiating a product,

$$\frac{d}{dx}(\int_a^x f_j(x)g_j(t)dt) = \frac{d}{dx}(f_j(x)\int_a^x g_j(t)dt) = f_j(x)g_j(x) + f_j'(x)\int_a^x g_j(t)dt$$

$$= f_j(x)g_j(x) + \int_a^x f_j'(x)g_j(t)dt .$$

The result follows on summation.   □

Corollary 2 Let $h > 0$, $g: [a - h, a + h] \to \mathbb{R}$ be continuous and define G_n by

$$G_n(x) = \frac{1}{(n-1)!}\int_a^x (x-t)^{n-1}g(t)dt \qquad (n = 1, 2, \dots) .$$

Then $G_n(a) = G_n'(a) = \cdots = G_n^{(n-1)}(a) = 0$ and $G_n^{(n)} = g$.

Proof: We prove the result by induction.

The Fundamental Theorem of Calculus gives the desired properties of G_1 .

Suppose the result is true for some $n \in \mathbb{N}$.   Then, since $(x - t)^n$ may be written as a sum of functions of the form $x^j t^{n-j}$, we may apply differentiation under the integral sign to G_{n+1} to obtain

$$G_{n+1}'(x) = \frac{1}{n!}(x-x)^n g(x) + \frac{1}{n!}\int_a^x n(x-t)^{n-1}g(t)dt = G_n(x) ,$$

and $G_{n+1}(a) = 0$. By the assumed result about G_n , $G_{n+1}^{(n+1)} = G_n^{(n)} = g$ and $G_{n+1}^{(j)}(a) = 0$ $(j = 1, ..., n)$, so the result also holds for $n+1$. $\square$

This Corollary gives a useful, but perhaps unexpected, method for calculating the result of n indefinite integration operations by performing a single integral. This yields another version of Taylor's Theorem where we obtain an exact formula for the remainder.

Corollary 3 Suppose that $f\colon [a - h, a + h] \to \mathbb{R}$ has derivatives of order up to and including n and that $f^{(n)}$ is continuous. Then if $a - h \leq x \leq a + h$,

$$f(x) = f(a) + (x - a)f'(a) + ... + (x - a)^{n-1}f^{(n-1)}(a)/(n - 1)! + R_n(x)$$

where
$$R_n(x) = \int_a^x \frac{(x - t)^{n-1}}{(n - 1)!} f^{(n)}(t)dt .$$

Proof: Define the function $R_n(x)$ by the formula given so, by Corollary 2, $R_n^{(n)} = f^{(n)}$ whence $R_n - f$ is a polynomial of degree at most $n - 1$. Let $f(x) - R_n(x) = \sum_{j=0}^{n-1} \alpha_j (x - a)^j$. Then, for $j = 0, 1, ..., n - 1$

$$f^{(j)}(a) - R_n^j(a) = f^{(j)}(a) = \alpha_j j!$$

Substituting these values of α_j gives the required result. $\square$

We shall postpone, for the moment, applications of these results and finish the task of relating the new concept of integration to the old, that is, show that all of the techniques used in the anti-differentiation view of the integral apply to our integral, so we have lost nothing, and gained much, in our new attack. The last two matters to be settled in this direction are integration by parts and by substitution.

Lemma 13.13 Integration by Parts. Suppose f and g have a continuous derivative. Then

$$\int_a^b fg' = (f(b)g(b) - f(a)g(a)) - \int_a^b f'g .$$

Proof: $(fg)' = f'g + fg'$, so since $(fg)'$ is continuous and therefore integrable,

$$\int_a^b (fg)' = \int_a^b fg' + f'g = f(b)g(b) - f(a)g(a) . \quad \square$$

The customary use of integration by parts to find $\int_a^b f\phi$ uses $g = \Phi$ where $\Phi' = \phi$.

Lemma 13.14 Integration by Substitution. Let f, ϕ and ϕ' be continuous where $f\colon [a, b] \to \mathbb{R}$ and $\phi\colon [\alpha, \beta] \to [a, b]$. Then if ϕ possesses an inverse function, ϕ^{-1} ,

$$\int_a^b f(x)dx = \int_{\phi^{-1}(a)}^{\phi^{-1}(b)} f(\phi(t))\phi'(t)dt \ .$$

Even if ϕ does not have an inverse function,

$$\int_{\phi(\alpha)}^{\phi(\beta)} f(x)dx = \int_\alpha^\beta f(\phi(t))\phi'(t)dt \ .$$

[Notice that these are exactly the results obtained from calculus by substituting $x = \phi(t)$, $dx = \phi'(t)dt$.]

Proof. We prove the second of these, since the first is a special case of it. Define

$$F(y) = \int_{\phi(\alpha)}^{\phi(y)} f(x)dx \quad \text{and} \quad G(y) = \int_\alpha^y f(\phi(t))\phi'(t)dt \ .$$

Then $F(y) = J(\phi(y))$ where $J(y) = \int_{\phi(\alpha)}^y f(x)dx$, so, using the Fundamental Theorem, which applies since all the integrands are continuous, we have, for all y, $F'(y) = J'(\phi(y))\phi'(y) = f(\phi(y))\phi'(y) = G'(y)$.   Since $F' - G' = 0$,   for all y, $F(y) - G(y) = F(\alpha) - G(\alpha) = 0$.   In particular, $F(\beta) = G(\beta)$,   the required result. □

13.2 Approximating the Value of an Integral

In this section we shall assume that f possesses as many derivatives as are needed for the calculation in hand and that these derivatives are continuous.

Suppose that $h > 0$ and that we wish to estimate the value of $\int_a^{a+h} f$.   An obvious first estimate is $h(f(a) + f(a + h))/2$, the area of the shaded region in Fig. 13.4, but for this to be of much use we need to know how accurate an estimate it is. For this we proceed as follows:

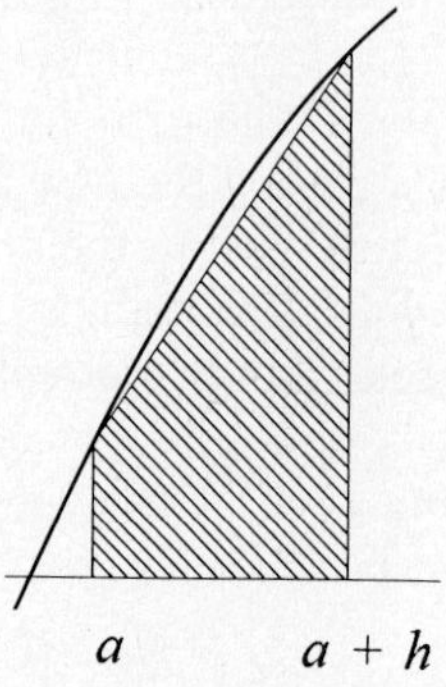

Let $F(x) = \int_a^x f$, so $F' = f$, $F'' = f'$, etc., and we have

$$F(a + h) = F(a) + hf(a) + h^2 f'(a)/2 + R_1 ,$$
$$\text{and } f(a + h) = f(a) + hf'(a) + R_2 ,$$

where R_1 and R_2 can be evaluated using Taylor's Theorem. Then

$$\int_a^{a+h} f = F(a+h) - F(a) = hf(a) + \frac{1}{2}h^2 f'(a) + R_1$$

$$= \frac{1}{2}h(f(a) + f(a+h)) + R_1 - \frac{1}{2}hR_2 ,$$

so the error in the estimate is $R_1 - \frac{1}{2}hR_2$.

At this point the ordinary form of remainder in Taylor's Theorem springs to mind, giving $R_1 = h^3 f''(\xi_1)/6$ and $R_2 = h^2 f''(\xi_2)/2$ for some points ξ_1 and ξ_2 between a and $a + h$.   Therefore

$$R_1 - \frac{1}{2}hR_2 = h^3(f''(\xi_1)/6 - f''(\xi_2)/4).$$

Fig. 13.4.

Since ξ_1 and ξ_2 need not be the same point, and, indeed, $f''(\xi_1)$ and $f''(\xi_2)$ need not have the same sign, we cannot cancel and the best this approach gives is

$$\left| R_1 - \tfrac{1}{2}hR_2 \right| \le h^3\left(\tfrac{1}{6}|f''(\xi_1)| + \tfrac{1}{4}|f''(\xi_2)|\right) \le \left(\tfrac{5}{12}\right)Mh^3$$

where $M = \sup\{f''(x): a \le x \le a + h\}$.

By the use of a spot of cunning, however, we can improve this. Write R_1 and R_2 in their integral forms (from Corollary 3 to Theorem 13.12):

$$R_1 = \tfrac{1}{2}\int_a^{a+h}(a+h-t)^2 f''(t)dt, \qquad R_2 = \int_a^{a+h}(a+h-t)f''(t)dt$$

so that $R_1 - \tfrac{1}{2}hR_2 = \tfrac{1}{2}\int_a^{a+h}(a+h-t)(a-t)f''(t)dt$.

Since, for $t \in [a, a+h]$, $a - t \le 0$ and $a + h - t \ge 0$,

$$\left| R_1 - \tfrac{1}{2}hR_2 \right| \le \tfrac{1}{2}\int_a^{a+h}|a+h-t||a-t||f''(t)|dt$$

$$\le \tfrac{1}{2}\int_a^{a+h}(a+h-t)(t-a)M\,dt$$

$$= Mh^3/12 \ .$$

Therefore, for a fixed function f and variable h, the error in this estimate decreases with h with order h^3 as $h \to 0$. We call this method the *Trapezium Rule*. It can be used to evaluate integrals to any desired accuracy, even when an 'antiderivative' is not available. Suppose we wish to find $\int_0^1 f$, and that we use the Trapezium Rule with n subintervals each of length $1/n$. Then, if $M = \sup\{f''(x): 0 \le x \le 1\}$ the error on each subinterval will not exceed $M/(12n^3)$ so the total error on adding the results for each subinterval will be at most $M/(12n^2)$, which obviously tends to 0 as $n \to \infty$.

It turns out that there is a more accurate approximation. Let $h > 0$ and consider the integral $\int_{a-h}^{a+h} f$. Letting $F(x) = \int_a^x f$ as before,

$$F(a + h) - F(a - h) = 2hf(a) + (h^3/3)f''(a) + R_1$$

where $\qquad R_1 = \tfrac{1}{24}\left(\int_a^{a+h}(a+h-t)^4 f^{(4)}(t)dt - \int_a^{a-h}(a-h-t)^4 f^{(4)}(t)dt\right)$,

obtained by using Taylor's Theorem on $[a, a + h]$ and $[a - h, a]$ separately. We wish to take a combination of $f(a - h)$, $f(a)$ and $f(a + h)$ which will give the terms in h and in h^3 above, with a remainder of order h^5. Now

$$f(a + h) + f(a - h) = 2\,f(a) + 2h^2 f''(a)/2 + R_2$$

where $R_2 = \tfrac{1}{6}\left(\int_a^{a+h}(a+h-t)^3 f^{(4)}(t)dt + \int_a^{a-h}(a-h-t)^3 f^{(4)}(t)dt\right)$.

Therefore $\tfrac{1}{3}h\{f(a-h)+4f(a)+f(a+h)\} = 2hf(a) + (\tfrac{1}{3}h^3)f''(a) + (\tfrac{1}{3}h)R_2$ so that

$$\int_{a-h}^{a+h} f - \tfrac{1}{3}h\{f(a-h)+4f(a)+f(a+h)\} = R_1 - \tfrac{1}{3}hR_2 \ .$$

Now

$$R_1 - \tfrac{1}{3}hR_2 = \int_a^{a+h} \left(\frac{(a+h-t)^4}{24} - \frac{(a+h-t)^3 h}{18} \right) f^{(4)}(t)\,dt$$

$$- \int_a^{a-h} \left(\frac{(a-h-t)^4}{24} + \frac{(a-h-t)^3 h}{18} \right) f^{(4)}(t)\,dt \ .$$

On noticing that for all $t \in [a, a+h]$

$$\frac{(a+h-t)^4}{24} - \frac{(a+h-t)^3 h}{18} = \frac{(a+h-t)^3 (3(a-t)-h)}{72} \leq 0$$

we see that

$$\left| \int_a^{a+h} \left(\frac{(a+h-t)^4}{24} - \frac{(a+h-t)^3 h}{18} \right) f^{(4)}(t)\,dt \right|$$

$$\leq \int_a^{a+h} \left(\frac{h(a+h-t)^3}{18} - \frac{(a+h-t)^4}{24} \right) |f^{(4)}(t)|\,dt \leq \tfrac{1}{180} M_1 h^5$$

where $M_1 = \sup\{|f^{(4)}(t)|: a \leq t \leq a+h\}$. We obtain a similar estimate for the other integral, so, combining, we see that $|R_1 - (h/3)R_2| \leq Mh^5/90$ where here $M = \sup\{|f^{(4)}(t)|: a-h \leq t \leq a+h\}$. This yields *Simpson's Rule*,

$$\left| \int_{a-h}^{a+h} f - \tfrac{1}{3}h\{f(a-h) + 4f(a) + f(a+h)\} \right| \leq \tfrac{1}{90} Mh^5 \ .$$

The result is interesting not only in its usefulness, as the 'error' tends to zero very rapidly as $h \to 0+$, but also as an illustration of how a little perseverance can extend the value of Taylor's Theorem. If we write the estimate for the integral as $2h\{f(a-h) + 4f(a) + f(a+h)\}/6$ it seems more plausible, being the length of the interval times a (complicated) average of the values of f at $a-h$, a and $a+h$.

13.3 Improper Integrals

There are still some issues to be tackled. If we view the integral as the area under a curve, there is the possibility that this may exist for some curves defined on the whole real line, or on $[0, \infty)$. In this connection, we notice that $\int_0^a e^{-t}\,dt = 1 - e^{-a} \to 1$ as $a \to \infty$, giving an obvious candidate for the area under the curve $\{(t, e^t): t \geq 0\}$. We make the definition which corresponds to this idea:

Definition Let $f: [a, \infty) \to \mathbb{R}$ and suppose that for all $b > a$, f is integrable on $[a, b]$. If $\int_a^b f$ tends to a limit as $b \to \infty$, we say that the **integral** $\int_a^\infty f$ **exists** (or converges) and define $\int_a^\infty f = \lim_{b \to \infty} \int_a^b f$. The analogous definition is made for $\int_{-\infty}^a f$. If, for some $a \in \mathbb{R}$, both $\int_{-\infty}^a f$ and $\int_a^\infty f$ exist we say that $\int_{-\infty}^\infty f$ exists and has the value $\int_{-\infty}^a f + \int_a^\infty f$.

The existence of these 'improper' integrals is a matter broadly similar to that of the convergence of infinite series, and we shall develop tests to show that certain integrals exist, these being essential when a usable formula for $\int_a^b f$ does not exist. These 'improper' integrals occur in diverse places in mathematics and there are some powerful techniques for evaluating them, though we shall have to content ourselves with the question of existence.

Notice that the definition of $\int_{-\infty}^\infty f$ is equivalent to the existence of the limit of $\int_a^b f$ as $b \to \infty$ and $a \to -\infty$ *independently*. This is a stronger assertion than the existence of $\lim_{b \to \infty} \int_{-b}^b f$, as may be seen by considering $f(t) = t$ $\forall t \in \mathbb{R}$; $\lim_{b \to \infty} \int_{-b}^b f = 0$ but neither $\int_0^\infty f$ nor $\int_{-\infty}^0 f$ exists. In this case, since $\int_0^a f$ exists for all $a \in \mathbb{R}$, it is easily seen that the non-existence of $\int_0^\infty f$ implies the non-existence of $\int_a^\infty f$ for all $a \in \mathbb{R}$.

It is worth noticing now that if $a \in \mathbb{R}$, $f \colon \mathbb{R} \to \mathbb{R}$ and both $\int_a^\infty f$ and $\int_{-\infty}^a f$ exist, then f is, by definition, integrable on $[a, b]$ for all $b \in \mathbb{R}$ so that $\int_b^\infty f$ and $\int_{-\infty}^b f$ will exist, as can be seen by noticing that for all x, $\int_a^x f = \int_a^b f + \int_b^x f$. The point a appearing in the definition of $\int_{-\infty}^\infty f$ is therefore immaterial.

We proceed by analogy with infinite series.

Lemma 13.15 Comparison Test. Suppose that $\forall x \geq a$ $f(x) \geq 0$ and f is integrable on $[a, x]$. If there is a constant K such that $\forall x \geq a$ $\int_a^x f \leq K$, then $\int_a^\infty f$ exists. If $\forall x \geq a$ $0 \leq f(x) \leq g(x)$ and $\int_a^\infty g$ exists, then $\int_a^\infty f$ exists.

Proof: Let $F(x) = \int_a^x f$. If $\forall x \geq a$ $f(x) \geq 0$ then F is increasing, since $y > x \Rightarrow F(y) - F(x) = \int_x^y f \geq 0$, and if F is bounded then $F(x) \to \sup F$ as $x \to \infty$. For the second part, if $\int_a^\infty g$ exists, then $\forall x \geq 0$ $0 \leq \int_a^x f \leq \int_a^x g \leq \int_a^\infty g$ so $\int_a^\infty f$ exists by the first part. $\square$

Example 13.5 $\int_{-\infty}^\infty e^{-x^2} dx$. Since e^{-x^2} is a continuous function of x, $\int_a^b e^{-x^2} dx$ exists for all finite a and b, so we need only consider the existence of $\int_1^\infty e^{-x^2} dx$ and $\int_{-\infty}^{-1} e^{-x^2} dx$. Now $x \geq 1 \Rightarrow x^2 \geq x \Rightarrow 0 \leq e^{-x^2} \leq e^{-x}$ so, since $\int_1^\infty e^{-x} dx = 1$, $\int_1^\infty e^{-x^2} dx$ exists by comparison. The existence of $\int_{-\infty}^{-1} e^{-x^2} dx$ is proved similarly. $\square$

Lemma 13.16 Absolute Convergence. Suppose that for all $x \geq a$, f is integrable on $[a, x]$ and that $\int_a^\infty |f|$ exists. Then $\int_a^\infty f$ exists.

Proof. Let $f_+(x) = \max(f(x), 0) = \frac{1}{2}(f(x) + |f(x)|)$, and $f_-(x) = \max(-f(x), 0)$. For all $x \geq a$, $f_\pm$ are integrable on $[a, x]$ and $0 \leq f_\pm(x) \leq |f(x)|$. By comparison $\int_a^\infty f_\pm$ both exist, hence so does $\int_a^\infty f = \int_a^\infty f_+ - \int_a^\infty f_-$. $\square$

Lemma 13.17 The Integral Test. Suppose that $f: [1, \infty) \to \mathbb{R}$ is non-negative and decreasing; then $\int_1^\infty f$ exists if and only if $\sum f(n)$ converges.

Proof. $\forall n \in \mathbb{N}$ $n \leq x \leq n + 1 \Rightarrow f(n) \geq f(x) \geq f(n + 1)$, so

$$f(n) \geq \int_n^{n+1} f \geq f(n + 1).$$

Letting $s_n = f(1) + f(2) + \ldots + f(n)$, we see that $\forall n \geq 2$ $s_{n-1} \geq \int_1^n f \geq s_n - f(1)$. This double inequality shows us that (s_n) is a bounded sequence if and only if $(\int_1^n f)$ is a bounded sequence. Thus if $\int_1^\infty f$ exists, $(\int_1^n f)$ is a bounded sequence, so (s_n) is increasing and bounded above, hence convergent. Conversely, if $\sum f(n)$ converges, $(\int_1^n f)$ is a bounded sequence, so if K is an upper bound for $(\int_1^n f)$, we see that if $x \geq 1$ $\exists n \in \mathbb{N}$ such that $n > x$ whence $\int_1^x f \leq \int_1^n f \leq K$. This shows that $\forall x \geq 1$ $\int_1^x f \leq K$ and $\int_1^\infty f$ exists by Lemma 13.15. $\square$

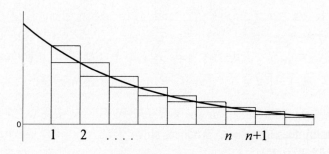

Fig. 13.5

As a convergence test for series, this gives us more or less the same information as the condensation test, although it is more transparent. We see, for example, that $\int_1^\infty (1/x^\alpha)\,dx$ exists if and only if $\alpha > 1$, either by direct calculation or by the knowledge that $\sum(1/n^\alpha)$ converges if and only if $\alpha > 1$. The technique of the integral test is, however, more useful than this, since it gives us estimates of the difference between $\sum_{n=1}^\infty f(n)$ and $\sum_{n=1}^N f(n)$ in some cases.

With the above notation, $\sum_{n=M}^{N-1} f(n) \geq \int_M^N f \geq \sum_{n=M+1}^{N} f(n)$ whence, letting $N \to \infty$, we have

$$\sum_{n=M}^{\infty} f(n) \geq \int_M^{\infty} f \geq \sum_{n=M+1}^{\infty} f(n)$$

(when the series and the integral converge).   Let $f(x) = 1/x^2$ so that if $s_n = 1/1^2 + 1/2^2 + ... + 1/n^2$ and $s = \sum_{n=1}^{\infty} 1/n^2$,

$$s - s_M = \sum_{n=M+1}^{\infty} \frac{1}{n^2} \leq \int_M^{\infty} f = \frac{1}{M} , \text{ and } s - s_M \geq \int_{M+1}^{\infty} f = \frac{1}{M+1}$$

so $1/(M + 1) \leq s - s_M \leq 1/M$, giving a very simple estimate of the difference between the sum to M terms and the sum to infinity.

In practice, the convergence of most improper integrals is established by comparison, with the occasional assistance of manipulative devices like integration by parts.

Example 13.6 $\int_0^{\infty} (\sin x)/x \, dx$ converges.

Solution: First notice that $(\sin x)/x \to 1$ as $x \to 0+$, so if we give the integrand the value 1 at $x = 0$, the resulting function is continuous, and $\int_0^a (\sin x)/x \, dx$ exists for all $a > 0$.   It is therefore enough to show the existence of $\int_0^{\infty} (\sin x)/x \, dx$.

At first sight, one might notice that $x > 0 \Rightarrow |(\sin x)/x | < 1/x$ and try the comparison test, but this is fruitless since $\int_1^{\infty} 1/x \, dx$ does not exist.   Let $F(y) = \int_1^y (\sin x)/x \, dx$.   Then, integrating by parts,

$$F(y) = \cos 1 - (\cos y) / y - \int_1^y (\cos x)/x^2 dx .$$

Since $\int_1^{\infty} 1/x^2 \, dx$ does exist and $x > 1 \Rightarrow |(\cos x)/x^2| < 1/x^2$ we see that $\int_1^{\infty} (\cos x)/x^2 dx$ exists.   Therefore the last term on the right of the expression for $F(y)$ tends to a limit as $y \to \infty$, whence, since $(\cos y)/y \to 0$ as $y \to \infty$, $F(y)$ tends to a limit as $y \to \infty$. ◻

There is another type of improper integral.   In the definition of the ordinary integral $\int_a^b f$ we demanded that f be a bounded function, yet we can ascribe a value to such integrals, in certain cases where f is unbounded, in much the same way as to $\int_a^{\infty} f$.

Definition Suppose that $f: (a, b] \to \mathbb{R}$ and that for all $x \in (a, b]$ f is integrable on $[x, b]$.   Then we say that the (improper) integral $\int_a^b f$ exists if $\int_x^b f$ tends to a

limit as $x \to a+$. The analogous definition applies if $\lim_{x \to b-} \int_a^x f$ exists. If
$f: [a, b] \backslash \{c\} \to \mathbb{R}$ and both $\int_a^c f$ and $\int_c^b f$ exist, we say the (improper) integral
$\int_a^b f$ exists and has the value $\int_a^b f = \int_a^c f + \int_c^b f$.

Example 13.7 $\int_0^1 x^{-\alpha} dx$ exists if and only if $\alpha < 1$. For

$$\int_a^1 x^{-\alpha} dx = (1 - a^{1-\alpha}) / (1 - \alpha) \to 1/(1 - \alpha) \quad \text{as} \quad a \to 0+$$

if $1 - \alpha > 0$, i.e. $\alpha < 1$, while the limit does not exist if $\alpha > 1$. The limit is also
non-existent if $\alpha = 1$ since $\int_a^1 x^{-1} dx = -\log a$. $\square$

The following two analogues of the results for $\int_a^\infty f$ are proved by simple
changes to the proofs of Lemmas 13.15 and 13.16.

Lemma 13.18 Comparison Test. Suppose that $\forall x \in (a, b]$ $f(x) \geq 0$ and f is
integrable on $[x, b]$. If $\{\int_x^b f : x \in (a, b]\}$ is bounded then $\int_a^b f$ exists. If
$\int_a^b g$ exists and $\forall x \in (a, b]$ $0 \leq f(x) \leq g(x)$ then $\int_a^b f$ exists. $\square$

Lemma 13.19 Absolute Convergence. Suppose that $\forall x \in (a, b]$ f is integrable
on $[x, b]$ and that $\int_a^b |f|$ exists. Then $\int_a^b f$ exists. $\square$

Example 13.8 The two types of integral may be combined. For example, if we
wished to examine $\int_0^\infty (1/\sqrt{(x|x - 1||x - 2|)}) \, dx$ we would investigate, separately, the
existence of the improper integrals on $(0, \frac{1}{2}], [\frac{1}{2}, 1), (1, \frac{3}{2}]$ $[\frac{3}{2}, 2)$, $(2, 3]$ and
$[3, \infty)$. On $(0, \frac{1}{2}]$ the function is continuous so the only issue is the behaviour
near 0 . Since $1/\sqrt{(|x - 1||x - 2|)}$ is continuous on $[0, \frac{1}{2}]$ it is bounded there; let
M be an upper bound. Then $\forall x \in (0, \frac{1}{2}]$ $1/\sqrt{(x|x - 1||x - 2|)} \leq Mx^{-1/2}$ so
$\int_0^{\frac{1}{2}}$ exists by comparison. The existence on the other four bounded intervals is
similarly proved, while on $[3, \infty)$ we see that $x \geq 3 \Rightarrow x(x - 1)(x - 2) \geq (2/9)x^3 \Rightarrow$
$1/\sqrt{(x|x - 1||x - 2|)} \leq \sqrt{(9/2)}x^{-3/2}$, so $\int_3^\infty$ exists by comparison with $\int_3^\infty x^{-3/2} dx$. $\square$

Example 13.9 The Gamma Function, $\Gamma: (0, \infty) \to \mathbb{R}$, is defined as follows:

$$\Gamma(x) = \int_0^\infty t^{x-1} e^{-t} dt .$$

The integral exists if both $\int_0^1$ and $\int_1^\infty$ exist. The former obviously exists if $x \geq 1$
for then the integrand is continuous on $[0, 1]$. If $x - 1 < 0$ the integrand is
unbounded near $t = 0$, so since $0 < t \leq 1$ implies $0 < t^{x-1} e^{-t} < t^{x-1}$, and
$\int_0^1 t^{x-1} dt$ exists, $\int_0^1 t^{x-1} e^{-t} dt$ exists for $x > 0$. (This integral diverges by the

comparison test for $x \le 0$ since $0 < t \le 1 \Rightarrow t^{x-1}e^{-t} \ge t^{x-1}e^{-1}$.) Since $t^{\alpha}e^{-\lambda t} \to 0$ as $t \to \infty$ for $\lambda > 0$, we see that $t^{x-1}e^{-t/2} \to 0$ as $t \to \infty$. From this it follows that $t^{x-1}e^{-t/2}$ is less than 1 for all sufficiently large t, say for $t \ge t_0$, and, since $t^{x-1}e^{-t/2}$ is continuous on $[1, t_0]$, it is bounded on that interval, and hence on the interval $[1, \infty)$. Therefore there is a constant T such that $t \ge 1 \Rightarrow t^{x-1}e^{-t/2} \le T$. Then $\forall t \ge 1$ $0 < t^{x-1}e^{-t} \le Te^{-t/2}$ so $\int_1^\infty t^{x-1}e^{-t}dt$ exists by comparison. Now, for $x > 0$, integrating by parts, we have

$$\Gamma(x+1) = \int_0^\infty t^x e^{-t} dt = \lim_{\substack{b \to \infty \\ a \to 0+}} \int_a^b t^x e^{-t} dt$$

$$= \lim_{\substack{b \to \infty \\ a \to 0+}} \left(-b^x e^{-b} + a^x e^{-a} + \int_a^b x t^{x-1} e^{-t} dt \right) = x\Gamma(x)$$

since $b^x e^{-b} \to 0$ as $b \to \infty$ and $a^x e^{-a} \to 0$ as $a \to 0+$ (using $x > 0$). Therefore $\forall x > 0$ $\Gamma(x+1) = x\Gamma(x)$. Moreover, $\Gamma(1) = 1$ so it is easily seen that $\forall n \in \mathbb{N}$ $\Gamma(n) = (n-1)!$ and we have found a function defined on $(0, \infty)$ whose value at the natural numbers is a factorial. At this point we recall the formula for the nth indefinite integral F_n of the continuous function f, that is, a function whose nth derivative if f. Rewriting the $(n-1)!$ as $\Gamma(n)$ we obtain

$$F_n(x) = \frac{1}{\Gamma(n)} \int_0^x (x-t)^{n-1} f(t) dt \ ,$$

an expression which has a meaning even if n is not an integer! This apparently whimsical observation can be put to good use in more sophisticated theories of differential equations.

Problems

1.   Let $f(x) = x^2$ for $0 \le x \le 1$ and let $D_n = \{0, 1/n, ..., (n-1)/n\}$ be the dissection of $[0, 1]$ into n subintervals of equal length. Calculate $S(D_n)$ and show that $\inf\{ S(D_n): n \in \mathbb{N} \} = \sup\{ s(D_n): n \in \mathbb{N} \} = 1/3$. Deduce that $\int_0^1 x^2 dx = 1/3$. [You may need to recall that $\sum_{k=1}^n k^2 = n(n+1)(2n+1)/6$.]

2.   Let $a < b < c$ and define $f: [a, c] \to \mathbb{R}$ by $f(x) = 1$ if $a \le x \le b$ and $f(x) = 0$ if $b < x \le c$. By calculating $\inf\{S(D)\}$ and $\sup\{s(D)\}$ directly, show that $\int_a^c f = b - a$.

3.   Suppose that $f: [a, b] \to \mathbb{R}$ is integrable and that $\forall x \in [a, b]$ $f(x) \ge 0$. By Lemma 13.10, then, $\int_a^b f \ge 0$. Show that if, for some $\alpha > 0$ and $\delta > 0$, there is an interval $(c - \delta, c + \delta) \subset [a, b]$ for which $\forall x \in (c - \delta, c + \delta)$ $f(x) \ge \alpha$ then $\int_a^b f > 0$.

4.   Let $f: [a, b] \to \mathbb{R}$ be *continuous* and satisfy $\forall x \in [a, b]$ $f(x) \geq 0$. Show, using Question 3, that if $\exists c \in [a, b]$ for which $f(c) > 0$ then $\int_a^b f > 0$. Deduce that if g is continuous and $\int_a^b |g| = 0$ then g is identically zero (i.e. $\forall x \in [a, b]$ $g(x) = 0$).

5.   Suppose that $f: [a, b] \to \mathbb{R}$ is continuous and that for all continuous functions $\phi: [a, b] \to \mathbb{R}$, $\int_a^b f\phi = 0$. Show that f is identically 0. (Use Question 4 to show that f^2 is identically 0.)

6.   Let $f: [a, b] \to \mathbb{R}$ be integrable and define F by $F(x) = \int_a^x f$. Prove that F is continuous, and deduce that there is a point $\xi \in [a, b]$ for which $\int_a^\xi f = \int_\xi^b f$.

7.   Let $f: [0, a] \to \mathbb{R}$ be continuous. Show that $\lim_{x \to 0+} \frac{1}{x} \int_0^x f(t)dt = f(0)$.

8.   By showing that the theorem for differentiation under the integral sign applies, show that if $y(x) = \int_0^x \sin(x - t) f(t) dt$, where f is a given continuous function, then y satisfies the differential equation $y'' + y = f$ and $y(0) = y'(0) = 0$.

9.   By showing that, for $x > 0$, $\int_1^x (1/t)dt = \log x$ (differentiate!), prove that $\forall n \in \mathbb{N}$ $1/(n + 1) \leq \log(1 + 1/n) \leq 1/n$. Let $s_n = 1 + 1/2 + 1/3 + \ldots + 1/n$ and $a_n = s_n - \log n$. Show that (a_n) is decreasing and that $\forall n \in \mathbb{N}$ $1/n \leq a_n \leq 1$. Deduce that there is a number $\gamma \in [0, 1]$ such that $a_n \to \gamma$ as $n \to \infty$, and that $s_n = \log n + \gamma + \varepsilon_n$ where $0 \leq \varepsilon_n \leq 1$ and $\varepsilon_n \to 0$ as $n \to \infty$.

10. Use the Trapezium Rule with no intermediate steps on $[n, n + 1]$ to show that $\int_n^{n+1} \log t\, dt = (\log n + \log(n + 1))/2 + \varepsilon_n$ where $|\varepsilon_n| \leq 1/(12n^2)$. By calculating $\int_1^n \log t\, dt$ exactly (integrate by parts!), show that

$$\log(n!) = (n + \tfrac{1}{2})\log n - n + \phi(n)$$

where $\phi(n)$ tends to a limit as $n \to \infty$. Deduce that $n!/((n/e)^n \sqrt{n})$ tends to a limit as $n \to \infty$.

11.* Let $I_n = \int_0^{\pi/2} (\sin t)^n dt$ $(n = 0, 1, 2, \ldots)$. Show that for all $n \geq 0$ $I_n \geq I_{n+1} > 0$ and that if $n \geq 2$ $I_n = ((n - 1)/n)I_{n-2}$. Deduce that $I_{2n} \geq I_{2n+1} \geq ((2n + 1)/(2n + 2))I_{2n}$ and hence that $I_{2n+1}/I_{2n} \to 1$ as $n \to \infty$. By using the expression $I_n = ((n - 1)/n)I_{n-2}$ show that

$$I_{2n+1} = \frac{(2n)(2n-2)\ldots 4.2}{(2n+1)(2n-1)\ldots 5.3} = \frac{(2^n n!)^2}{(2n+1)!} \quad \text{and } I_{2n} = \frac{(2n)!\pi}{(2^n n!)^2 2} .$$

Deduce that $\lim_{n\to\infty} \frac{1}{\sqrt{(2n+1)}} \frac{(2^n n!)^2}{(2n)!} = \sqrt{\frac{\pi}{2}}$.

Use this to show that $\lim_{n\to\infty} n!/(\sqrt{n}(n/e)^n) = \sqrt{(2\pi)}$ (cf. Question 10); hence show that the ratio of $n!$ to $\sqrt{(2\pi n)}(n/e)^n$ tends to 1 as $n\to\infty$.

12. Show that the following improper integrals exist:

$$\int_0^\infty \frac{dx}{1+x^2}, \quad \int_0^\infty e^{-ax}\sin(bx)\,dx \quad (a>0), \quad \int_1^\infty \frac{\sin x}{x^\alpha}\,dx \quad (\alpha>0),$$

$$\int_0^1 \log x\,dx, \quad \int_0^1 \frac{dx}{\sqrt{(1-x)}}, \quad \int_0^1 \frac{dx}{\sqrt{(x(1-x))}}, \quad \int_{-\infty}^\infty e^{-|x|}dx .$$

13. Suppose that $f: [0, \infty) \to \mathbb{R}$ and that $\int_0^\infty |f|$ exists. Show that, for all $\xi \in \mathbb{R}$, $\int_0^\infty f(x)\sin(\xi x)\,dx$ and $\int_0^\infty f(x)\cos(\xi x)\,dx$ both exist.

14. Show that $\int_0^\infty (\cos x)/\sqrt{x}\,dx$ exists and, by changing variable, deduce that $\int_0^\infty \cos(t^2)\,dt$ exists. This is noteworthy since $\cos(t^2) \not\to 0$ as $t\to\infty$.

15. For $\xi \in \mathbb{R}$ define $F(\xi) = \int_0^\infty (\sin(\xi x))/x\,dx$. Show that if $\xi > 0$ then $F(\xi) = F(1)$ and if $\xi < 0$ then $F(\xi) = -F(1)$. It follows that if we can show that $F(1) > 0$, F must be discontinuous at 0 . (See the next problem!)

16.* Show that $\int_0^\infty (\sin x)/x\,dx = \sum_{n=1}^\infty \int_{2(n-1)\pi}^{2n\pi} (\sin x)/x\,dx$. By splitting the range of integration into two, or otherwise, show that for all $n \in \mathbb{N}$ $\int_{2(n-1)\pi}^{2n\pi} (\sin x)/x\,dx > 0$ and hence that $\int_0^\infty (\sin x)/x\,dx > 0$.

17. Which of the following improper integrals exist?

$$\int_0^\infty \frac{dx}{x^2-1}, \quad \int_0^\infty \frac{dx}{(x+1)\sqrt{|x^2-1|}}, \quad \int_0^\infty te^{-t}dt .$$

For which real α does $\int_0^\infty x^\alpha dx$ exist?

18. Show that $\sum_{n\geq 2} 1/(n(\log n)^2)$ converges and show that if s_n denotes the sum up to the term $1/(n(\log n)^2)$ inclusive, and s is the sum to infinity, then $1/\log(n + 1) < s - s_n < 1/\log n$. Deduce that for the partial sum s_n to be 'correct' to two decimal places' (i.e. within $1/200$ of the sum to infinity) we

must let n be at least $e^{200} - 1$.

19.* Suppose that $f: [a, b] \to \mathbb{R}$ is bounded and that for all $x \in (a, b]$ f is integrable on $[x, b]$. Show that for all $\varepsilon > 0$ there is a dissection D of $[a, b]$ with $S(D) - s(D) < \varepsilon$, and hence that f is integrable. Deduce that $\int_0^1 \sin(1/x)\,dx$ exists in the ordinary sense, that is, the integral is not improper.

20.* (i)   Prove that if $c < d$ then $\int_c^d \sin(\lambda x)\,dx \to 0$ as $\lambda \to \infty$.

(ii) $g: [a, b] \to \mathbb{R}$ is said to be a **step function** if there are points $c_0, ..., c_n$ with $a = c_0 < c_1 < ... < c_n = b$ such that g is constant on the intervals (c_{i-1}, c_i) and the value of $g(c_i)$ is that taken by g on one of the adjacent intervals. Prove that if g is a step function then

$$\forall \varepsilon > 0 \; \exists \Lambda \text{ such that } \forall \lambda \geq \Lambda \; \left| \int_a^b g(x)\sin(\lambda x)\,dx \right| < \tfrac{1}{2}\varepsilon .$$

(iii)   Let $f: [a, b] \to \mathbb{R}$ be integrable. Show directly from the definition of integrability that there is a step function g on $[a, b]$ for which

$\forall x \in [a, b]$ $g(x) \leq f(x)$ and $\int_a^b (f - g) < \tfrac{1}{2}\varepsilon$. (What is a lower sum?)

Deduce that there is a step function g such that $\int_a^b |f - g| < \tfrac{1}{2}\varepsilon$.

(iv)   Assemble all this into a proof that

$$\int_a^b f(x)\sin(\lambda x)\,dx \to 0 \text{ as } \lambda \to \infty. \; .$$

21.* Suppose that f is integrable on $[a, b]$ and that $\forall x \in [a, b]$ $f(x) > 0$. We wish to show $\int_a^b f > 0$, which is rather more awkward than one might expect.

Assume the result is false, i.e. that $\int_a^b f = 0$.

(i)   Show that if $[c, d] \subset [a, b]$ then $\int_c^d f = 0$.

(ii)   Prove that there is a dissection D_1 of $[a, b]$ with $S(D_1) \leq b - a$ and deduce that there is a subinterval $[a_1, b_1]$ of $[a, b]$ such that $a_1 < b_1$ and $\sup_{[a_1, b_1]} f \leq 1$.

(iii)   By noticing that $\int_{a_1}^{b_1} f = 0$, repeat the process to show by induction that there is a sequence of intervals $[a_n, b_n]$ such that $\forall n \in \mathbb{N}$ $a_n < b_n$, $[a_{n+1}, b_{n+1}] \subset [a_n, b_n]$ and $\sup_{[a_n, b_n]} f \leq 1/n$.

(iv)   Show that (a_n) is increasing and bounded above, and let $x = \lim_{n \to \infty} a_n$.

Show that $\forall n \in \mathbb{N}$ $x \in [a_n, b_n]$, and deduce that $f(x) = 0$. Use this contradiction to show that $\int_a^b f > 0$.

14

Functions of Several Variables

"And thick and fast they came at last, and more, and more, and more."

Lewis Carroll

14.1 Continuity

There are evidently many mathematical quantities which depend on more than one variable so we must augment our analysis to include the consideration of functions of several variables. Much of what we have established for functions of one variable remains true in the new situation, but some features change. To oversimplify, the $\varepsilon - \delta$ techniques remain much the same, the extra variables adding a certain long-windedness, but there is an underlying change arising from the fact that the geometry of n-dimensional space is rather different from that of one dimension.

The first issue is to decide what we mean by limits and continuity. If A and B are sets we denote by $A_\times B$ the set of all ordered pairs (x, y) where $x \in A$ and $y \in B$, that is, $A_\times B = \{(x, y): x \in A, y \in B\}$. $A_\times B_\times C$ is the set of triples (x, y, z) with $x \in A$, $y \in B$, $z \in C$, and so on. We denote by $\mathbb{R}^n$ ($= \mathbb{R}_\times \mathbb{R}_\times .... \times \mathbb{R}$ with n occurrences of $\mathbb{R}$) the set of all n-tuples $(x_1, ..., x_n)$ of real numbers. Where helpful, we shall abbreviate $(x_1, ..., x_n)$ as $\mathbf{x}$. To define a limit as $\mathbf{x}$ tends to $\mathbf{a}$ we need an idea of distance, which we generalise from that used in two- and three-dimensional geometry. The definitions of limit and continuity are then obtained from the idea that $f(\mathbf{x}) \to L$ as $\mathbf{x} \to \mathbf{a}$ if $|f(\mathbf{x}) - L|$ can be made as small as desired by choosing $\mathbf{x}$ 'near enough' to $\mathbf{a}$.

Definitions Let $\mathbf{x} = (x_1, ..., x_n) \in \mathbb{R}^n$. The **norm** of $\mathbf{x}$, denoted by $\|\mathbf{x}\|$, is the quantity $\|\mathbf{x}\| = \sqrt{\Sigma_{i=1}^n |x_i|^2}$. The distance apart of two points $\mathbf{x}$ and $\mathbf{y}$ is said to be $\|\mathbf{x} - \mathbf{y}\|$, where $\mathbf{x} - \mathbf{y} = (x_1 - y_1, ..., x_n - y_n)$.

Let $\mathbf{a} = (a_1, ..., a_n)$ and suppose that $f: A \to \mathbb{R}$ where A is some subset of $\mathbb{R}^n$ containing $\{\mathbf{x} \in \mathbb{R}^n : 0 < \|\mathbf{x} - \mathbf{a}\| < R\}$ for some $R > 0$. We say $f(\mathbf{x}) \to L$ as $\mathbf{x} \to \mathbf{a}$ if and only if

$$\forall \varepsilon > 0 \quad \exists \delta > 0 \text{ such that } \forall \mathbf{x} \in \mathbb{R}^n \quad 0 < \|\mathbf{x} - \mathbf{a}\| < \delta \Rightarrow |f(\mathbf{x}) - L| < \varepsilon .$$

Suppose that $A \subset \mathbb{R}^n$ and that $f: A \to \mathbb{R}$. Then if $\mathbf{a} \in A$ we say that f is **continuous at** $\mathbf{a}$ if and only if

$$\forall \varepsilon > 0 \quad \exists \delta > 0 \text{ such that } \forall \mathbf{x} \in A \quad \|\mathbf{x} - \mathbf{a}\| < \delta \Rightarrow |f(\mathbf{x}) - f(\mathbf{a})| < \varepsilon .$$

f is said to be **continuous** if it is continuous at each point of A. If $\mathbf{a}$ is 'interior to' A in the sense that there is a positive h for which $\{\mathbf{x}: \|\mathbf{x} - \mathbf{a}\| < h\} \subset A$, then continuity at $\mathbf{a}$ is exactly the same as the property that $f(\mathbf{x}) \to f(\mathbf{a})$ as $\mathbf{x} \to \mathbf{a}$.

If, however, $\mathbf{a} \in A$ but $\mathbf{a}$ is not interior to A then continuity at $\mathbf{a}$ does not imply that $f(\mathbf{x}) \to f(\mathbf{a})$ as $\mathbf{x} \to \mathbf{a}$, since the definition of continuity only pays heed to those points $\mathbf{x}$ near $\mathbf{a}$ which happen to lie in A. This is akin to the definition of a continuous function on the interval $[0, 1]$ where we make a slightly less stringent demand for continuity at the endpoints. Since the geometry of the 'edge' of a set in $\mathbb{R}^n$ is more complicated than that in one dimension, we cannot just formulate what we wish in terms of one-sided continuity. It is easy to see that if A is an interval in $\mathbb{R}^1$, the new definition of a continuous function from A to $\mathbb{R}$ coincides with our old one, though the new one also allows us to consider less regular sets than intervals.

Example 14.1 Let $f: \mathbb{R}^2 \to \mathbb{R}$ be given by $f(x, y) = xy$. (We shall label the variables x and y in place of x_1 and x_2 to accord with tradition.) Then $f(x, y) \to 1$ as $(x, y) \to (1, 1)$.

Solution: Let $\varepsilon > 0$. Let $\delta = \min(1, \varepsilon/3)$.

$$\text{Then } \|(x, y) - (1, 1)\| < \delta \implies \sqrt{((x-1)^2 + (y-1)^2)} < \delta$$
$$\implies |x-1| < \delta \text{ and } |y-1| < \delta$$
$$\implies |x| < 2 \text{ and } |y| < 2 \text{ (since } \delta \leq 1) \;.$$

Therefore $|(x, y) - (1, 1)| < \delta \implies |f(x, y) - 1| = |xy - 1|$
$$\leq |(x-1)y| + |y-1|$$
$$< 2\delta + \delta \leq \varepsilon \;.$$

Since $\varepsilon > 0$ was arbitrary we have shown the result.

From this it follows that f is continuous at $(1, 1)$ and a similar argument shows f is continuous everywhere. □

Example 14.2 More generally, let $g, h: A \to \mathbb{R}$ be continuous, where A is an interval in $\mathbb{R}$. Define $f: A_\times A \to \mathbb{R}$ by $f(x, y) = g(x)h(y)$; f is continuous.

Solution: To show this, let $(a, b) \in A_\times A$ and let $\varepsilon > 0$. Since g and h are continuous,

$$\exists \delta_1 > 0 \text{ such that } \forall x \in A \;\; |x - a| < \delta_1 \implies |g(x) - g(a)| < \varepsilon \text{ and}$$
$$\exists \delta_2 > 0 \text{ such that } \forall y \in A \;\; |y - b| < \delta_2 \implies |h(y) - h(b)| < \varepsilon \;.$$

Also, $\exists \delta_3 > 0$ such that $\forall y \in A \;\; |y - b| < \delta_3 \implies |h(y) - h(b)| < 1$.

Then $\|(x, y) - (a, b)\| < \delta = \min(\delta_1, \delta_2, \delta_3) \implies |x - a| \leq \|(x, y) - (a, b)\| < \delta$ and $|y - b| \leq \|(x, y) - (a, b)\| < \delta$, so

$$\|(x, y) - (a, b)\| < \delta \implies |f(x, y) - f(a, b)| = |g(x)h(y) - g(a)h(b)|$$
$$\leq |g(x) - g(a)||h(y)|$$
$$+ |h(y) - h(b)||g(a)|$$
$$< \varepsilon(|h(y)| + |g(a)|)$$
$$< \varepsilon(1 + |h(b)| + |g(a)|)$$

Since the factor multiplying ε is a constant and the above is true for all $\varepsilon > 0$, we could have substituted $\varepsilon/(1 + |h(b)| + |g(a)|)$ for ε throughout; this shows that f is continuous at (a, b) and, since (a, b) was typical, f is continuous. $\square$

Lemma 14.1 Let $f, g: A \to \mathbb{R}$ be continuous, where $A \subset \mathbb{R}^n$ and let $\lambda \in \mathbb{R}$. Then $f + g$, fg, λf and $|f|$ are all continuous, and $1/f$ is continuous on the set $A \backslash \{x: f(x) = 0\}$. $\square$

We leave the proof of the Lemma as an exercise.   From this and the example preceding it, we see that functions which can be defined by adding, multiplying and dividing functions of one variable are continuous except where their denominator is zero.   Thus, for example, $(x_1 x_2 + x_3)/(x_1^2 + x_2^2 + x_3^2)$ is a continuous function of (x_1, x_2, x_3) except possibly at $(0, 0, 0)$.

Suppose that $f: \mathbb{R}^n \to \mathbb{R}$, so that $f(x_1, ..., x_n)$ is a function of the n independent variables $x_1, ..., x_n$.   By fixing some of these variables we can obtain a related function dependent on fewer variables, which raises the question of how we relate the continuity of the various functions so obtained.   Suppose $a_2, ..., a_n \in \mathbb{R}$ and $g_1: \mathbb{R} \to \mathbb{R}$ is defined by $g_1(x) = f(x, a_2, ..., a_n)$.   Then if f is continuous at $(a_1, ..., a_n)$, g_1 is continuous at a_1.   To see this, let $\varepsilon > 0$;

$$\exists \delta > 0 \text{ such that } \|x - a\| < \delta \Rightarrow |f(x) - f(a)| < \varepsilon$$

whence $|x - a_1| < \delta \Rightarrow \|(x, a_2, ..., a_n) - (a_1, ..., a_n)\| < \delta \Rightarrow |g_1(x) - g_1(a_1)| < \varepsilon$.
Similarly g_i, given by $g_i(x) = f(a_1, ..., a_{i-1}, x, a_{i+1}, ..., a_n)$, is continuous at a_i.   This result we can paraphrase as saying that if f is continuous in the n variables $x_1, ..., x_n$ jointly, then f is continuous in each variable separately (i.e. fixing all but one of the variables).   The converse result is false.

Example 14.3 Define $f: \mathbb{R}^2 \to \mathbb{R}$ by $f(x, y) = 2xy/(x^2 + y^2)$ (for $(x, y) \neq (0, 0)$) and $f(0, 0) = 0$. f is continuous at all points other than $(0, 0)$, since at all such points, $x^2 + y^2 \neq 0$.   Now for $x \neq 0$,

$$f(x, x) - f(0, 0) = 1 \text{ while } f(x, -x) - f(0, 0) = -1,$$

so, whatever $\delta > 0$ we choose, there are points (x, y) with $\|(x, y)\| < \delta$ for which $|f(x, y) - f(0, 0)| = 1$ so f is not continuous at $(0, 0)$, nor could a different choice of $f(0, 0)$ produce a continuous function.
The result here is more easily visualised by changing to polar co-ordinates and letting $x = r \cos\theta$, $y = r \sin\theta$ so $f(x, y) - f(0, 0) = 2 \cos\theta\sin\theta$.   This does not tend to 0 as $r = \|(x, y\| \to 0$ (unless $\cos\theta \sin\theta = 0$).
However, $f(x, 0) = 0$ and $f(0, y) = 0$ so the functions obtained from f by fixing one of the variables are continuous.   We conclude that 'joint' continuity with respect to several variables is stronger than continuity with respect to each variable separately.   This difference can be significant and occasionally awkward.   The reader may take heart from the knowledge that, in the development of analysis, even such an eminent mathematician as Cauchy failed to notice this! $\square$

We shall prove an analogue of Theorem 10.5 which shows that if $f: [a, b] \to \mathbb{R}$ is continuous, it is bounded.   For this we need to find a suitable type of set to form the domain of our function.

Definitions Let $(\mathbf{x}_n)$ be a sequence of points of $\mathbb{R}^m$. We say that $\mathbf{x}_n \to \mathbf{x}$ as $n \to \infty$ if and only if $\forall\, \varepsilon > 0 \;\exists\, N$ such that $\forall n \geq N \;\|\mathbf{x}_n - \mathbf{x}\| < \varepsilon$. Let $A \subset \mathbb{R}^m$. We say A is **closed** if, whenever $(\mathbf{x}_n)$ is a sequence of elements of A and $\mathbf{x}_n \to \mathbf{x}$ as $n \to \infty$, then $\mathbf{x} \in A$.

Example 14.4 $\mathbb{R}^m$ itself is closed. Also $[a, b]$ is a closed subset of $\mathbb{R}^1$ (since if $\forall n$, $x_n \in [a, b]$ and $x_n \to x$ as $n \to \infty$, then $x \in [a, b]$). The set $(0, 1)$ is not closed since $\forall n \in \mathbb{N} \; 1/n \in (0, 1)$ and $1/n \to 0$ but $0 \notin (0, 1)$. $\square$

Lemma 14.2 Let A be a subset of $\mathbb{R}^m$ and $f\colon A \to \mathbb{R}$ be continuous. If $\forall n \in \mathbb{N} \; \mathbf{x}_n \in A$ and $\mathbf{x}_n \to \mathbf{a} \in A$ as $n \to \infty$ then $f(\mathbf{x}_n) \to f(\mathbf{a})$ as $n \to \infty$. $\square$

This result, whose proof is a direct adaptation of that for sequences and functions defined in $\mathbb{R}$, leads us to a useful technique for spotting closed sets.

Let $f\colon \mathbb{R}^m \to \mathbb{R}$ be continuous and $A = \{\mathbf{x}\colon f(\mathbf{x}) \in [a, b]\} = f^{-1}([a, b])$; A is closed. Notice that f has the whole of $\mathbb{R}^m$ as its domain. The proof is easy: let $\mathbf{x} \in \mathbb{R}^m$ and suppose that $\forall n \in \mathbb{N} \; \mathbf{x}_n \in A$ and $\mathbf{x}_n \to \mathbf{x}$. Then $f(\mathbf{x}_n) \to f(\mathbf{x})$ as $n \to \infty$ so, since $a \leq f(\mathbf{x}_n) \leq b$ for all n, $a \leq f(\mathbf{x}) \leq b$ and $\mathbf{x} \in A$. A is thus closed.

The same argument shows that $f^{-1}([a, \infty))$ and $f^{-1}((-\infty, a])$ are closed, and a slight change shows that the intersection of two closed sets, or of any collection of closed sets, is closed. Thus we can see easily that, for example, the 'disc' $\{\mathbf{x} \in \mathbb{R}^m\colon \|\mathbf{x}\| \leq 1\}$ is closed, since it is $\{\mathbf{x}\colon f(\mathbf{x}) \in [0, 1]\}$ where $f(\mathbf{x}) = \sqrt{(x_1^2 + \dots x_m^2)}$ and f is continuous. This is worth stating as a Lemma.

Lemma 14.3 Let $f\colon \mathbb{R}^m \to \mathbb{R}$ be continuous. The sets $f^{-1}([a, b])$, $f^{-1}([a, \infty))$ and $f^{-1}((-\infty, a])$ are closed, and the intersection of a collection of closed sets is closed. $\square$

A useful intuitive view of a closed set is one which includes the points 'at its edges'; $\{\mathbf{x} \in \mathbb{R}^m\colon \|\mathbf{x}\| < 1\}$ is not closed but $\{\mathbf{x} \in \mathbb{R}^m\colon \|\mathbf{x}\| \leq 1\}$ is closed.

To proceed further, we need a short discussion on sequences.

Definition Let (a_n) be a sequence of real numbers. A **subsequence** of (a_n) is a sequence consisting of some of the terms of (a_n) in their original order, that is, a sequence of the form $(a_{n_k})_{k=1}^{\infty}$ where $\forall k \in \mathbb{N} \; n_{k+1} > n_k$ and $n_k \in \mathbb{N}$. Thus $(a_2, a_4, a_6, \dots)$ is a subsequence of (a_n). Notice that $n_1 \geq 1$, and $n_2 > n_1$ so $n_2 \geq 2$ and, in general, $n_k \geq k$.

Lemma 14.4 Every sequence of real numbers has either an increasing subsequence or a decreasing one.

Proof. Let (a_n) be a sequence of real numbers. We first try to find an increasing subsequence. If we just proceed naïvely, choosing a_{n_1} and seeking $n_2 > n_1$ satisfying $a_{n_2} \geq a_{n_1}$, we shall become unable to proceed if any of our suffices n_j has the property that $\forall n > n_j \; a_n < a_{n_j}$. This suggests that we look at the set of such points, so let $S = \{n \in \mathbb{N} : \forall m > n, \; a_m < a_n\}$, the set of suffices such that the corresponding term is greater than all subsequent terms.

If S is finite (including the case $S = \varnothing$) then choose n_1 to be greater than all the elements of S. Thus $n_1 \notin S$. By definition, then, $\exists n_2 > n_1$ such that $a_{n_2} \geq a_{n_1}$. Since $n_2 > n_1$, $n_2 \notin S$ and $\exists n_3 > n_2$ such that $a_{n_3} \geq a_{n_2}$. Proceeding in this way we obtain an increasing subsequence (a_{n_k}) of (a_n).

If S is infinite let $n_1 \in S$. For $n_k \in S$ let n_{k+1} be chosen to belong to S and satisfy $n_{k+1} > n_k$. This is possible for all k since S is infinite. This gives a subsequence (a_{n_k}) with the property that $\forall k \in \mathbb{N} \; a_{n_{k+1}} < a_{n_k}$ (since $n_k \in S$), a strictly decreasing sequence. $\square$

Lemma 14.5 A subsequence of a convergent sequence converges to the same limit as the whole sequence.

Proof. Let $x_n \to x$ as $n \to \infty$, where (x_n) is a sequence of real numbers. Let (x_{n_k}) be a subsequence of (x_n), so $\forall k \in \mathbb{N} \; n_k \geq k$.

Let $\varepsilon > 0$. $\exists N$ such that $\forall n \geq N \; |x_n - x| < \varepsilon$, whence, for this N,
$$k \geq N \Rightarrow n_k \geq k \geq N \Rightarrow |x_{n_k} - x| < \varepsilon \; .$$

Thus $x_{n_k} \to x$ as $k \to \infty$.

This result is equally true of a sequence in $\mathbb{R}^m$. $\square$

Definition A sequence (x_n) in $\mathbb{R}^m$ is said to be **bounded** if $(\|x_n\|)$ is a bounded sequence in $\mathbb{R}$. A subset $A \subset \mathbb{R}^m$ is said to be **bounded** if $\{\|x\| : x \in A\}$ is a bounded subset of $\mathbb{R}$.

Theorem 14.6 Bolzano-Weierstrass Theorem. Every bounded sequence in $\mathbb{R}^m$ has a convergent subsequence.

Proof. We first take the simpler case of a sequence of real numbers. Let (a_n) be a bounded sequence of real numbers. Then there is a subsequence (a_{n_k}) which is either increasing or decreasing. Since (a_n) is bounded, so is (a_{n_k}), hence (a_{n_k}), being bounded and either increasing or decreasing, is convergent.

Now let (x_n) be a bounded sequence in $\mathbb{R}^m$; suppose that $\forall n \in \mathbb{N} \; \|x_n\| \leq K$. For each n, let $x_n = (x_1^{(n)}, \dots, x_m^{(n)})$. The sequence $(x_1^{(n)})$ (as n varies) is a sequence of real numbers, and since $\forall n, \; |x_1^{(n)}| \leq \|x_n\| \leq K$ it is a bounded sequence. Choose a subsequence $(x_1^{n_1(k)})$ which is convergent, which is possible by the last paragraph. Now $|x_2^{n_1(k)}| \le \|x_{n_1(k)}\| \le K$ whence $(x_2^{n_1(k)})$ is a bounded sequence of real numbers, so it has a convergent subsequence $(x_2^{n_2(k)})$. Since $(x_1^{n_2(k)})$ is a subsequence of the convergent sequence $(x_1^{n_1(k)})$, it also converges. Proceeding in this way we obtain after m steps a sequence $(n_m(k))_{k=1}^{\infty}$ of integers such that for $j = 1, ..., m$ each sequence $(x_j^{n_m(k)})$ converges, to x_j say. Since $\|x^{n_m(k)} - x\| \to 0$ as $k \to \infty$, where $x = (x_1,..., x_m)$, the result is proved. $\square$

Theorem 14.7 Uniform Continuity. Let A be a closed, bounded subset of $\mathbb{R}^m$ and $f: A \to \mathbb{R}$ be continuous. Then f is uniformly continuous, that is,
$$\forall \varepsilon > 0 \;\; \exists \delta > 0 \;\; \text{such as} \;\; \forall x, y \in A \;\; \|x - y\| < \delta \Rightarrow |f(x) - f(y)| < \varepsilon.$$

Proof. Suppose the result were false, so that $\exists \varepsilon > 0$ such that $\forall \delta > 0$ $\exists x, y \in A$ such that $\|x - y\| < \delta$ and $|f(x) - f(y)| \ge \varepsilon$. Let $\varepsilon > 0$ have the stated property. For each $n \in \mathbb{N}$ since $1/n > 0$, $\exists x_n, y_n \in A$ such that $\|x_n - y_n\| < 1/n$ and $|f(x_n) - f(y_n)| \ge \varepsilon$. Since A is bounded, (x_n) has a convergent subsequence, (x_{n_k}); let $x_{n_k} \to x$ as $k \to \infty$. Since A is closed, $x \in A$. Also, letting $x_{n_k} = (x_1^{n_k}, ..., x_m^{n_k})$ and $y_{n_k} = (y_1^{n_k}, ..., y_m^{n_k})$ we have, for each i and k, $|x_i^{n_k} - y_i^{n_k}| \le \|x_{n_k} - y_{n_k}\| < \frac{1}{n_k}$ so that $y_i^{n_k} \to x_i = \lim_{k \to \infty} x_i^{n_k}$ and it follows that $y_{n_k} \to x$ as $k \to \infty$. But f is continuous at x, so $\exists \delta_1 > 0$ such that $\|y - x\| < \delta_1 \Rightarrow |f(y) - f(x)| < \varepsilon/2$. Choose k so large that $\|x_{n_k} - x\| < \delta_1$ and $\|y_{n_k} - x\| < \delta_1$, then
$$|f(x_{n_k}) - f(y_{n_k})| \le |f(x_{n_k}) - f(x)| + |f(x) - f(y_{n_k})| < \varepsilon/2 + \varepsilon/2 \ .$$
This is a contradiction, since by definition $|f(x_{n_k}) - f(y_{n_k})| \ge \varepsilon$. $\square$

In common with the corresponding result for functions of one variable, Theorem 13.6, the significant result here is that the δ above can be chosen to be independent of x and y. Let $A = [a, b] \times [c, d]$ and $f: A \to \mathbb{R}$ be continuous. A is closed and bounded; to see the closure, notice that $p_i: \mathbb{R}^2 \to \mathbb{R}$ given by $p_i(x_1, x_2) = x_i$ is continuous and $A = p_1^{-1}([a, b]) \cap p_2^{-1}([c, d])$ by Lemma 14.3. Therefore the Theorem applies and f is uniformly continuous. Let $\varepsilon > 0$. Then we can choose $\delta > 0$ such that $\forall x, y \in A \;\; \|x - y\| < \delta \Rightarrow |f(x) - f(y)| < \varepsilon$. In particular, if $|x_1 - x_2| < \delta/2$ and $|y_1 - y_2| < \delta/2$ so that $\|(x_1, y_1) - (x_2, y_2)\| < \delta$, then $|f(x_1, y_1) - f(x_2, y_2)| < \varepsilon$, so that we can use the same value of δ for all $y \in [c, d]$ to ensure that $|f(x_1, y) - f(x_2, y)| < \varepsilon$. One use of this is:

Theorem 14.8 Let $f: [a, b] \times [c, d] \to \mathbb{R}$ be continuous and define $F: [a, b] \to \mathbb{R}$ by $F(x) = \int_c^d f(x, y)\,dy$. Then F is continuous.

Proof: Notice first that, for each $x \in [a, b]$, $y \mapsto f(x, y)$ is continuous, hence integrable, so the expression for F is meaningful.

Let $\varepsilon > 0$. By the uniform continuity of f on $[a, b]_\times [c, d]$, $\exists \delta > 0$ such that $\forall x, x' \in [a, b]_\times [c, d]$ $\|\mathbf{x} - \mathbf{x}'\| < \delta \Rightarrow |f(\mathbf{x}) - f(\mathbf{x}')| < \varepsilon/(d - c)$. Therefore let $x, x' \in [a, b]$ with $|x - x'| < \delta$. Then $\forall y \in [c, d]$ $|f(x, y) - f(x', y)| < \varepsilon/(d - c)$, so that

$$|F(x) - F(x')| \le \int_c^d |f(x, y) - f(x', y)| \, dy < \varepsilon.$$

Since $\varepsilon > 0$ was arbitrary we have shown that $\forall \varepsilon > 0$ $\exists \delta > 0$ such that $\forall x, x' \in [a, b]$ $|x - x'| < \delta \Rightarrow |F(x) - F(x')| < \varepsilon$ so F is (uniformly) continuous. $\square$

Obviously, the analogous result, obtaining a function of y by integrating with respect to x, is true. More generally, if f is a continuous function of the m variables $x_1, ..., x_m$ in $[a_1, b_1]_\times ... _\times[a_m, b_m]$, the function obtained by integrating with respect to one of these is continuous with respect to the remaining variables.

That a continuous function on a closed, bounded set is bounded is now easy:

Theorem 14.9 Let A be a closed, bounded subset of $\mathbb{R}^m$ and $f : A \to \mathbb{R}$ be continuous. Then f is a bounded function and f attains its supremum and infimum at points of A.

Proof: By Theorem 14.7, f is uniformly continuous, so $\exists \delta > 0$ such that $\forall \mathbf{x}, \mathbf{y} \in A$, $\|\mathbf{x} - \mathbf{y}\| < \delta \Rightarrow |f(\mathbf{x}) - f(\mathbf{y})| < 1 \Rightarrow |f(\mathbf{x})| < |f(\mathbf{y})| + 1$.

Since A is bounded, there is an M for which $\mathbf{x} \in A \Rightarrow \|\mathbf{x}\| \le M$ so $\mathbf{x} = (x_1, ..., x_m) \in A \Rightarrow |x_i| \le M$ $i = 1, 2, ..., m$. Thus A is a subset of the 'cube' $[-M, M]_\times[-M, M]_\times ... _\times[-M, M]$. Let $a_0 = -M < a_1 < \cdots < a_r = M$ be such that, for each i, $a_i - a_{i-1} < \delta/m$, so if $x \in [-M, M]$, $x \in [a_{i-1}, a_i]$, for some i. Therefore $A = A_1 \cup ... \cup A_s$ where each A_i is the intersection of A with a set of the form $[a_{i_1}-1, a_{i_1}]_\times ... _\times[a_{i_m}-1 - a_{i_m}]$ and we omit the set from the list if it is empty. (See Fig. 14.1.)

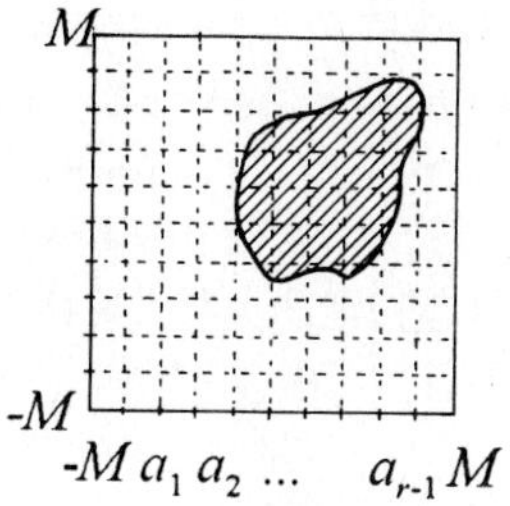

Fig. 14.1

Choose $\mathbf{z}_i = (z_1^{(i)}, ..., z_m^{(i)}) \in A_i$ for each i. If $\mathbf{x} = (x_1, ..., x_m) \in A_i$ then $|z_j^{(i)} - x_j| < \delta/m$ for each j, $\|\mathbf{x} - \mathbf{z}_i\| < \delta$ and so $|f(\mathbf{x})| < |f(\mathbf{z}_i)| + 1$. Thus

$\forall \mathbf{x} \in A$, $|f(\mathbf{x})| < \max_{1 \le i \le s} (|f(\mathbf{z}_i)| + 1)$, since $\mathbf{x}$ must belong to some A_i. It follows that f is bounded.

To show the supremum is attained, let $a = \sup\{f(\mathbf{x}): \mathbf{x} \in A\}$. Then if a is not attained and we set $g(\mathbf{x}) = 1/(a - f(\mathbf{x}))$ g is continuous on A and thus bounded. This gives a contradiction in exactly the same way as in the case of a function of one variable. $\square$

14.2 Differentiation

As with continuity, we may consider the differentiation of a function of several variables either by fixing all but one of them and differentiating the resulting function of one variable or by considering a differentiation process where all the variables are treated simultaneously. As with continuity again, these two processes differ.

Definitions Suppose that A is a subset of $\mathbb{R}^m$, $f: A \to \mathbb{R}$ and that $\mathbf{a} \in A$. If the function g_i defined by $g_i(x) = f(a_1, ..., a_{i-1}, x, a_{i+1}, ..., a_m)$ is differentiable at a_i, then we say the **partial derivative** $\dfrac{\partial f}{\partial x_i}$ exists at $\mathbf{a} = (a_1, ..., a_m)$ and that its value is $g_i'(a_i)$. f is said to be **differentiable** at $\mathbf{a}$ if for some $h > 0$ $\{\mathbf{x} \in \mathbb{R}^m : \|\mathbf{x} - \mathbf{a}\| < h\} \subset A$ and, for some constants $\beta_1, ..., \beta_m$,

$$f(\mathbf{x}) = f(\mathbf{a}) + \sum_{i=1}^{m} \beta_i (x_i - a_i) + \theta(\mathbf{x})$$

where $\theta(\mathbf{x})/\|\mathbf{x} - \mathbf{a}\| \to 0$ as $\mathbf{x} \to \mathbf{a}$.

Partial derivatives are the type of derivative we have met before, of functions of one variable, and we can apply our existing theorems to them. Notice that there is no need for all partial derivatives to exist: if $f(x_1, x_2) = x_1 + |x_2|$ then $\partial f/\partial x_1$ exists at $(0, 0)$ but $\partial f/\partial x_2$ does not.

Differentiability of f as a whole is about the approximation of $f(\mathbf{x}) - f(\mathbf{a})$ by a linear function of $\mathbf{x} - \mathbf{a}$, the error θ being 'small' in the sense stated. This is often expressed in terms of 'differentials' as $df = \sum_{i=1}^{m} \beta_i(dx_i)$ where, as we shall see, β_i turns out to equal $\partial f/\partial x$ at $\mathbf{a}$. The more precise statement is the one we have made about approximating $f(\mathbf{x}) - f(\mathbf{a})$ by the linear function $\sum_{i=1}^{m} \beta_i(x_i - a_i)$.

Example 14.5 Let $f(x_1, x_2) = x_1 x_2$. It is easy to see that at the point (a_1, a_2), $\partial f/\partial x_1 = a_2$ and $\partial f/\partial x_2 = a_1$, so f possesses partial derivatives. Also

$$f(x_1, x_2) - f(a_1, a_2) = x_1 x_2 - a_1 a_2 = (x_1 - a_1)a_2 + (x_2 - a_2)a_1 + \theta(x_1, x_2)$$

where $\theta(x_1, x_2) = (x_1 - a_1)(x_2, - a_2)$. At this point we recall that if $u, v \in \mathbb{R}$ $|2uv| \le u^2 + v^2$ so that $2|(x_1 - a_1)(x_2 - a_2)| \le (x_1 - a_1)^2 + (x_2 - a_2)^2$ and $|\theta(x_1, x_2)|/\|\mathbf{x} - \mathbf{a}\| \le \frac{1}{2}\sqrt{((x_1 - a_1)^2 + (x_2 - a_2)^2)}$ whence $\theta(x_1, x_2)/\|\mathbf{x} - \mathbf{a}\| \to 0$ as $\mathbf{x} \to \mathbf{a}$. f is differentiable at (a_1, a_2) and the constants β_1, β_2 equal $(\partial f/\partial x_1)(a_1, a_2)$ and $(\partial f/\partial x_2)(a_1, a_2)$ respectively.

In practice, writing $x_1 - a_1 = r \cos\theta$, $x_2 - a_2 = r \sin\theta$ is illuminating in the above calculation. □

Lemma 14.10 Let A be a subset of $\mathbb{R}^m$ and $\{x: \|x - a\| < h\} \subset A$, where $h > 0$. If $f: A \to \mathbb{R}$ is differentiable at a then all the partial derivatives at a exist and $\frac{\partial f}{\partial x_i}(a_1, ..., a_m) = \beta_i$ $(i = 1, 2, ..., m)$ where β_i is the number occurring in the definition of differentiability.

Conversely, if all the partial derivatives of f exist at all points of A, and are continuous at a, then f is differentiable at a.

Proof: Suppose that f is differentiable at a and that

$$f(x) = f(a) + \sum_{i=1}^{m} \beta_i (x_i - a_i) + \theta(x)$$

where $\theta(x)/\|x - a\| \to 0$ as $x \to a$. Then let $x = (a_1, ..., a_{i-1}, t, a_{i+1}, ..., a_m)$ and $g_i(t) = f(a_1, ..., a_{i-1}, t, a_{i+1}, ..., a_m)$. Since $\|x - a\| = |t - a_i|$ we have

$$g_i(t) = g_i(a_i) + \beta_i(t - a_i) + \theta_i(t)$$

where $\theta_i(t) = \theta(x)$ and $\theta_i(t)/|t - a_i| \to 0$ as $t \to a_i$. From this we see that g_i is differentiable at a_i, and that $\beta_i = g_i'(a_i) = \frac{\partial f}{\partial x_i}(a_1, ..., a_m)$.

Now suppose that all the partial derivatives exist, and that they are continuous at a. To deduce information about $f(x) - f(a)$ we need to re-express this in a form where in each part only one variable changes, so if $\|x - a\| < h$

$$f(x_1, ..., x_m) - f(a_1, ..., a_m) = \sum_{i=1}^{m} (f(x_1, ..., x_i, a_{i+1}, ..., a_m) - f(x_1, ..., x_{i-1}, a_i, ..., a_m))$$

$$= \sum_{i=1}^{m} \frac{\partial f}{\partial x_i}(x_1, ..., x_{i-1}, \xi_i, a_{i+1}, ..., a_m)(x_i - a_i), \tag{1}$$

obtained by applying the one-variable Mean Value Theorem to the function $t \mapsto f(x_1, ..., x_{i-1}, t, a_{i+1}, ..., a_m)$; ξ_i will be between a_i and x_i (and will depend on $x_1, ..., x_i$). In this notation, we have, for $\|x - a\| < h$, that $f(x) = f(a) + \sum_{i=1}^{m} \frac{\partial f}{\partial x_i}(a)(x_i - a_i) + \theta(x)$ where

$$\theta(x) = \sum_{i=1}^{m} \left(\frac{\partial f}{\partial x_1}(x_1, ..., x_{i-1}, \xi_i, a_{i+1}, ..., a_m) - \frac{\partial f}{\partial x_i}(a_1, ..., a_m) \right)(x_i - a_i)$$

Let $\varepsilon > 0$. Then, by continuity of $\frac{\partial f}{\partial x_i}$ at a, $\exists \delta > 0$ such that $\|y - a\| < \delta \Rightarrow$ for all i, $|\frac{\partial f}{\partial x_i}(y) - \frac{\partial f}{\partial x_i}(a)| < \varepsilon/m$. Choose $\|x - a\| < \delta$, so, with ξ_i as in (1),

$$\left\| (x_1, ..., x_{i-1}, \xi_i, a_{i+1}, ..., a_m) - (a_1, ..., a_m) \right\| \leq \|x - a\| < \delta \quad \text{and}$$

$$|\theta(\mathbf{x})| \;\le\; \sum_{i=1}^{m} \left| \frac{\partial f}{\partial x_i}(x_1, ..., x_{i-1}, \xi_i, a_{i+1}, ..., a_m) - \frac{\partial f}{\partial x_i}(a_1, ..., a_m) \right| |x_i - a_i|$$

$$< \;\; \varepsilon/m \sum_{i=1}^{m} |x_i - a_i| \;\le\; \varepsilon\|\mathbf{x} - \mathbf{a}\| \qquad (\text{since } |x_i - a_i| \le \|\mathbf{x} - \mathbf{a}\|) \; .$$

So $\;\; 0 < \|\mathbf{x} - \mathbf{a}\| < \delta \Rightarrow |\theta(x)|/\|\mathbf{x} - \mathbf{a}\| < \varepsilon$, and, since $\;\; \varepsilon > 0 \;\;$ was arbitrary, $\theta(x)/\|\mathbf{x} - \mathbf{a}\| \to 0$ as $\mathbf{x} \to \mathbf{a}$. $\;\; \square$

The condition for differentiability may be put in a version which is co-ordinate-free by observing that $\;\; \sum_{i=1}^{m} \beta_i (x_i - a_i) \;\;$ is the form of a typical linear function from $\mathbb{R}^m$ to $\mathbb{R}$ evaluated at $\mathbf{x} - \mathbf{a}$; we shall not pursue this. Notice that differentiability is a stronger condition than the existence of the partial derivatives. It is differentiability that we require in order to establish the chain rule for functions of several variables, but we have a few items to tidy up first.

Example 14.6 Let $f(x, y) = xy/\surd(x^2 + y^2) \; ((x, y) \ne (0, 0))$ and $f(0, 0) = 0$. Then $(\partial f/\partial x)(0, 0) = \lim_{x \to 0} (f(x, 0) - f(0, 0))/x = 0 \;\;$ and $\;\; (\partial f/\partial y)(0, 0) = 0$. $\;\; f$ is not, however, differentiable at $(0, 0)$, since

$$f(x, y) - f(0, 0) - \beta_1 x - \beta_2 y = xy/\surd(x^2 + y^2)$$

where $\;\; \beta_1 = (\partial f/\partial x)(0, 0) \;\;$ and $\;\; \beta_2 = (\partial f/\partial y)(0, 0) \;\;$ and $\;\; xy/\surd(x^2 + y^2) \not\to 0$ as $(x, y) \to (0, 0)$. $\;\; \square$

The existence of partial derivatives of f does not imply that f is continuous. If $\frac{\partial f}{\partial x_1}$ exists at a point $\mathbf{a}$, this means that, *as a function of x_1 only, f is differentiable* and so f is continuous with respect to the variable x_1. Therefore the existence of all the partial derivatives guarantees only that f is continuous with respect to each variable separately. Consideration of the function defined by $f(x, y) = xy/(x^2 + y^2)$ for $(x, y) \ne (0, 0)$, $f(0, 0) = 0$, gives a function which is not continuous at $(0, 0)$ but $(\partial f/\partial x)$ and $(\partial f/\partial y)$ exist at all points.

If f is differentiable at $\mathbf{a}$, then f is continuous at $\mathbf{a}$. By definition of differentiability, $f(\mathbf{x}) = f(\mathbf{a}) + \sum_{i=1}^{m} \beta_i (x_i - a_i) + \theta(\mathbf{x})$ where $\theta(\mathbf{x})/\|\mathbf{x} - \mathbf{a}\| \to 0$ as $\mathbf{x} \to \mathbf{a}$. Since for each i, $|x_i - a_i| \le \|\mathbf{x} - \mathbf{a}\|$, $\sum_{i=1}^{m} \beta_i (x_i - a_i) \to 0$ as $\mathbf{x} \to \mathbf{a}$, and also $\theta(\mathbf{x}) = (\theta(\mathbf{x})/\|\mathbf{x} - \mathbf{a}\|)\|\mathbf{x} - \mathbf{a}\| \to 0$ so f is continuous at $\mathbf{a}$.

Theorem 14.11 The Chain Rule. Suppose that $\mathbf{b} \in B \subset \mathbb{R}^m$ and that $f: B \to \mathbb{R}$ is differentiable at $\mathbf{b}$. Let $g_1, ..., g_m: A \to \mathbb{R}$ be m functions, differentiable at $\mathbf{a} \in A \subset \mathbb{R}^n$ and with $g_i(\mathbf{a}) = b_i \;\; (i = 1, 2, ..., m)$. Then the function $h: A \to \mathbb{R}$ where $h(x_1, ..., x_n) = f(g_1(x_1, ..., x_n), ..., g_m(x_1, ..., x_n))$ is differentiable at $\mathbf{a}$ and

$$\frac{\partial h}{\partial x_j}(\mathbf{a}) = \sum_{i=1}^{m} \frac{\partial f}{\partial y_i}(\mathbf{g}(\mathbf{a})) \frac{\partial g_i}{\partial x_j}(\mathbf{a})$$

(where $\mathbf{g}(\mathbf{a}) = (g_1(\mathbf{a}), ..., g_m(\mathbf{a})))$. Here f is regarded as a function of $(y_1, ..., y_m)$.

Proof: By the assumption of differentiability,

$$f(\mathbf{y}) = f(\mathbf{b}) + \sum_{i=1}^{m} \beta_i (y_i - b_i) + \theta(\mathbf{y})$$

and $g_i(\mathbf{x}) = g_i(\mathbf{a}) + \sum_{j=1}^{n} \alpha_{ij}(x_j - a_j) + \phi_i(\mathbf{x})$

where $\theta(\mathbf{y})/\|\mathbf{y} - \mathbf{b}\| \to 0$ as $\mathbf{y} \to \mathbf{b}$ and $\phi_i(\mathbf{x})/\|\mathbf{x} - \mathbf{a}\| \to 0$ as $\mathbf{x} \to \mathbf{a}$ $(i = 1, ..., m)$. Then

$$h(\mathbf{x}) = f(\mathbf{g}(\mathbf{x})) = f(\mathbf{g}(\mathbf{a})) + \sum_{i=1}^{m} \beta_i (g_i(\mathbf{x}) - g_i(\mathbf{a})) + \theta(\mathbf{g}(\mathbf{x}))$$

$$= f(\mathbf{g}(\mathbf{a})) + \sum_{i=1}^{m} \beta_i \left(\sum_{j=1}^{n} \alpha_{ij}(x_j - a_j) \right) + \sum_{i=1}^{m} \beta_i \phi_i(\mathbf{x}) + \theta(\mathbf{g}(\mathbf{x}))$$

$$= f(\mathbf{g}(\mathbf{a})) + \sum_{j=1}^{n} \gamma_j (x_j - a_j) + \psi(\mathbf{x}) , \qquad (1)$$

where $\gamma_j = \sum_{i=1}^{m} \beta_i \alpha_{ij}$ and $\psi(\mathbf{x}) = \sum_{i=1}^{m} \beta_i \phi_i(\mathbf{x}) + \theta(\mathbf{g}(\mathbf{x}))$. We have to show $\psi(\mathbf{x})/\|\mathbf{x} - \mathbf{a}\| \to 0$ as $\mathbf{x} \to \mathbf{a}$. From what we already know of ϕ_i, we need only prove $\theta(\mathbf{g}(\mathbf{x}))/\|\mathbf{x} - \mathbf{a}\| \to 0$. Let $M = (1 + \max_i \sum_j |\alpha_{ij}|)\sqrt{m}$. $\exists \delta_1 > 0$ such that $0 < \|\mathbf{y} - \mathbf{b}\| < \delta_1 \Rightarrow |\theta(y)|/\|\mathbf{y} - \mathbf{b}\| < \varepsilon/M$, since $\theta(y)/\|\mathbf{y} - \mathbf{b}\| \to 0$, so

$$\|\mathbf{y} - \mathbf{b}\| < \delta_1 \Rightarrow |\theta(\mathbf{y})| \le \varepsilon \|\mathbf{y} - \mathbf{b}\| / M . \qquad (2)$$

Since $\phi_i(\mathbf{x})/\|\mathbf{x} - \mathbf{a}\| \to 0$ as $\mathbf{x} \to \mathbf{a}$,

$\exists \delta_2 > 0$ such that $0 < \|\mathbf{x} - \mathbf{a}\| < \delta_2 \Rightarrow |\phi_i(\mathbf{x})|/\|\mathbf{x} - \mathbf{a}\| < 1$ $(i = 1, 2, ..., m)$

$$\Rightarrow |g_i(\mathbf{x}) - g_i(\mathbf{a})| \le \sum_{j=1}^{n} |\alpha_{ij}||x_j - a_j| + |\phi_i(\mathbf{x})|$$

$$< \left(1 + \sum_{j=1}^{n} |\alpha_{ij}|\right) \|\mathbf{x} - \mathbf{a}\|$$

$$\Rightarrow \|\mathbf{g}(\mathbf{x}) - \mathbf{g}(\mathbf{a})\| \le M \|\mathbf{x} - \mathbf{a}\| . \qquad (3)$$

Thus $0 < \|\mathbf{x} - \mathbf{a}\| < \min(\delta_2, \delta_1/M) \Rightarrow \|\mathbf{g}(\mathbf{x}) - \mathbf{g}(\mathbf{a})\| < \delta_1$ by (3)

$$\Rightarrow |\theta(\mathbf{g}(\mathbf{x}))| < \varepsilon \|\mathbf{g}(\mathbf{x}) - \mathbf{g}(\mathbf{a})\| / M \le \varepsilon \|\mathbf{x} - \mathbf{a}\|$$

by (2) and (3) .

Therefore $\theta(\mathbf{g}(\mathbf{x}))/\|\mathbf{x} - \mathbf{a}\| \to 0$ as $\mathbf{x} \to \mathbf{a}$ so h is differentiable, by (1).

Finally, we see that γ_j, which is $\dfrac{\partial h}{\partial x_j}(\mathbf{a})$, equals $\sum_{i=1}^{m} \beta_i \alpha_{ij}$, that is,

$$\frac{\partial h}{\partial x_j}(\mathbf{a}) = \sum_{i=1}^{m} \frac{\partial f}{\partial y_i}(\mathbf{g}(\mathbf{a})) \frac{\partial g_i}{\partial x_j}(\mathbf{a}) . \quad \square$$

The Chain Rule for functions of several variables is easiest remembered in the form where f is a function of the variables $y_1, ..., y_m$ and each of $y_1, ..., y_m$ is a function of $x_1, ..., x_n$ giving, in the standard form,

$$\frac{\partial f}{\partial x_j} = \sum_{i=1}^{m} \frac{\partial f}{\partial y_i} \cdot \frac{\partial y_i}{\partial x_j} ,$$

the rate of change of f with x_j being the sum of the rate of change of f with each co-ordinate times the rate of change of that co-ordinate with x_j.   We have not chosen to state the theorem in this way because the symbol f is used here with two distinct meanings, and the same is true of y_i. This is clearer if we take an example.

Let $f(y_1, y_2) = y_1 y_2$ and let $g_1(x_1, x_2) = x_1 + x_2$, $g_2(x_1, x_2) = x_1 - x_2$. Then let $h(x_1, x_2) = f(g_1(x_1, x_2), g_2(x_1, x_2)) = x_1^2 - x_2^2$. The first thing to notice is that h is not the same function as f (e.g. $h(1, 2) \neq f(1, 2)$). The confusion usually arises in applications where the value of f represents some physical quantity and h represents the same quantity as a function of different variables, so that by '$\frac{\partial f}{\partial x_1}$' we may really mean $\frac{\partial h}{\partial x_1}$.

The second issue with the traditional notations is that the y_1 in $\frac{\partial f}{\partial y_1}$ is spurious; $\frac{\partial f}{\partial y_1}$ is a function of two variables, being that function obtained from f by fixing the second variable and differentiating with respect to the first.   Since these variables are 'dummy' variables the fact that we have to attach a name to them is unfortunate. $\frac{\partial f}{\partial y_1}$ above is the function whose value at (a, b) is b.   The Newtonian notation of f_{y_1} for $\frac{\partial f}{\partial y_1}$ is no better in this respect, while the various attempts to introduce a more rational notation have not penetrated far into practice so we shall just have to make the best of the situation.   The principal analytical advice is to be quite clear which functions are distinct and, where possible, to use different symbols to distinguish them.

Example 14.7 Let f be a function of the two variables (x, y) and define $F(x, y) = f(y, x)$.   Show that $\frac{\partial F}{\partial x}(a, b) = \frac{\partial f}{\partial y}(b, a)$.   (This is, of course, deliberately awkward.)

Solution: Let $g_1(x, y) = y$ and $g_2(x, y) = x$ so that $F(x, y) = f(g_1(x, y), g_2(x, y))$ and $\frac{\partial F}{\partial x}(a, b) = \frac{\partial f}{\partial x}(g(a, b))\frac{\partial g_1}{\partial x}(a, b) + \frac{\partial f}{\partial y}(g(a, b))\frac{\partial g_2}{\partial x}(a, b)$

$$= \frac{\partial f}{\partial x}(b, a) \ 0 + \frac{\partial f}{\partial y}(b, a) \ 1 = \frac{\partial f}{\partial y}(b, a) .$$

(If it helps, you can write $\frac{\partial f}{\partial g_1}$ in place of $\frac{\partial f}{\partial x}$ and $\frac{\partial f}{\partial g_2}$ in place of $\frac{\partial f}{\partial y}$.) □

Armed with the Chain Rule, it is now easy to apply earlier results.

Theorem 14.12. **Mean Value Theorem.** Let $f: A \to \mathbb{R}$ be continuous, $\mathbf{a}, \mathbf{b} \in A \subset \mathbb{R}^m$ and suppose that f is differentiable at all points on the straight line joining $\mathbf{a}$ to $\mathbf{b}$. Then, for some $\theta \in (0, 1)$

$$f(\mathbf{b}) - f(\mathbf{a}) = \sum_{i=1}^{m} \frac{\partial f}{\partial x_i}(\mathbf{a} + \theta(\mathbf{b} - \mathbf{a}))(b_i - a_i) .$$

Proof: Define $F(t) = f(\mathbf{a} + t(\mathbf{b} - \mathbf{a}))$ for $t \in [0, 1]$. F is continuous on $[0, 1]$ and differentiable on $(0, 1)$ and $F'(t) = \sum_{i=1}^{m} \frac{\partial f}{\partial x_i}(\mathbf{a} + t(\mathbf{b} - \mathbf{a}))(b_i - a_i)$ by the Chain Rule. By the ordinary Mean Value Theorem $\exists \theta \in (0, 1)$ such that $F(1) - F(0) = F'(\theta)$. □

Definition Let $A \subset \mathbb{R}^m$ and $f: A \to \mathbb{R}$. If f is differentiable then each of the partial derivatives $\frac{\partial f}{\partial x_i}$ is a function from A to $\mathbb{R}$ and it may in turn be differentiable. We define $\frac{\partial^2 f}{\partial x_j \partial x_i}$ to be $\frac{\partial}{\partial x_j}(\frac{\partial f}{\partial x_i})$, the partial derivative of $\frac{\partial f}{\partial x_i}$ with respect to the jth variable, when this exists. F is said to be **twice differentiable** if f and each of its partial derivatives $\frac{\partial f}{\partial x_i}$ are differentiable. The analogous definitions apply to higher order derivatives.

The 'mixed' partial derivatives $\frac{\partial^2 f}{\partial x_1 \partial x_2}$ and $\frac{\partial^2 f}{\partial x_2 \partial x_1}$ are not necessarily equal, since the two limiting processes involved are taken in the opposite order in the two cases. Under very mild additional assumptions, however, they are equal, as we shall show shortly.

14.3 Results Involving Interchange of Limits
We shall devote a little effort to some results, all of which are essentially about reversing the order of limiting processes.

Theorem 14.13 Differentiation under the Integral Sign. Suppose that $f: [a, b] \times [c, d] \to \mathbb{R}$, and that $\frac{\partial f}{\partial x}$ exists and is continuous (with respect to both variables jointly). Then, with $F(x) = \int_c^d f(x, t)\, dt$, F is differentiable and $F'(x) = \int_c^d \frac{\partial f}{\partial x}(x, t)\, dt$.

Proof: We have to show that the derivative of the integral and the integral of the derivative are equal. Fix $x_0 \in (a, b)$; if x_0 is an endpoint, the usual minor changes are needed. Then, applying the Mean Value Theorem to $x \mapsto f(x, t)$ we see that, for each t, there is some ξ_t between x and x_0 satisfying $f(x, t) - f(x_0, t) = (x - x_0)\frac{\partial f}{\partial x}(\xi_t, t)$ and therefore

$$F(x) - F(x_0) = \int_c^d (x - x_0)\frac{\partial f}{\partial x}(\xi_t, t)\, dt .$$

Let $\varepsilon > 0$. Then, by the uniform continuity of $\frac{\partial f}{\partial x}$ on $[a, b] \times [c, d]$, $\exists \delta > 0$ such that $\|(x, y) - (x', y')\| < \delta \Rightarrow |\frac{\partial f}{\partial x}(x, y) - \frac{\partial f}{\partial x}(x', y')| < \varepsilon/(d - c)$. Let $0 < |x - x_0| < \delta$. Then

$$\left| \frac{F(x) - F(x_0)}{x - x_0} - \int_c^d \frac{\partial f}{\partial x}(x_0, t)\,dt \right| \le \int_c^d \left| \frac{\partial f}{\partial x}(\xi_t, t) - \frac{\partial f}{\partial x}(x_0, t) \right| dt < \varepsilon \ ,$$

since for each t, ξ_t is between x_0 and x, so $\|(\xi_t, t) - (x_0, t)\| < |x - x_0| < \delta$, whence $\forall t \in [c, d]$ $|\frac{\partial f}{\partial x}(\xi_t, t) - \frac{\partial f}{\partial x}(x_0, t)| < \varepsilon/(d - c)$. This establishes that $F'(x_0) = \int_c^d \frac{\partial f}{\partial x}(x_0, t)\,dt$. $\square$

Corollary With f as above, if $\phi, \psi: [a, b] \to [c, d]$ are differentiable and $F(x) = \int_{\phi(x)}^{\psi(x)} f(x, t)\,dt$ then

$$F'(x) = \int_{\phi(x)}^{\psi(x)} \frac{\partial f}{\partial x}(x, t)\,dt + f(x, \psi(x))\,\psi'(x) - f(x, \phi(x))\,\phi'(x) \ .$$

Proof. Let $G(x, y, z) = \int_y^z f(x, t)\,dt$. Then $\frac{\partial G}{\partial x}$ exists by the Theorem, while $\frac{\partial G}{\partial y} = -f(x, y)$, and $\frac{\partial G}{\partial z} = f(x, z)$ by the Fundamental Theorem of Calculus. All three partial derivatives are continuous, so we may apply the Chain Rule to $F(x) = G(x, \phi(x), \psi(x))$. $\square$

Theorem 14.14 Let $f: [a, b] \times [c, d] \to \mathbb{R}$ be continuous. Then

$$\int_a^b \left\{ \int_c^d f(x, y)\,dy \right\} dx = \int_c^d \left\{ \int_a^b f(x, y)\,dx \right\} dy \ .$$

Proof. Define $g(x, z) = \int_c^z f(x, y)\,dy$. Then $\frac{\partial g}{\partial z} = f$, which is continuous. Now let $G(z) = \int_a^b g(x, z)\,dx$, so by differentiation under the integral sign, $G'(z) = \int_a^b f(x, z)\,dx$. Since G' is continuous (Theorem 14.8),

$$\int_c^d \left\{ \int_a^b f(x, z)\,dx \right\} dz = \int_c^d G' = G(d) - G(c) = G(d) = \int_a^b \left\{ \int_c^d f(x, y)\,dy \right\} dx \ . \quad \square$$

The result about inverting the order of integration remains true under more general circumstances, but it may fail for improper integrals. To see this, consider inverting the order of integration in the improper integrals $\int_1^\infty (\int_1^\infty (x - y)(x + y)^{-3} dx)\,dy$ or $\int_0^1 (\int_0^1 (x - y)(x + y)^{-3} dx)\,dy$.

Theorem 14.15 Suppose that f is defined in the neighbourhood of the point (a, b) in $\mathbb{R}^2$, that is, on the set $\{(x, y) \in \mathbb{R}^2 : \|(x, y) - (a, b)\| < h\}$ for some $h > 0$, that

$\frac{\partial f}{\partial x}$ and $\frac{\partial f}{\partial y}$ are continuous on this set and that one of $\frac{\partial^2 f}{\partial x \partial y}$ and $\frac{\partial^2 f}{\partial y \partial x}$ exists and is (jointly) continuous on this set. Then both $\frac{\partial^2 f}{\partial x \partial y}$ and $\frac{\partial^2 f}{\partial y \partial x}$ exist and are equal on this set.

Proof: Suppose that $\frac{\partial^2 f}{\partial x \partial y}$ is continuous in the neighbourhood of (a, b). Regarding y as fixed, we see, by differentiating under the integral sign, that

$$\int_b^y \frac{\partial^2 f}{\partial x \partial y}(x, t)\,dt = \frac{\partial}{\partial x}\int_b^y \frac{\partial f}{\partial y}(x, t)\,dt = \frac{\partial}{\partial x}\left(f(x, y) - f(x, b)\right)$$

$$= \frac{\partial f}{\partial x}(x, y) - \frac{\partial f}{\partial x}(x, b) .$$

(This uses the continuity of $\frac{\partial f}{\partial y}$ and the Fundamental Theorem of Calculus.) Since this is true of all y sufficiently close to b, and since, for fixed x, the left-hand side is differentiable with respect to y, the right-hand side is differentiable with respect to y and

$$\frac{\partial^2 f}{\partial x \partial y}(x, y) = \frac{\partial}{\partial y}\int_b^y \frac{\partial^2 f}{\partial x \partial y}(x, t)\,dt = \frac{\partial^2 f}{\partial y \partial x}(x, y) - 0 . \quad \Box$$

There are several possible minor variations on Theorem 14.15, which may easily be made to apply to functions of more than two variables, and to higher order partial derivatives. For most purposes it is sufficient to know that if the two mixed derivatives are (jointly) continuous they are equal.

Theorem 14.16 Taylor's Theorem. Suppose that $a \in \mathbb{R}^m$ and that $f: A \to \mathbb{R}$ where $A = \{x \in \mathbb{R}^m : \|x - a\| < \delta\}$ for some $\delta > 0$. Then if f possesses all partial derivatives of orders up to and including n and if all these partial derivatives are continuous, we have, for $\|h\| < \delta$,

$$f(a+h) = f(a) + \sum_{i=1}^m h_i \frac{\partial f}{\partial x_i}(a) + \cdots + \sum_{i_1, i_2, \ldots, i_{n-1}=1}^m \frac{h_{i_1} \cdots h_{i_{n-1}}}{(n-1)!} \frac{\partial^{n-1} f}{\partial x_{i_1} \cdots \partial x_{i_n}}(a) + R_n$$

where, for some $\theta \in (0, 1)$,

$$R_n = \sum_{i_1, i_2, \ldots, i_n=1}^m \frac{h_{i_1} \cdots h_{i_n}}{n!} \frac{\partial^n f}{\partial x_{i_1} \cdots \partial x_{i_n}}(a + \theta h) .$$

Proof: Because all the various partial derivatives are continuous, all up to order $n - 1$ are differentiable as functions of m variables, so the Chain Rule, applied n times, shows that if $F(t) = f(a + th)$ $(0 \le t \le 1)$ then F is n-times differentiable. The Chain Rule shows that $F'(t) = \sum_{i=1}^m h_i \frac{\partial f}{\partial x_i}$, $F''(t) = \sum_{i_1, i_2=1}^m h_{i_1} h_{i_2} \frac{\partial^2 f}{\partial x_{i_1} \partial x_{i_2}}$, etc.

The rest is messy but routine. $\square$

Example 14.8 Suppose that $f: \mathbb{R}^2 \to \mathbb{R}$ is differentiable. If f has a local maximum at (a, b) then the functions $x \mapsto f(x, b)$ and $y \mapsto f(a, y)$ have local maxima at $x = a$ and $y = b$ respectively so $\frac{\partial f}{\partial x}$ and $\frac{\partial f}{\partial y}$ are both zero at (a, b).

This, of course, is true if f has a local minimum, so a *necessary* condition for a local maximum or minimum is that both (or all if there are more than two variables) the first order partial derivatives should be zero at that point.

Now suppose $\frac{\partial f}{\partial x}(a, b) = \frac{\partial f}{\partial y}(a, b) = 0$. To find sufficient conditions for a maximum, minimum etc., we presume that f has partial derivatives of order two and that these are continuous. Then, for $\|\mathbf{h}\|$ small enough,

$$f(a + h_1, b + h_2) = f(a, b) + \tfrac{1}{2}\left(h_1^2 \frac{\partial^2 f}{\partial x^2}(\mathbf{x}) + 2h_1 h_2 \frac{\partial^2 f}{\partial x \partial y}(\mathbf{x}) + h_2^2 \frac{\partial^2 f}{\partial y^2}(\mathbf{x})\right)$$

for some point $\mathbf{x} = (a + \theta h_1, b + \theta h_2)$ with $0 < \theta < 1$. If we choose $\|\mathbf{h}\|$ sufficiently small, the term on the right will be close to

$$f(a, b) + (\alpha h_1^2 + 2\beta h_1 h_2 + \gamma h_2^2)/2$$

where $\alpha = \frac{\partial^2 f}{\partial x^2}$, $\beta = \frac{\partial^2 f}{\partial x \partial y}$ and $\gamma = \frac{\partial^2 f}{\partial y^2}$, all evaluated at (a, b), so we investigate expressions of the form $\alpha h_1^2 + 2\beta h_1 h_2 + \gamma h_2^2$. Such expressions are called **quadratic forms** and their algebraic theory is elegant and complete; we shall merely poach some fragments from it. Notice that $\alpha h_1^2 + 2\beta h_1 h_2 + \gamma h_2^2 = \alpha(h_1 + (\beta/\alpha)h_2)^2 + ((\alpha \gamma - \beta^2)/\alpha)h_2^2$ so that

(i)   if $\alpha\gamma - \beta^2 > 0$ and $\alpha > 0$ then $\alpha h_1^2 + 2\beta h_1 h_2 + \gamma h_2^2 > 0$ unless $h_1 = h_2 = 0$, whence $f(a, b) + (\alpha h_1^2 + 2\beta h_1 h_2 + \gamma h_2^2)/2$ has a **minimum** at $h_1 = h_2 = 0$.

(ii) if $\alpha\gamma - \beta^2 > 0$ and $\alpha < 0$ then $\alpha h_1^2 + 2\beta h_1 h_2 + \gamma h_2^2 < 0$ unless $h_1 = h_2 = 0$ so $f(a, b) + (\alpha h_1^2 + 2\beta h_1 h_2 + \gamma h_2^2)/2$ has a **maximum** at $h_1 = h_2 = 0$.

(iii) if $\alpha\gamma - \beta^2 < 0$ then $f(a, b) + (\alpha h_1^2 + 2\beta h_1 h_2 + \gamma h_2^2)/2$ has neither a maximum nor a minimum. If $\alpha \neq 0$ this may be seen by considering the cases $(h_1, h_2) = (\lambda, 0)$ and $(-\lambda\beta/\alpha, \lambda)$, while re-expressing by taking a $\gamma(h_2 + (\beta/\gamma)h_1)^2$ term shows the same situation if $\gamma \neq 0$. The case $\alpha = \gamma = 0$, $\beta \neq 0$ is easy.

Since $\alpha\gamma - \beta^2 > 0$ implies $\alpha \neq 0$, we have considered all cases where $\alpha\gamma - \beta^2 \neq 0$. In case (iii), f is said to have a **saddle point**.

These are the cases we can deal with, since the expression for $f(a + h_1, b + h_2)$ involves $\frac{\partial^2 f}{\partial x^2}$ etc. evaluated not at (a, b) but at $(a + \theta h_1, b + \theta h_2)$. If $\frac{\partial^2 f}{\partial x^2}$ is continuous and $\alpha = \frac{\partial^2 f}{\partial x^2}(a, b) \neq 0$, then for $\|\mathbf{h}\|$ small enough,

$\frac{\partial^2 f}{\partial x^2}(a + \theta h_1, b + \theta h_2)$ will have the same sign as α; similarly, if $\alpha\gamma - \beta^2 \neq 0$

$(\frac{\partial^2 f}{\partial x^2})(\frac{\partial^2 f}{\partial y^2}) - (\frac{\partial^2 f}{\partial x \partial y})^2$ will have constant sign near (a, b). Since our arguments

about the sign of $\alpha h_1^2 + 2\beta h_1 h_2 + \gamma h_2^2$ depend only on the sign of α and $\alpha\gamma - \beta^2$, the conclusions about the sign of $f(a + h_1, b + h_2) - f(a, b)$ are valid.

For the remaining cases, $\alpha\gamma - \beta^2 = 0$, and we see, for example, that $\alpha h_1^2 + 2\beta h_1 h_2 + \gamma h_2^2$ is always non-negative if $\alpha > 0$, but that there are non-zero values of h_1, h_2 for which the expression is zero. For these values of h_1, h_2, if we substitute the appropriate derivatives evaluated at $(a + \theta h_1, a + \theta h_2)$ we cannot be sure of the sign of $f(a + h_1, b + h_2) - f(a, b)$.

The results for a function of three or more variables are similar but a little more complicated. $\square$

At this point the techniques we have used are becoming somewhat strained, largely because in handling functions of several variables there is a degree of technicality and complication which can obscure matters. This is particularly true if one tries to consider whether a function whose domain is a subset of $\mathbb{R}^2$ and whose image is in $\mathbb{R}^2$ has an inverse function, where we need to treat simultaneously the two real-valued functions corresponding to the first and second co-ordinates of the image. This sort of issue is best tackled using a rather more abstract and sophisticated viewpoint which eliminates some of the technicalities, so we shall not pursue it here. A good source is Apostol (1974).

14.4 Solving Differential Equations

Ordinary differential equations (that is, differential equations involving only one independent variable) are important in many aspects of mathematical modelling, from biology to economics and physics. They are also quite difficult to solve! An elementary treatment of the subject introduces various ingenious techniques for finding explicit solutions of particular classes of differential equations - for example,

for solving
$$\frac{d^2 y}{dx^2} + a\frac{dy}{dx} + by = f(x) \tag{1}$$

where a and b are given constants and f is a given function. The technique of these calculus-based methods is to construct explicit solutions, often by interesting and ingenious means. These methods do not, however, deal with all cases and if we modified equation (1) so that the constants a and b were replaced by functions we would have an example where the calculus-based methods would be useful only in certain special cases.

What can we do in these awkward cases? If we expect that, perhaps because the functions involved are awkward, we will not find an explicit solution, is there any progress to be made? We shall use our analysis to show that in fairly general cases the differential equation *does* have a solution. Now the practical person may not be too impressed by knowing that there is a solution if he or she cannot find it, but the techniques used will actually produce a means of approximating this elusive solution. Taking another perspective, we shall also produce conditions under which our differential equation has a *unique* solution, which will lead us to a much-used technique for constructing all solutions out of a known number of specific solutions. We shall begin with a result which contains all the hard work.

Lemma 14.16 Suppose that $F: [a, b] \times [a, b] \to \mathbb{R}$ is a continuous function of two variables, and that $f: [a, b] \to \mathbb{R}$ is also continuous. Then the equation

$$z(x) = f(x) + \int_a^x F(x, t) z(t) \, dt \qquad (a \le x \le b) \tag{1}$$

has a unique continuous solution $z: [a, b] \to \mathbb{R}$.

Comment: Notice that this result is so general that we should be able to apply it in many specific cases. It also has the property that, because of the generality, we are unlikely to find a neat formula for the solution.

Proof: Define a sequence of functions z_n by

$$z_0(x) = f(x), \quad z_{n+1}(x) = f(x) + \int_a^x F(x, t) z_n(t) \, dt \qquad (a \le x \le b) \tag{2}$$

Now z_0 is continuous, and (by Problem 16) if z_n is continuous so is z_{n+1} so it follows that for all $n \in \mathbb{N} \cup \{0\}$, z_n is continuous. We need to show that (z_n) tends to a limit as $n \to \infty$.

Let $M = \sup\{ |F(x, t)| : a \le x \le b \}$, and define d_n by

$$d_n = \sup\{ |z_n(t) - z_{n-1}(t)| : a \le t \le b \}.$$

Then for all $x \in [a, b]$ and $n \in \mathbb{N}$,

$$\left| z_{n+1}(x) - z_n(x) \right| = \left| \int_a^x F(x, t)(z_n(t) - z_{n-1}(t)) \, dt \right| \qquad \text{by (2)}$$

$$\le \int_a^x |F(x, t)| |z_n(t) - z_{n-1}(t)| \, dt$$

$$\le \int_a^x M d_n \, dt \le M(b - a) d_n \ .$$

As this holds for all $x \in [a, b]$, $\forall n \in \mathbb{N}$

$$d_{n+1} \le M(b - a) d_n \ . \tag{3}$$

Therefore $d_2 \le M(b - a)d_1$, $d_3 \le M(b - a)d_2 \le (M(b - a))^2 d_1$ and so on.

To simplify matters we shall consider the case where $M(b - a) < 1$ first. Then from (3), $d_n \le (M(b - a))^{n-1} d_1$ and, by comparison, the series Σd_n converges. By observing that for all $x \in [a, b]$ $|z_n(x) - z_{n-1}(x)| \le d_n$ we see in turn that $\Sigma(z_n(x) - z_{n-1}(x))$ also converges. Now

$$\sum_{n=1}^N (z_n(x) - z_{n-1}(x)) = z_N(x) - z_0(x) \ ,$$

so the convergence of the series implies that the sequence $(z_n(x))$ tends to a limit as $n \to \infty$. Let $z(x) = \lim_{n \to \infty} z_n(x)$.

We need to show that z is continuous. Choose $x_0 \in [a, b]$ and $\varepsilon > 0$. For all $x \in [a, b]$

$$|z(x) - z_N(x)| = \left| \sum_{n=N+1}^\infty (z_n(x) - z_{n-1}(x)) \right| \le \sum_{n=N+1}^\infty d_n \ ,$$

so choose N so that $\sum_{n=N+1}^{\infty} d_n < \varepsilon/3$. By the continuity of z_n, $\exists \delta > 0$ such that $|x - x_0| < \delta \Rightarrow |z_N(x) - x_N(x_0)| < \varepsilon/3$. Then

$$|x - x_0| < \delta \Rightarrow |z(x) - z(x_0)| \leq |z(x) - z_N(x)| + |z_N(x) - z_N(x_0)| + |z_N(x_0) - z(x_0)|$$

$$\leq 2\sum_{n=N+1}^{\infty} d_n + \varepsilon/3 < \varepsilon \ .$$

Since $\varepsilon > 0$ was arbitrary we have shown z is continuous at x_0, and since x_0 was typical, z is continuous.

Because z is continuous, $\int_a^x F(x,t)z(t)\,dt$ is continuous, and we might hope that this is the limit of $\int_a^x F(x,t)z_n(t)\,dt$ as $n \to \infty$. As this conclusion is equivalent to interchanging the operations of integrating and letting $n \to \infty$, we need to prove its truth. Choose N so that $\sum_{n=N+1}^{\infty} d_n < \varepsilon$, so that, as above

$$\forall t \in [a,b] \text{ and } \forall m \geq N \quad |z(t) - z_m(t)| \leq \sum_{n=m+1}^{\infty} |z_n(t) - z_{n-1}(t)|$$

$$\leq \sum_{n=m+1}^{\infty} d_n < \varepsilon \ . \tag{4}$$

Then, if $x \in [a, b]$ and $m \geq N$,

$$\left| z_{m+1}(x) - f(x) - \int_a^x F(x,t)z(t)\,dt \right| \leq \left| \int_a^x F(x,t)(z_m(t) - z(t))\,dt \right|$$

$$\leq \int_a^x M|z_m(t) - z(t)|\,dt$$

$$\leq \int_a^x M\varepsilon \, dt \leq M(b-a)\varepsilon \ .$$

Since $\varepsilon > 0$ is arbitrary here, $z_{m+1}(x) \to f(x) + \int_a^x F(x,t)z(t)\,dt$ as $m \to \infty$, and so, by the definition of z as the limit of (z_m), z satisfies (1). We have shown that a continuous solution of (1) exists.

To show that the solution is unique, suppose that z and w are continuous solutions of (1), and let $d = \sup\{|z(t) - w(t)| : t \in [a, b]\}$. Then, for all $x \in [a, b]$

$$|z(x) - w(x)| \leq \int_a^x |F(x,t)||z(t) - w(t)|\,dt \leq M(b-a)d$$

Because this is true for all $x \in [a, b]$, $d \leq M(b-a)d$ which is impossible with $M(b-a) < 1$ unless $d = 0$. Therefore $d = 0$ and $z = w$, proving the uniqueness.

Finally, suppose that we no longer presume that $M(b-a) < 1$. Choose h to satisfy $0 < h < 1/M$, and apply the result above on the interval $[a, a+h]$, obtaining a unique solution of (1) valid on $[a, a+h]$. Now set

$$f_1(x) = f(x) + \int_a^{a+h} F(x,t)z(t)\,dt$$

for the solution z just found. By the result above the equation

$$z(x) = f_1(x) + \int_{a+h}^x F(x,t)z(t)\,dt$$

has a unique continuous solution z on $[a+h, a+2h]$, which extends the function z already found to $[a, a+2h]$. Proceeding in this way, extending in steps of less than $1/M$, we extend z to the whole of $[a, b]$. □

Lemma 14.17 actually tells us a little more than we have said. We know a unique solution z exists, and that z_n, as defined in the proof, tends to z as $n \to \infty$. We can therefore ensure that z_n is as close to the true solution as we wish if n is large enough. In this case, we can even say how large is enough. We saw in equation (4) above that

$$\forall t \in [a, b] \quad |z(t) - z_m(t)| \leq \sum_{n=m+1}^{\infty} d_n \ .$$

Now, considering the simpler case where $M(b - a) < 1$, we have for all $n \geq m + 1$, $d_n \leq M(b - a)d_{n-1} \leq \ldots \leq (M(b - a))^{n-m}d_m$, so

$$\sum_{n=m+1}^{\infty} d_m \leq \sum_{n=m+1}^{\infty} (M(b - a))^{n-m} d_m = \frac{M(b-a)}{1-M(b-a)} d_m$$

(on summing the geometric series). We can calculate z_1, z_2 in turn and from these find d_m. Then once we find an m such that $\dfrac{M(b-a)}{1-M(b-a)} d_m < \varepsilon$, we know that the approximation z_m has the property that for all $t \in [a, b]$, $z(t)$ differs from the true solution $z(t)$ by less than ε.

We can relate all this immediately to the general second-order, "linear" differential equation.

Theorem 14.18 Suppose that p, q and $g: [a, b] \to \mathbb{R}$ are continuous, and that $\alpha, \beta \in \mathbb{R}$. Then the differential equation problem

$$\frac{d^2y}{dx^2} + p(x)\frac{dy}{dx} + q(x)y(x) = g(x) \qquad (a \leq x \leq b)$$

$$y(a) = \alpha, \ y'(a) = \beta$$

has a unique solution y.

Proof. If y satisfies the differential equation then it must be twice differentiable, hence both y and y' are continuous. From this, and the differential equation, we see that y'' is continuous. Therefore, letting z denote y'', z is continuous,

$$y(x) = \alpha + \beta(x - a) + \int_a^x (x - t)z(t)\,dt \quad (a \leq x \leq b) \tag{1}$$

(by Corollary 2 to Theorem 13.12) and z satisfies

$$z(x) = f(x) + \int_a^x F(x, t)z(t)\,dt \quad (a \leq x \leq b) \tag{2}$$

where $f(x) = g(x) - \beta p(x) - (\alpha + \beta(x - a))q(x)$ and $F(x, t) = -(p(x) + (x - t)q(x))$.

Moreover, if z is continuous and satisfies (2), then the Fundamental Theorem of Calculus shows us that y given by (1) is twice differentiable and satisfies the original differential equation problem. It follows that the differential equation has a unique solution y if and only if equation (2) has a unique continuous solution z. By Lemma 14.17 this is true, and we have finished. $\square$

Suppose now that p and q are continuous on $[a, b]$ and we let y_1 and y_2 be the (unique) solutions of

$$\frac{d^2y}{dx^2} + p(x)\frac{dy}{dx} + q(x)y(x) = 0 \quad (a \leq x \leq b) \tag{*}$$

with $$y_1(a) = 1, \qquad y_1'(a) = 0 ,$$
$$y_2(a) = 0, \qquad y_2'(a) = 1 .$$

Then if y is another solution of (*), let $\alpha = y(a)$ and $\beta = y'(a)$. We now notice that the function y_3 where

$$y_3(x) = \alpha y_1(x) + \beta y_2(x)$$

satisfies (*) (because y_1 and y_2 do), and $y_3(a) = \alpha$, $y_3'(a) = \beta$. Therefore, by the uniqueness part of Theorem 14.18, y and y_3 must be equal, that is,

$$y(x) = \alpha y_1(x) + \beta y_2(x) .$$

In other words each solution of (*) can be expressed in the form of a constant times y_1 plus a constant times y_2. This can be extended slightly (see Problem 22) to yield the result that if we can find two independent solutions y_1 and y_2 of (*) then all other solutions are of the form $\alpha y_1 + \beta y_2$ for constants α and β.

The results about differential equations can be extended in many directions, but we shall be content with what we have achieved, which we could not have done without some form of analysis. However, our arguments have become somewhat lengthy, and this length tends to obscure the structure of the proofs. This can be remedied by placing the analysis in a more elegant framework, taking account of algebraic and other mathematical structures available. Good examples are in the books by Brown and Page, Porter and Stirling, Sutherland, and Conway.

Problems

1. In each case below, define $f(x, y)$ to have the value stated for $(x, y) \neq (0, 0)$ and to have the value 0 at $(0, 0)$. Determine whether f is continuous at $(0, 0)$:

$$x^2 y/(x^2 + y^2), \quad x/\sqrt{(x^2 + y^2)}, \quad (x + y)^2/\sqrt{(x^2 + y^2)}, \quad (x^2 - y^4)/(x^2 + y^4) .$$

2. Let $f, g_1, g_2 \colon \mathbb{R}^2 \to \mathbb{R}$ be three continuous functions and define F by $F(x, y) = f(g_1(x, y), g_2(x, y))$. Show that F is continuous. Prove also that if $h \colon \mathbb{R} \to \mathbb{R}$ is continuous so is H, where $H(x, y) = h(f(x, y))$.

3. Let f and g be defined as follows:

$$f(x, y) = \begin{cases} \dfrac{\sin(x + y)}{x + y} & (x + y \neq 0), \\ 1 & (x + y = 0), \end{cases} \qquad g(x, y) = \begin{cases} \dfrac{\sin x - \sin y}{x - y} & (x \neq y) \\ \cos x & (x = y) . \end{cases}$$

Show that both functions are continuous at $(0, 0)$.

4. Let $f \colon \mathbb{R}^m \to \mathbb{R}$ be continuous at $\mathbf{a}$. For $\mathbf{b} \in \mathbb{R}^m$, define $F \colon \mathbb{R} \to \mathbb{R}$ by $F(t) = F(\mathbf{a} + t\mathbf{b})$ and show that F is continuous at 0. The converse result is false; to see this define $f \colon \mathbb{R}^2 \to \mathbb{R}$ by $f(x, y) = \begin{cases} 1 & x^2 < y < 2x^2, \\ 0 & \text{otherwise.} \end{cases}$

f is not continuous at $(0, 0)$ but $\lim_{t \to 0} f(t\mathbf{b}) = 0$ for each vector $\mathbf{b} \in \mathbb{R}^2$.

5. Which of the following subsets of $\mathbb{R}^2$ are closed? $\{x \in \mathbb{R}^2 : \|x\| \leq 1\}$, $\{(x_1, x_2) : x_1 \geq 0\}$, $\{(x_1, x_2) : x_1^2 + x_2^2 < 1$ or $(x_1^2 + x_2^2 = 1$ and $x_1 \geq 0)\}$.

6. Suppose that (x_n) is a sequence of real numbers with the property that $\forall \varepsilon > 0 \; \exists N$ such that $\forall m, n \geq N \; |x_m - x_n| < \varepsilon$; such a sequence is called a **Cauchy sequence**. Prove that if (x_n) has a convergent subsequence then (x_n) is itself convergent. By showing that (x_n) is a bounded sequence, deduce that it converges. (Thus every Cauchy sequence in $\mathbb{R}$ converges.)

7. Let $A = \{x \in \mathbb{R}^m : \|x - a\| \leq r\}$ and suppose that $f : A \to \mathbb{R}$ is continuous. If $b, c \in A$ and $f(b) < \gamma < f(c)$, prove that there is a point $x \in A$ for which $f(x) = \gamma$. (Hint: Define $F : [0, 1] \to \mathbb{R}$ such that $F(0) = f(b)$ and $F(1) = f(c)$?)

8. Calculate the partial derivatives $\dfrac{\partial f}{\partial x}$ and $\dfrac{\partial f}{\partial y}$ of the following functions at $(0, 0)$ and decide whether f is differentiable at $(0, 0)$. In all cases, $f(0, 0) = 0$ and for other values of (x, y), $f(x, y)$ is given by the formula stated: $x^2 y/(x^2 + y^2)$, $xy/\sqrt{(x^2 + y^2)}$, $x\sqrt{(x^2 + y^2)}$, $x^2 y^2/(x^2 + y^2)$.

9. Let $f : \mathbb{R}^m \to \mathbb{R}$ and $a \in \mathbb{R}^m$. If $b \in \mathbb{R}^m$ and $\|b\| = 1$ we say that the **directional derivative of** f at a in the direction b is $F'(0)$ where $F(t) = f(a + tb)$, when this exists. (So $\dfrac{\partial f}{\partial x_1}$ is the directional derivative in the direction $(1, 0, ..., 0)$.) Show that if f is differentiable, the directional derivative in the direction $b = (b_1, b_2, ..., b_m)$ is $\sum_{i=1}^{m} b_i \dfrac{\partial f}{\partial x_i}(a)$.

10. Let $f(x, y) = xy(x^2 - y^2)/(x^2 + y^2)$ if $(x, y) \neq (0, 0)$ and $f(0, 0) = 0$. Calculate the first- and second-order partial derivatives and show that $\dfrac{\partial^2 f}{\partial x \partial y}$ and $\dfrac{\partial^2 f}{\partial y \partial x}$ are unequal at $(0, 0)$.

11. Define $g_1, g_2 : \mathbb{R}^2 \to \mathbb{R}$ by $g_1(x_1, x_2) = x_2$, $g_2(x_1, x_2) = x_1$ and set $G(x_1, x_2) = g_1(g_1(x_1, x_2), g_2(x_1, x_2))$. Use the Chain Rule (carefully!) to show that $\dfrac{\partial G}{\partial x_1} = 1$ and $\dfrac{\partial G}{\partial x_2} = 0$.

12. Let $f : \mathbb{R}^3 \to \mathbb{R}$ be differentiable and let $y, z : \mathbb{R} \to \mathbb{R}$ also be differentiable. Calculate $F'(x)$ where $F(x) = f(x, y(x), z(x))$. (If this is confusing, let $x(t) = t$ and find $(d/dt)(f(x(t), y(t), z(t)))$.)

13. The internal energy, u, of a gas can be expressed as a function of the pressure, volume and temperature of unit mass of the gas, as $u(p, v, t)$. The pressure p may be expressed as a function of volume and temperature,

$p = p(v, t)$.   Thus the internal energy can be expressed in terms of v and t as $U(v, t) = u(p(v, t), v, t)$.   Show that $\dfrac{\partial U}{\partial t} = \dfrac{\partial u}{\partial p}\dfrac{\partial p}{\partial t} + \dfrac{\partial u}{\partial t}$.

14. f is twice differentiable and $\forall x, y \in \mathbb{R}$ $f(x, y) = f(y, x)$.   Show that

$$\frac{\partial f}{\partial x}(a, b) = \frac{\partial f}{\partial y}(b, a), \quad \frac{\partial^2 f}{\partial x^2}(a, b) = \frac{\partial^2 f}{\partial y^2}(b, a) \quad \text{and} \quad \frac{\partial^2 f}{\partial x \partial y}(a, b) = \frac{\partial^2 f}{\partial y \partial x}(b, a) .$$

15.   Let $A = \{x \in \mathbb{R}^m : \|x\| \le 1\}$ and $f : A \to \mathbb{R}$ be continuous and be differentiable at all points a satisfying $\|a\| < 1$.   Prove that if f attains a maximum or minimum at an internal point of A (i.e. at a point a with $\|a\| < 1$), then $\dfrac{\partial f}{\partial x_1} = \dfrac{\partial f}{\partial x_2} = \cdots = \dfrac{\partial f}{\partial x_m} = 0$ there.   Hence prove that if $f(x)$ is constant on the set $\{x : \|x\| = 1\}$ then there is a point y with $\|y\| < 1$ for which $\dfrac{\partial f}{\partial x_i}(y) = 0$ $(i = 1, 2, ..., m)$.

16.   Let $f : [a, b] \times [c, d] \to \mathbb{R}$ be continuous and define $F : [a, b] \times [c, d] \to \mathbb{R}$ by $F(x, y) = \int_c^y f(x, t)\, dt$.   Prove that F is continuous.   (Modify Theorem 14.7 and consider small changes in x and y separately.)

17.   Show that if $f(x, y)$ is continuous in the two variables separately and $\dfrac{\partial f}{\partial y}$ exists and is jointly continuous, then f is jointly continuous.   (Hint: Use $f(x, y) - f(x, y_0) = \int_{y_0}^y \dfrac{\partial f}{\partial y}(x, t)\, dt$.)

18.   Let $f : [a, b] \times [a, b] \to \mathbb{R}$ be continuous, and define $F, G : [a, b] \to \mathbb{R}$ by $F(z) = \int_a^z \left(\int_a^x f(x, y)\, dy \right) dx$, $G(z) = \int_a^z \left(\int_y^z f(x, y)\, dx \right) dy$.   By showing that the result of the inner integration in both cases is (jointly) continuous, and differentiating under the integral sign, deduce that $F' = G'$ and hence that $F = G$.   Sketch the sets over which integration is taken in the (x, y)-plane.

19.   Given that $\displaystyle \int_0^{\pi/2} \frac{dx}{a^2 \sin^2 x + b^2 \cos^2 x^2} = \frac{\pi}{2ab}$,   show by differentiating under the integral sign that $\displaystyle \int_0^{\pi/2} \frac{\sin^2 x \, dx}{(a^2 \sin^2 x + b^2 \cos^2 x)^2} = \frac{\pi}{4a^3 b}$ and find

$\displaystyle \int_0^{\pi/2} \frac{dx}{(a^2 \sin^2 x + b^2 \cos^2 x)^2}$.

20.   Decide whether the functions given by the formulae below have a maximum, minimum or saddle point at $(0, 0)$: xy, $x^2 + y^2$, $x^2 + xy + y^2$, $x^4 + xy + y^2$.

21.*   Let $A = \{x \in \mathbb{R}^2 : \|x\| < r\}$ and let $f_1, f_2 : A \to \mathbb{R}$ have continuous first-order partial derivatives. By considering $F_1(t) = f_1(a + t(b - a))$ show that if $f_1(a) = f_1(b)$ where a and b are two distinct points of A, then there is a point $x_1 \in A$ for which $(b_1 - a_1)\frac{\partial f_1}{\partial x_1}(x_1) + (b_2 - a_2)\frac{\partial f_1}{\partial x_2}(x_1) = 0$.

If, in addition, $f_2(a) = f_2(b)$ show that there is a second point $x_2 \in A$ for which $G(x_1, x_2) = 0$ where $G(x, y) = \frac{\partial f_1}{\partial x_1}(x)\frac{\partial f_2}{\partial x_2}(y) - \frac{\partial f_1}{\partial x_2}(x)\frac{\partial f_2}{\partial x_1}(y)$.

Deduce that if $G(0, 0) \neq 0$, then for r sufficiently small, the function $x \mapsto (f_1(x), f_2(x))$ is injective on A. (Notice that the condition $G(0, 0) \neq 0$ is equivalent to the statement that the matrix $\begin{bmatrix} \frac{\partial f_1}{\partial x_1} & \frac{\partial f_1}{\partial x_2} \\ \frac{\partial f_2}{\partial x_1} & \frac{\partial f_2}{\partial x_2} \end{bmatrix}$ of partial derivatives is non-singular at 0.)

22.*   Suppose that p and q are continuous functions on $[a, b]$ and that y_1 and y_2 satisfy

$$\frac{d^2 y}{dx^2} + p(x)\frac{dy}{dx} + q(x)y(x) = 0 \quad (a \leq x \leq b) \; . \tag{*}$$

Let $y_j(a) = \alpha_j$ and $y_j'(a) = \beta_j$ $(j = 1, 2)$.

(i)   Show that if there is a constant λ such that $\alpha_1 = \lambda\alpha_2$ and $\beta_1 = \lambda\beta_2$, then $y_1 = \lambda y_2$. In this case one of the solutions is a constant multiple of the other (and $\alpha_1\beta_2 - \alpha_2\beta_1 = 0$).

(ii)   Show that if $\alpha_1\beta_2 - \alpha_2\beta_1 \neq 0$ then y_1 and y_2 are independent in the sense that neither of them is a constant multiple of the other.

(iii)   By noticing that if $\alpha_1\beta_2 - \alpha_2\beta_1 \neq 0$ then, given two real numbers α and β, there are constants λ and μ such that

$$\lambda\alpha_1 + \mu\alpha_2 = \alpha$$
$$\lambda\beta_1 + \mu\beta_2 = \beta \; ,$$

deduce that if $\alpha_1\beta_2 - \alpha_2\beta_1 \neq 0$ and y is a solution of (*), then there are constants λ and μ such that $y = \lambda y_1 + \mu y_2$. (Use Part (iii) and Theorem 14.18.)

Appendix

The Expression of an Integer as a Decimal
Theorem 6.1 Let x be a non-negative integer. Then there are integers $n \geq 0$, and numbers $a_0, a_1, ..., a_n$ belonging to $\{0, 1, ..., 9\}$ such that

$$x = \sum_{k=0}^{n} a_k 10^k .$$

If $x > 0$ and we demand that $a_n \neq 0$ then n and the numbers $a_0, a_1, ..., a_n$ are unique.

Proof: Clearly if $x = 0$ we can express x in the form $\sum_{k=0}^{n} a_k 10^k$ if and only if $a_0 = a_1 = ... = a_n = 0$. (Notice that this expression is not unique in that we can choose any value of $n \geq 0$, but it *is* unique if we fix n.)

Let $P(n)$ be the statement "if $0 \leq x < 10^{n+1}$ then x can be expressed in a unique way in the form $\sum_{k=0}^{n} a_k 10^k$ with $a_k \in \{0, 1, ..., 9\}$ for all k".

$P(0)$ is true for if $0 \leq x < 10$, then $x = a_0$ where $a_0 \in \{0, 1, ..., 9\}$ and this is unique.

Suppose $n \geq 0$ and that $P(n)$ is true. Let $0 \leq x < 10^{n+2}$, where x is an integer. Then $0 \leq x/10^{n+1} < 10$ so if we set $a_{n+1} = [x/10^{n+1}]$ then $a_{n+1} \in \{0, 1, ..., 9\}$. Also $0 \leq x/10^{n+1} - a_{n+1} < 1$ so that $0 \leq x - a_{n+1}10^{n+1} < 10^{n+1}$. By the statement $P(n)$ then, there is a unique collection of integers $a_0, a_1, ...,$ $a_n \in \{0, 1, ..., 9\}$ for which $x - a_{n+1}10^{n+1} = \sum_{k=0}^{n} a_k 10^k$. Hence we deduce that

$$x = \sum_{k=0}^{n+1} a_k 10^k .$$

To see the uniqueness, suppose that $b_0, ..., b_{n+1} \in \{0, 1, ..., 9\}$ and $x = \sum_{k=0}^{n+1} b_k 10^k$. Then $x = b_{n+1} 10^{n+1} + \sum_{k=0}^{n} b_k 10^k$ where

$$0 \leq \sum_{k=0}^{n} b_k 10^k \leq \sum_{k=0}^{n} 9.10^k = 9\frac{10^{n+1}-1}{10-1} = 10^{n+1} - 1 < 10^{n+1} .$$

Therefore $0 \leq x - b_{n+1}10^{n+1} < 10^{n+1}$, giving $0 \leq x/10^{n+1} - b_{n+1} < 1$ and $b_{n+1} \leq x/10^{n+1} < b_{n+1} + 1$. This shows us that $b_{n+1} = [x/10^{n+1}]$ so $b_{n+1} = a_{n+1}$ where a_{n+1} is as defined earlier in the paragraph. Then we have $0 \leq x - a_{n+1}10^{n+1} = \sum_{k=0}^{n} b_k 10^k < 10^{n+1}$. Then, by the uniqueness part of the statement $P(n)$ we see that, since $\sum_{k=0}^{n} b_k 10^k = \sum_{k=0}^{n} a_k 10^k$, $b_k = a_k$, for $k = 0, 1, ..., n$.

Therefore $b_k = a_k$ for $k = 0, 1, ..., n + 1$ and we have shown that if $0 \leq x < 10^{n+2}$ then x can be expressed in a unique way in the form $\sum_{k=0}^{n+1} a_k 10^k$ with $a_k \in \{0, 1, .., 9\}$ for all k. (The uniqueness is because any other admissible expression, $\sum_{k=0}^{n+1} b_k 10^k$ has to have the same coefficients b_k as the a_k we have already found.) We have shown $P(n) \Rightarrow P(n+1)$.

By induction, for all $n \in \mathbb{N}$ $P(n)$ is true.

We have not quite finished! Let $x > 0$. Then since x is a natural number $10^x > x(10 - 1) = 9x > x$ (by Bernoulli's inequality), so there certainly exist values of m for which $x < 10^{m+1}$ (e.g. $m = x$), so that x *can* be expressed in the form $\sum_{k=0}^{n} a_k 10^k$ for various values of n. If we wish $a_n \neq 0$ then $a_n \geq 1$ so that

$$10^n \leq \sum_{k=0}^{n} a_k 10^k \leq \sum_{k=0}^{n} 9.10^k = 10^{n+1} - 1 .$$

There is exactly one value of n for which this is true. It is obvious that there is no more than one and on noticing that $10^0 \leq x$ and $x < 10^{m+1}$ we see that $10^n \leq x < 10^{n+1} - 1$ for some n between 0 and m. In fact, $n = \min\{m \in \mathbb{N} \cup \{0\} : x < 10^{m+1}\}$.

Therefore n is uniquely determined by x if $x > 0$ and we demand that $a_n \neq 0$. This finishes the proof. $\square$

References

(The dates given are the first edition; several of the books below have later editions.)

Ahlfors, L.V., *Complex Analysis*, McGraw Hill, 1966.

Apostol, T.M., *Mathematical Analysis*, Addison Wesley (Chapter 11), 1974.

Brown, A.L. and Page, A., *Elements of Functional Analysis*, Van Nostrand Rheinhold, 1970.

Conway, J.B., *Functions of One Complex Variable*, Springer Verlag, 1978.

Dunning-Davies, J., *Mathematical Methods for Mathematicians, Physical Scientists and Engineers*, Ellis Horwood, 1982.

Kline, M., *Mathematical Thought form Ancient to Modern Times*, Oxford University Press, 1990.

Körner, T.W., *Fourier Analysis*, Cambridge University Press, 1988.

May, R.M., 'Simple mathematical models with very complicated dynamics', *Nature*, **261**, 10 June 1976, 459-467.

Porter, D., and Stirling, D.S.G., *Integral Equations*, Cambridge University Press, 1990.

Rose, H.E., *A Course in Number Theory*, Oxford University Press, 1994.

Rudin, W., *Real and Complex Analysis*, McGraw Hill, 1970.

Sutherland, W.A., *Introduction to Metric and Topological Spaces*, Oxford University Press, 1975.

Taylor, A.E., *Introduction to Functional Analysis*, Wiley, 1958.

Young, N. J., *Introduction Hilbert Space*, Cambridge University Press, 1988.

Further Reading

Analysis extends in many directions, not all of them immediately accessible on completing this book. Further classical analysis along the lines of this book, in more advanced and condensed for, may be found in Apostol's work quoted. For extensions to "modern" analysis, see Rudin's *Real and Complex Analysis*, which blends many strands together. Sutherland provides a gentle introduction to one area of modern analysis, and Brown and Page a very readable treatment of functional analysis. Young's *Introduction to Hilbert Space* is also an excellent introduction to its subject.

Analysis may be extended to deal with functions of a complex variable, which is a subject different in flavour from "real" analysis, partly because it considers only "well-behaved" functions. The resulting subject is extremely elegant, and the books of Ahlfors (an all-time classic) and Conway are excellent expositions of it.

For a topic which uses analysis in a particular area, with many interesting applications (and an individual style) try Körner's *Fourier Analysis*.

For more about number theory, as well as putting the fragments used in Chapter 4 into a wider context see Rose's book quoted above.

Finally, for a comprehensive history of mathematics, on a level deeper than anecdotes about mathematicians, one can do no better than Kline's monumental *Mathematical Thought from Ancient to Modern Times*, (or some of his other books), which shows clearly how analysis arose and where it is used.

Hints and Solutions to Selected Problems

Chapter 2

1. (i) $\sqrt{(1+2x)} = 1 - \sqrt{x} \Rightarrow x = 0$ or $x = 4$; 0 is the only solution.

(ii) $\sqrt{(1 + x)} = 1 + \sqrt{(1 - x)} \Rightarrow x = \pm(\sqrt{3})/2$. Substituting $x = (\sqrt{3})/2$, we see that $\sqrt{(1 + (\sqrt{3})/2)} = (1 + \sqrt{3})/2$ (by the hint, since $((1 + \sqrt{3})/2)^2 = 1 + (\sqrt{3})/2$, and $1 + \sqrt{(1 - (\sqrt{3})/2)} = 1 + (\sqrt{3} - 1)/2 = (1 + \sqrt{3})/2$, also by the hint. As both sides are equal $(\sqrt{3})/2$ is a solution. $-(\sqrt{3})/2$ is not a solution.

(iii) $x = \pi/4 + n\pi$, where n is an integer (so that there are infinitely many solutions in this case).

2. $\Rightarrow, \Rightarrow, \Leftrightarrow, \Leftrightarrow, \Rightarrow, \Leftarrow$ respectively.

3. $x^4 = y^4 \Leftrightarrow x^4 - y^4 = 0 \Leftrightarrow (x - y)(x + y)(x^2 + y^2) = 0 \Leftrightarrow x = y$ or $x = -y$ or $x = y = 0 \Leftrightarrow x = y$ or $x = -y$.

5(ii) $a = \dfrac{x}{1+x^2} = \dfrac{y}{1+y^2} \Leftrightarrow y = x$ or $y = 1/x$ by part (i). Therefore if $y \neq x$ we need x and $1/x$ both to exist and be different, Therefore $x \neq 0$, 1, or -1. the corresponding values of a are 0, 1/2, -1/2.

6. $\dfrac{x}{1+2x^2} = \dfrac{y}{1+2y^2} \Leftrightarrow x + 2xy^2 = y + 2x^2y \Leftrightarrow (x - y)(1 - 2xy) = 0$.

Chapter 3

1(i) The formula is true for $n = 1$. Suppose it is true for $n = k$. Then
$1^3 + 2^3 + ... + k^3 + (k + 1)^3 = k^2(k + 1)^2/4 + (k + 1)^3 = (k + 1)^2(k^2 + 4(k + 1))/4 = (k + 1)^2(k + 2)^2/4 = (k + 1)^2((k + 1) + 1)^2/4$. Therefore if the formula is true for $n = k$ it is true for $n = k + 1$.
By induction, for all n $1^3 + 2^3 + ... + n^3 = n^2(n + 1)^2/4$.

(iii) Let $P(n)$ be the statement $\sum_{r=1}^{n} \frac{1}{r(r+1)} = 1 - \frac{1}{n+1}$. $P(1)$ is true. Suppose that $P(k)$ is true. Then $\sum_{r=1}^{k+1} \frac{1}{r(r+1)} = \sum_{r=1}^{k} \frac{1}{r(r+1)} + \frac{1}{(k+1)(k+2)}$
$= 1 - \frac{1}{k+1} + \frac{1}{(k+1)(k+2)} = 1 - \frac{k+2-1}{(k+1)(k+2)} = 1 - \frac{1}{k+2}$
Therefore $P(k) \Rightarrow P(k + 1)$. By induction, for all natural numbers n $P(n)$ is true.

2. (i) Let $P(n)$ be "$1^2 + 3^2 + ... + (2n - 1)^2 = n(4n^2 - 1)/3$". (Notice that there are n terms on the left.) Since $1^2 = 1(4.1^2 - 1)/3$, $P(1)$ is true.
Suppose $P(k)$ is true. Then
$1^2 + 3^2 + ... + (2(k + 1) - 1)^2 = 1^2 + 3^2 + ... + (2k - 1)^2 + (2k + 1)^2$
$= k(4k^2 - 1)/3 + (2k + 1)^2 = (2k + 1)(k(2k - 1) + 3(2k + 1))/3$
$= (2k + 1)(2k^2 + 5k + 3)/3 = (k + 1)(2k + 1)(2k + 3)/3 = (k + 1)(4(k + 1)^2 - 1)$
Therefore $P(k) \Rightarrow P(k+1)$. By induction, for all n $P(n)$ is true.

(ii) Let $P(n)$ be "$1^2 - 2^2 + 3^2 + ... + (2n - 1)^2 - (2n)^2 = -n(2n + 1)$". (Again, there are $2n$ terms on the left.) Then $P(1)$ is $1^2 - 2^2 = -1.3$ which is true.
Suppose $P(k)$ is true. Then (putting $k + 1$ in place of k on the left)
$1^2 - 2^2 + 3^2 + ... + (2k + 1)^2 - (2k + 2)^2 = -k(2k + 1) + (2k + 1)^2 - (2k + 2)^2$
$= -2k^2 - k - 4k - 3 = -(k + 1)(2k + 3) = -(k + 1)(2(k+1) + 1)$.

Therefore $P(k) \Rightarrow P(k+1)$. The result follows by induction.

3.   (i) Let $P(n)$ be $(1-\frac{1}{2})(1-\frac{1}{3})...(1-\frac{1}{n+1}) = \frac{1}{n+1}$. $P(1)$ is true. Suppose $P(k)$ is true. Then $(1-\frac{1}{2})(1-\frac{1}{3})...(1-\frac{1}{k+2}) = \frac{1}{k+1}(1-\frac{1}{k+2}) = \frac{1}{k+1}\frac{k+1}{k+2} = \frac{1}{k+2}$. Therefore $P(k) \Rightarrow P(k+1)$ and the result follows by induction.

  (iii) Do by induction or by mutliplying the left hand sides of (i) and (ii).

4.   (i) Let $P(n)$ be "10^n leaves a remainder of 1 on division by 9". Since $10 = 9 + 1$, $P(1)$ is true.

Suppose $P(k)$ is true. Then 10^k leaves a remainder of 1 on division by 9, so $10^k = 9m + 1$ for some integer m. $\therefore 10^{k+1} = 90m + 10 = 9(10m + 1) + 1$, so 10^{k+1} leaves a remainder of 1 on division by 9. $P(k) \Rightarrow P(k+1)$ and the result follows by induction.

  (ii) $4^2 = 16$ leaves a remainder of 6 on division by 10. $\therefore$ The result is true for $n = 1$. Suppose it is true for $n = k$. Then $4^{2k} = 10m + 6$ for some integer m. $\therefore 4^{2(k+1)} = 16.4^{2k} = 160m + 96 = 10(16m + 9) + 6$, which leaves a remainder of 6 on division by 10. $\therefore P(k) \Rightarrow P(k+1)$ and the result follows by induction.

5.   Let $P(n)$ be "$10^n + (-1)^{n-1}$ is a multiple of 11". $P(1)$ is true because $10^0 + (-1)^0 = 11$ is a multiple of 11. Suppose that for some k $P(k)$ is true, that is, $10^k + (-1)^{k-1}$ is a multiple of 11. Then $10^k = 11m - (-1)^{k-1}$ for some integer m whence $10^{k+1} = 110m - (-1)^{k-1}10 = 110m - (-1)^{k-1}11 - (-1)^k$, and so $10^{k+1} + (-1)^k = 11(10m - (-1)^{k-1})$, a multiple of 11.

6.   The auxiliary equation $y^2 = 4y - 3$ has solutions $y = 1$ and $y = 3$, so $a_n = A1^n + B3^n$. Solving $a_1 = 1$, $a_2 = 2$ gives $A = 1/2$, $B = 1/6$. Therefore $a_n = (1 + 3^{n-1})/2$. (Example 3.7.)

7. $a_n = 2^n - 3^{n-1}$.

8.   The auxiliary equation has only one solution, $x = 1/2$. Therefore $a_n = (A + nB)(1/2)^n$ and to ensure that a_1 and a^2 are correct we need $A = 0$, $B = 2$, giving $a_n = n2^{-(n-1)}$.

10. $(a_m 10^m + a_{m-1}10^{m-1} + ...+ a_0) - (a_m + ... + a_0) = \sum_{n=1}^{m} a_n (10^n - 1)$ which is divisible by 9 since $10^n - 1$ is for each $n \in \mathbb{N}$.

11. $(a_m 10^m + a_{m-1}10^{m-1} + ...+ a_0) - (a_0 - a_1 + ... + (-1)^m a_m) = \sum_{n=1}^{m} a_n (10^n + (-1)^{n-1})$.

13. Since $b < 0$ (for $b < -a^2/4$), $-b > 0$. Also $(a/2)^2 + (\sqrt{(-a^2/4 - b)})^2 = -b$ so there is a θ satisfying the equations for θ. $\sin \theta \neq 0$ because $-a^2/4 - b > 0$. Then A and B exist because the determinant $\begin{vmatrix} \cos\theta & \sin\theta \\ \cos 2\theta & \sin 2\theta \end{vmatrix} = \cos\theta \sin 2\theta - \cos 2\theta \sin\theta = \sin \theta \neq 0$. (By direct calculation if you don't know about determinants.) This choice of A and B ensures that $a_n = (\sqrt{(-b)})^n (A\cos n\theta + B\sin n\theta)$ for $n = 1$ and $n = 2$. Then since $a = 2\sqrt{(-b)}\cos \theta$ we have $A(a\cos n\theta - \sqrt{(-b)}\cos(n-1)\theta) = A\sqrt{(-b)}(2\cos n\theta \cos \theta - \cos(n-1)\theta) = A\sqrt{(-b)}(2\cos n\theta \cos \theta - \cos n\theta \cos \theta - \sin n\theta \sin \theta) = A\sqrt{(-b)}\cos(n+1)\theta$. This gives the correct coefficient of A, and a similar calculation gives the coefficient of B.

14. $2^n = (1 + 1)^n = \sum_{k=0}^{n} \binom{n}{k} 1^k 1^{n-k} = \sum_{k=0}^{n} \binom{n}{k}$. Similarly using

$(1+(-1))^n = \sum_{k=0}^{n} \binom{n}{k}(-1)^{n-k}$ we see that if n is even $\binom{n}{0} - \binom{n}{1} + ... + \binom{n}{n} = 0$. Adding these gives $2(\binom{n}{0} + \binom{n}{2} + ... + \binom{n}{n}) = 2^n$ if n is even. If n is odd $-\binom{n}{0} + \binom{n}{1} - ... - \binom{n}{n-1} + \binom{n}{n} = 0$ and subtracting this from the first equation gives $2(\binom{n}{0} + \binom{n}{2} + ... + \binom{n}{n}) = 2^n$ for n odd. It follows that if $n \le 2k + 1$ $b_n = \binom{n}{0} + \binom{n}{2} + ... + \binom{n}{r}$ where $r = n$ if n is even and $r = n - 1$ if n is odd (for the omitted terms are zero). this shows that for $n \le 2k + 1$, $b_n = 2^{n-1}$. However $b_{2k+2} = \binom{2k+2}{0} + \binom{2k+2}{2} + ... + \binom{2k+2}{2k} = 2^{2k+1} - \binom{2k+2}{2k+2} = 2^{2k+1} - 1$.

Chapter 4

1. Suppose that $n \in \mathbb{Z}$, n^2 is a multiple of 3 and that n is *not* a multiple of 3. The for some integer m, $n = 3m + 1$ or $n = 3m + 2$, whence $n^2 = 3(3m^2 + 2m) + 1$ or $3(3m^2 + 4m + 1) + 1$, neither of which is a multiple of 3. theis contradiction proves the result. Suppose that $\sqrt{3}$ is rational. Then there are integers p and q, with no common factor greater than 1 for which $\sqrt{3} = p/q$. Then $p^2 = 3q^2$, so p^2 and hence p, is a multiple of 3. Therefore, for some integer p', $p = 3p'$ and $q^2 = 3(p')^2$. It follows that q^2 and therefore q is a multiple of 3. We have shown p and q have a common factor of 3, a contradiction. This completes the proof.

3. Suppose that $\sqrt{6}$ is rational. Then there are integers p and q, with no common factor greater than 1 for which $\sqrt{6} = p/q$. Therefore $p^2 = 6q^2$ which shows that p^2 is a multiple of 2)and a multiple of 3). Therefore by Q1 above and Example 4.2, p is a multiple of 2. Let $p = 2p'$, where p' is an integer, and we see that $2(p')^2 = 3q^2$. Therefore q^2 is a multiple of 2 (for if not, $3q^2$ would not be), and hence so is q. We have shown p and q have a common factor of 2; contradiction.

4. Let $a \ne 0$ be rational and b be irrational. Suppose that ab is rational. Then $b = (ab)(1/a)$ is the product of two rational numbers, hence rational. This is a contradiction, since b is irrational.

5. Let x^2 be irrational. If x were rational, then x^2 would be rational, so by contradiction, x must be irrational. Now $(\sqrt{2} + \sqrt{3})^2 = 2 + 2\sqrt{6} + 3 = 5 + 2\sqrt{6}$. Since $2\sqrt{6}$ is irrational (Q2 and Q3), so is $5 + 2\sqrt{6}$ (Ex.4.4). Therefore $(\sqrt{2} + \sqrt{3})^2$ is irrational and hence so is $(\sqrt{2} + \sqrt{3})^2$.

6. Suppose that a, $b \in \mathbb{Q}$, both non-zero, and $a\sqrt{2} + b\sqrt{3}$ is rational. Then $(a\sqrt{2} + b\sqrt{3})^2 = 2a^2 + 3b^2 + 2ab\sqrt{6}$ is rational. Since $2a^2 + 3b^2$ is rational, we deduce that $2ab\sqrt{6}$ is rational. Then since $2ab$ is non-zero and rational, it follows that $\sqrt{6} \in \mathbb{Q}$, a contradiction. (Notice that we can say more; if one of a or b is zero and the other non-zero and rational, then Q4 shows $a\sqrt{2} + b\sqrt{3}$ is irrational.)

7. The possible remainders of $x^2 + y^2$ on division by 7 are all the (eight) sums of two of 0, 1, 2, 4, that is, all possible remainders. But the only combination leaving a remainder of 0 arises from $0 + 0$, so that if $x^2 + y^2$ leaves a remainder of 0 on division by 7 then x^2 and y^2 do separately. If x^2 is divisible by 7 then so is x (by Theorem 4.4, because 7 is prime), hence x and

y are both divisible by 7 , showing that $x^2 + y^2$ is divisble by 49. Let $x = 7X$ and $y = 7Y$; then $(x^2 + y^2)/49 = (X^2 + Y^2)$.

8. Suppose that $p^3 + pq^2 + q^3 = 0$. Then if $p = 2p'$ $q^3 = -8(p') - 2p'q^2$, showing that q^3 and hence q is a multiple of 2 . . Similarly if q is a multiple of 2 so is p . If p and q are both odd, then $p^3 + pq^2 + q^3$ is the sum of three odd numbers, hence odd, hence not equal to the even number 0 . Therefore p and q are both divisible by 2 (even). Now suppose that x satisfies $x^3 + x + 1 = 0$. If $x \in \mathbb{Q}$ then for some integers integers p and q , with no common factor greater than 1 , $x = p/q$, and hence $p^3 + pq^2 + q^3 = 0$. But we have just shown that this requires p and q to have a common factor of 2 ; contradiction.

10. Suppose that p is a prime and $\sqrt{p}$ is rational. Then for some integers integers a and b , with no common factor greater than 1 , $\sqrt{p} = a/b$, so that $a^2 = pb^2$. Then p divides $a^2 = a.a$, so by Theorem 4.4 p divides a . Then $a = pa'$ for some integer a' , and $b^2 = p(a')^2$, so that b^2 is also a multiple of p, whence so is b , This a contradiction, as we said that a and b had no common factor greater than 1 .

Chapter 5

1. If $z < 0$ then $-z > 0$. Then if also $x < y$, $-xz = x(-z) < y (-z) = -yz$. Hence $xz > yz$ by Ex. 5.2.

2. Let $0 < x < y$ and $0 < z < t$ then by Ex. 5.6, $0 < 1/t < 1/z$, whence by Ex.5.4 $0 < x/t < y/z$. (If you didn't see it, notice the similarity with Ex. 5.3.)

3. Let $0 < x < y < 1$. The $0 < 1 - y < 1 - x$ so $0 < 1/(1 - x) < 1/(1 - y)$. The use Ex 5.4.

5. Let $0 < x < y$. The $0 < x^n < y^n$ is true for $n = 1$. Use Ex. 5.4 to show that if it is true for n then it is true for $n + 1$, and show the result by induction. Checking the cases with equality is simple. Now suppose that $x \geq 0$ and $y \geq 0$ and that $x^n > y^n$. Then suppose $x > y$ is false, that is $x \leq y$. Then $x^n \leq y^n$.

6. $x^2 + xy + y^2 = (x + y/2)^2 + 3y^2/4$, which is non-negative (≥ 0) since all squares are. Also $x^2 + xy + y^2 = 0 \Rightarrow (x + y/2)^2 + 3y^2/4 = 0 \Rightarrow (x + y/2)^2 = y^2 = 0 \Rightarrow (x + y/2) = y = 0 \Rightarrow x = y = 0$. Last part: use $y^3 - x^3 = (x^2 + xy + y^2)(x - y)$.

7. For $0 < x < 1$, $1 + x > 0$, so prove that $1 - x^2 = (1 - x)(1 + x) < 1 < (1 - x + x^2)(1 - x) = 1 - x^3$, and deduce the result from this.

9. $x + y < 1 + xy \Leftrightarrow 1 - x - y + xy = (1 - x)(1 - y) > 0$.

11. For all $x \in \mathbb{R}$ $x^2 - 2x + 1 = (x - 1)^2 \geq 0$ (the square of a real number), hence $x^2 + 1 \geq 2x$. If $x > 0$ divide both sides by x to obtain $x + 1/x \geq 2$. If $x < 0$ then we obtain $x + 1/x \leq -2$.

13. $2 \leq x \leq 3 \Rightarrow - 1/2 \leq x - 5/2 \leq 1/2 \Rightarrow 0 \leq (x - 5/2)^2 = x^2 -5x + 25/4 \leq 1/4 \Rightarrow$ $-1/4 \leq x^2 - 5x + 6 \leq 0$ (use Ex 5.4). $-3 \leq x \leq 3 \Rightarrow -2 \leq x + 1 \leq 4 \Rightarrow$ $0 \leq (x + 1)^2 \leq 16 \Rightarrow -1 \leq x^2 + 2x \leq 15$.

14. $x^2 - 5x + 6 \leq 0 \Rightarrow (x - 2)(x - 3) \leq 0 \Rightarrow \{x - 2 \geq 0$ and $x - 3 \leq 0\}$ or $\{x - 2 < 0$ and $x - 3 \geq 0\} \Rightarrow 2 \leq x \leq 3$ or $\{x < 2$ and $x \geq 3$, which is impossible$\} \Rightarrow$ $2 \leq x \leq 3$. $x^2 - 5x + 6 > 0 \Rightarrow (x - 2)(x - 3) > 0 \Rightarrow \{x - 2 > 0$ and $x - 3 > 0\}$ or $\{x - 2 < 0$ and $x - 3 < 0\} \Rightarrow x > 3$ or $x < 2$.

15. Mulitply the x out of the denominator. $1 \leq x \leq 3 \Rightarrow x^2 > 1 \Rightarrow x^2 - 1 > 0 \Rightarrow$ $x - 1/x > 0$. Also $1 \leq x \leq 3 \Rightarrow -1/3 \leq x - 4/3 \leq 5/3 \Rightarrow (x - 4/3)^2 =$

$x^2 - (8/3)x + 16/9 \le 25/9 \Rightarrow x^2 - 1 \le (8/3)x \Rightarrow x - 1/x \le 8/3$.
$2 \le x \le 4 \Rightarrow x^2 - 4 \ge 0 \Rightarrow x - 4/x \ge 0$. Also $2 \le x \le 4 \Rightarrow 1/2 \le x - 3/2 \le 5/2 \Rightarrow$
$1/4 \le (x - 3/2)^2 \le 25/4$ (notice that one - the left hand inequality is correct
because $1/.2 > 0$). Therefore $2 \le x \le 4 \Rightarrow x^2 - 3x + 9/4 \le 25/4 \Rightarrow x^2 - 4 \le 3x \Rightarrow$
$x - 4/x \le 3$.

17. $1 \le x \le 3 \Rightarrow 0 \le x - 1 \le 2 \Rightarrow 0 \le x^2 - 2x + 1 \Rightarrow 2x \le x^2 + 1 \Rightarrow 2 \le x + 1/x$.
Also $1 \le x \le 3 \Rightarrow -2/3 \le x - 5/3 \le 4/3 \Rightarrow x^2 - (10/3)x + 25/9 \le 16/9 \Rightarrow$
$x^2 + 1 \le (10/3)x \Rightarrow x + 1/x \le 10/3$.
$0 \le (x - \sqrt{2})^2 \Rightarrow (2\sqrt{2})x \le x^2 + 2 \Rightarrow 2\sqrt{2} \le x + 2/x$. Also $1 \le x \le 3 \Rightarrow$
$-5/6 \le x - 11/6 \le 7/6 \Rightarrow x^2 - (11/3)x + 121/36 \le 49/36 \Rightarrow x^2 + 2 \le (11/3)x$.

18. Suppose that $0 \le u < v$. Then if $u^{1/n} \ge v^{1/n}$, it follows that $u = (u^{1/n})^n \ge (v^{1/n})^n$
$= v$, a contradiction. Therefore $u^{1/n} < v^{1/n}$. The if $x \ge 1$, $x^{1/n} \ge 1$ and so, by
Bernoulli's Inequality $x = (x^{1/n})^n > n(x^{1/n} - 1)$ hence $x^{1/n} < 1 + x/n$.

19. By Q18 with $x = n$, $1 < n^{1/n} \le 2$. Using $x = \sqrt{n}$ gives $1 \le \sqrt{(n^{1/n})} \le 1 + 1/\sqrt{n}$,
and then squaring finishes the problem.

21. Suppose that $3^n \ge n^3$ for all $n \ge N$. We need to spot a valuye of N which
would allow us to deduce the "induction step" that $3^{n+1} \ge (n + 1)^3 = n^3 + 3n^2 +$
$3n + 1$.) Then $3^{n+1} \ge 3n^3 = n^3 + 2n^3 \ge n^3 + 2Nn^2 = n^3 + 3n^2 + (2N - 3)n^2$.
Now we need $(2N - 3)n^2 > (2N - 3)Nn$ and we will want a $3n$ term so we
need $(2N - 3)N > 3$. $N = 3$ will do. The we see that for all $n \ge 3$ $3^n \ge n^3 \Rightarrow$
$3^{n+1} \ge n^3 + 3n^2 + 3n^2 \ge n^3 + 3n^2 + 9n > (n + 1)^3$. Since $3^3 \ge 3^3$, induction
now shows that $3^n \ge n^3$ for all $n \ge 3$. It is easy to check the result for $n = 1$
and $n = 2$.

23. $|x| \le a$ and $|y| \le b \Rightarrow$ (since $|x| \ge 0$ and $|y| \ge 0$) $|x||y| \le ab$ hence $|xy| \le ab$.

25. $a \le x \le b \Rightarrow -b \le -x \le -a$. Now $b \le |b|$ and $-a \le |a|$ so that we deduce that
$x \le |b|$ and $-x \le |a|$. Then $|x| = \max(x, -x) \le \max(|b|, |a|)$.

27. The result is obvious for $n = 1$ and is just the triangle inequality for $n = 2$.
For $n \ge 2$ the result holds for n then $|x_1 + x_2 + ... + x_{n+1}| \le$
$|x_1 + x_2 + ... + x_n| + |x_{n+1}|$ by the triangle inequality, and the rest follows by
induction.

28. (i) $|x| < \delta \Rightarrow |2x^2 + x| = |x||2x + 1| \le \delta|2x + 1| \le \delta(2|x| + 1) < \delta(2\delta + 1)$
$$\le 3\delta \quad (\text{if } \delta \le 1)$$
$$\le 1 \quad (\text{if } \delta \le 1/3)$$
Therefore $\delta = \min(1, 1/3) = 1/3$ will suffice.
(ii) $|x| < \delta \Rightarrow |3x^3 + 5x^2 - 3x| = |x||3x^2 + 5x - 3| < \delta(|3x^2| + |5x| + |-3|) <$
$\delta(3\delta^2 + 5\delta + 3) \le^{*1} 11\delta \le^{*2} 1$ (Notes:^{*1} if $\delta \le 1$, ^{*2} if $\delta \le 1/11$). So choose δ
$= \min(1, 1/11) = 1/11$.
(iii) $|x| < \delta \Rightarrow |x^3 - 2x^2| = |x||x^2 - 2x| \le \delta(|x|^2 + 2|x|) < \delta(\delta^2 + 2\delta) \le^{*1} 3\delta \le^{*2} 1$
(Notes: ^{*1} if $\delta \le 1$, ^{*2} if $\delta \le 1/3$). Therefore $\delta = 1/3$ will do.
(iv) $|x - 1| < \delta \Rightarrow |x^3 - 3x^2 + 2| = |x - 1||x^2 - 2x - 2| \le \delta(|x|^2 + 2|x| + 2) <$
$\delta(\delta^2 + 2\delta + 2) \le^{*1} 4\delta \le^{*2} 1$ (provided ^{*1} $\delta \le 1$ and ^{*2} $\delta \le 1/4$). So let
$\delta = 1/4$.
(v) $|x + 2| < \delta \Rightarrow |x^3 - 2x + 4| = |x + 2||x^2 - 2x + 2| \le \delta(|x|^2 + 2|x| + 2) \le^{*1} 5\delta \le^{*2} 1$
(provided ^{*1} $\delta \le 1$ and ^{*2} $\delta \le 1/5$). Choose $\delta = 1/5$.

29. (For simplicity we re-use some of the calculation in Q28.) In all cases $\varepsilon > 0$.

(i) $|x| < \delta \Rightarrow |2x^2 + x| = |x||2x + 1| \le \delta(2|x| + 1) < \delta(2\delta + 1) \le^{*1} 3\delta \le^{*2} \varepsilon$. (Notes:
$*1$ if $\delta \le 1$, $*2$ if $\delta \le \varepsilon/3$). Therefore if we choose $\delta = \min(1/\varepsilon/3)$ we have
$|x| < \delta \Rightarrow |2x^2 + x| = \varepsilon$.

(ii) $|x| < \delta \Rightarrow |3x^3 + 5x^2 - 3x| = |x||3x^2 + 5x - 3| < \delta(3\delta^2 + 5\delta + 3) \le^{*1} 11\delta \le^{*2} \varepsilon$
(provided $*1$ $\delta \le 1$ and $*2$ $\delta \le \varepsilon/11$). Therefore $\delta = \min(1, \varepsilon/11)$ will suffice.

(iii) $|x| < \delta \Rightarrow |x^3 - 2x^2| = |x||x^2 - 2x| \le \delta(|x|^2 + 2|x|) < \delta(\delta^2 + 2\delta) \le^{*1} 3\delta \le^{*2} \varepsilon$
(provided $*1$ $\delta \le 1$ and $*2$ $\delta \le \varepsilon/3$). Therefore set $\delta = \min(1, \varepsilon/11)$.

(iv) $|x - 1| < \delta \Rightarrow |x^3 - 3x^2 + 2| = |x - 1||x^2 - 2x - 2| \le \delta(|x|^2 + 2|x| + 2) <$
$\delta(\delta^2 + 2\delta + 2) \le^{*1} 4\delta \le^{*2} \varepsilon$ (provided $*1$ $\delta \le 1$ and $*2$ $\delta \le \varepsilon/4$). Therefore let
$\delta = \min(1/\varepsilon/4)$.

(v) $|x + 2| < \delta \Rightarrow |x^3 - 2x + 4| = |x + 2||x^2 - 2x + 2| \le \delta(|x|^2 + 2|x| + 2) \le^{*1} 5\delta \le^{*2} \varepsilon$
(provided $*1$ $\delta \le 1$ and $*2$ $\delta \le \varepsilon/5$). Choose $\delta = \min(1, \varepsilon/5)$.

30. $|x| < 1 \Rightarrow -1 < x < 1 \Rightarrow 1 < 2 + x < 3 \Rightarrow 1 < |2 + x|$. So set $\delta_1 = 1$, giving
$|x| < \delta_1 \Rightarrow 1/|x + 2| \le 1$. Also,

$$|x| < \delta_2 \Rightarrow \left|\frac{x+1}{x+2} - \frac{1}{2}\right| = \left|\frac{x}{(x+2)2}\right| < \frac{\delta_2}{|x+2|2} \le^{*1} \frac{\delta_2}{2} \le^{*2} \varepsilon \quad \text{(provided } *1 \quad |x| < \delta_1$$

which is true if $\delta_2 \le \delta_1$ and $*2$ $\delta_2 \le 2\varepsilon$). Therefore choose $\delta_2 = \min(\delta_1, 2\varepsilon) = \min(1, 2\varepsilon)$.

31. (i) We are going to have $|x|$ on the denominator. Use $|x - 1| < 1/2 \Rightarrow |x| > 1/2$
$\Rightarrow 1/|x| < 2$. Now $|x - 1| < \delta \Rightarrow |1/x - 1| = |1 - x|/|x| < \delta/|x| <^{*1} 2\delta \le^{*2} \varepsilon$
(provided $*1$ $\delta \le 1/2$ to ensure that $1/|x| < 2$ and $*2$ $\delta \le \varepsilon/2$). Choose $\delta = \min(1/2, \varepsilon/2)$.

(ii) Here the denominator is $x^2 + 1 \ge 1$, so $1/(x^2 + 1) \le 1$.
$|x - 1| < \delta \Rightarrow |(x + 1)/(x^2 + 1) - 1| = |x||1 - x|/|x^2 + 1| \le \delta|x| = \delta(|x - 1| + 1) <$
$\delta(\delta + 1) <^{1*} 2\delta <^{*2} \varepsilon$ (provided $*1$ $\delta \le 1$ and $*2$ $\delta \le \varepsilon/2$). Let
$\delta = \min(1, \varepsilon/2)$.

(iii) Here the denominator will be $|x^2 + 2x| = |(x + 1)^2 - 1|$. So $|x + 1| < 1/2 \Rightarrow$
$(x + 1)^2 < 1/4 \Rightarrow (x + 1)^2 - 1 < -3/4 \Rightarrow |x^2 + 2x| > 3/4 \Rightarrow 1/|x^2 + 2x| < 4/3$.
Then $|x + 1| < \delta \Rightarrow |(x + 1)/(x^2 + 2x)| < \delta/|x^2 + 2x| <^{*1} 4\delta/3 \le^{*2} \varepsilon$ (provided
$*1$ $\delta \le 1/2$ and 2 $\delta \le 3\varepsilon/4$). Let $\delta = \min(1/2, 3\varepsilon/4)$.

34. $\sqrt{(n + 1)} - \sqrt{n} = (n + 1 - n)/(\sqrt{(n + 1)} + \sqrt{n}) = 1/(\sqrt{(n + 1)} + \sqrt{n})$. Then use

$$\frac{1}{2\sqrt{n}} \ge \sqrt{(n+1)} - \sqrt{n} \ge \frac{1}{2\sqrt{(n+1)}}.$$ For (ii) use $1/\sqrt{k} \le 2(\sqrt{k} - \sqrt{(k - 1)})$ obtained

by substituting k for $n + 1$ above. Sum this from $k = 1$ to n. The other
side of the inequality involving $\sqrt{(n + 1)} - \sqrt{n}$ gives the other part.

Chapter 6

2. Let x have a terminating decimal, so $x = m/10^k$, where m and k are
integers and $k \ge 0$. Then, since the factorisation of $10k$ into primes is $2^k 5^k$
cancelling any common factors between numerator and denominator will leave
a denominator of the form $2^a 5^b$ where a and b are non-negative integers.
For the converse, let $c = \max(a, b)$ and multiply numerator and denominator
of $p/2^a 5^b$ by $2^{c-a} 5^{b-a}$.

3. Since $b_1 = [10y]$, $b_1 \le 10p/q < b_1 + 1$, whence $qb_1 \le 10p < qb_1 + q$, so that
$10p = qb_1 + r_1$ where $0 \le r_1 < q$. Since p, q, b_1 and r_1 are integers, b_1 is
the quotient and r_1 the remainder on long division of $10p$ by q. The rest is
similar.

5.* The sequence a_N, a_{N+1}, a_{N+2}, ...repeats at intervals of k terms, so that $a_{N+k} 10^{-(N+k)} = a_N 10^{-(N+k)} = (a_N 10^{-N})10^{-k}$. Then

$$\sum_{n=N}^{N+mk-1} a_n 10^{-n} = \left(\sum_{n=N}^{N+k-1} a_n 10^{-n}\right) + \left(\sum_{n=N+k}^{N+2k-1} a_n 10^{-n}\right) + ... + \left(\sum_{n=N+(m-1)k}^{N+mk-1} a_n 10^{-n}\right)$$

$$= \left(\sum_{n=N}^{N+k-1} a_n 10^{-n}\right) + \left(\sum_{n=N}^{N+k-1} a_n 10^{-n-k}\right) + ... + \left(\sum_{n=N}^{N+k-1} a_n 10^{-n-(m-1)k}\right)$$

$$= \left(\sum_{n=1}^{N+k-1} a_n 10^{-n}\right)\left(1 + 10^{-k} + 10^{-2k} + ... + 10^{-(m-1)k}\right)$$

$$= \left(\sum_{n=1}^{N+k-1} a_n 10^{-n}\right)\frac{1-(10^{-k})^m}{1-10^{-k}}$$

Letting m tend to ∞, so that $(10^{-k})^m = 10^{-km} \to 0$, gives the expression for $\sum_{n=N}^{\infty} a_n 10^{-n}$, which is rational because numerator and denominator are integers. $\sum_{n=1}^{\infty} a_n 10^{-n}$ differs from this by a rational number.

6.* Let p be a prime, so that $1/p$ is rational and by Q5 $\dfrac{1}{p} = \dfrac{m}{10^{N-1}(10^k -1)}$ for some positive integers m, N and k. Then $m = 10^{N-1}(10^k - 1)/p$; since m is an integer, p divides into the numerator, and since it does not divide the first factor, it must divide the second, by Theorem 4.4.

Chapter 7

1. (i) $|a_n - 1| = 2/(n + 1) \le 2/(N + 1)$ (if $n \ge N$) so that $|a_n - 1| < \varepsilon$ if $2/(N + 1) < \varepsilon$ which is true if $N + 1 > 2/\varepsilon$, which in turn is true if $N \ge 2/\varepsilon$. A formal proof is: Let $\varepsilon > 0$. Then we can choose $N \in \mathbb{N}$ with $N \ge 2/\varepsilon$ (by the Archimedean property, which we shall not mention again). Then $\forall n \ge N$ $|a_n - 1| = |(n - 1)/(n + 1) - 1| = 2/(n + 1) \le 2/(N + 1) < 2/N \le \varepsilon$. Since $\varepsilon > 0$ was arbitrary, we have shown that $a_n \to 1$ as $n \to \infty$.
    (ii) $\forall n \in \mathbb{N}$ $-2/n^2 \le a_n - 1 \le 1/n$ so $|a_n - 1| \le \max(2/n^2, 1/n) = 1/n$ (if $n \ge 2$). Let $\varepsilon > 0$, then choose $N > \max(1/\varepsilon, 2)$. Then $\forall n \ge N, n \ge 2$ and so $|a_n - 1| \le \max(2/n^2, 1/n) = 1/n \le 1/N < \varepsilon$. Since $\varepsilon > 0$ was arbitrary, we have shown that $a_n \to 1$ as $n \to \infty$.
    (iii) $|a_n - 0| = 1/n$ or $1/n^2$ so, since $n^2 \ge n$ for $n \in \mathbb{N}$ $|a_n - 0| \le 1/n$. Let $\varepsilon > 0$ and choose $N > 1/\varepsilon$. Then $\forall n \ge N$ $|a_n - 0| \le 1/n \le 1/N < \varepsilon$. Since $\varepsilon > 0$ was arbitrary, we have shown that $a_n \to 0$ as $n \to \infty$.

2. Let $a_n \to L$ as $n \to \infty$. Then let $\varepsilon > 0$. $\exists N$ such that $\forall n \ge N$ $|a_n - L| < \varepsilon$. Then $\forall n \ge N$, $n + 1 \ge N$ so $|b_n - L| = |a_{n+1} - L| < \varepsilon$. Since $\varepsilon > 0$ was arbitrary, we have shown that $b_n \to L$ as $n \to \infty$.
    Now suppose that $b_n \to L$ as $n \to \infty$. Let $\varepsilon > 0$. $\exists N$ such that $\forall n \ge N$ $|b_n - L| < \varepsilon$. Then $\forall n \ge N + 1$, $n - 1 \ge N$ so $|a_n - L| = |b_{n-1} - L| < \varepsilon$. therefore, since ε was arbitrary, $a_n \to L$ as $n \to \infty$.

3   Let $\varepsilon > 0$. Since $a_n \to L$ as $n \to \infty$ and $c_n \to L$ as $n \to \infty$ there exist natural numbers N_1 and N_2 for which $\forall n \ge N_1$ $L - \varepsilon < a_n < L + \varepsilon$ and $\forall n \ge N_2$ $L - \varepsilon < c_n < L + \varepsilon$. Then $\forall n \ge N = \max(N_1, N_2)$ $n \ge N_1$ and $n \ge N_2$ so that $L - \varepsilon < a_n \le b_n \le c_n < L + \varepsilon$, hence $L - \varepsilon < b_n < L + \varepsilon$. It follows that $b_n \to L$ as $n \to \infty$.

4. The limits are, respectively, 2, 1/2, 0, 0, none, 1/2. (In all cases, divide the

numerator and denominator by the highest power of n or in the case of the fourth, the largest term, which is 5^n . In this case note that $(2/5)^n \to 0$ as $n \to \infty$.

7. Suppose that (a_n) is increasing and not bounded above. Let $R \in \mathbb{R}$. Then R is not an upper bound for (a_n) , so $\exists N$ such that $a_N > R$. Then $\forall n \geq N$ $a_n \geq a_N > R$ and therefore, since R was arbitrary,, $a_n \to \infty$ as $n \to \infty$. Then if (b_n) is increasing, either (b_n) is bounded above or not bounded above.

8. (i) $a_{n+1} = (a_n - 1)^2 + 1$ so $1 < a_n < 2 \Rightarrow 1 < a_{n+1} < 2$. By induction $\forall n \in \mathbb{N}$ $1 < a_n < 2$. (ii) $\forall n \in \mathbb{N}$ $a_{n+1} - a_n = (a_n - 1)(a_n - 2) < 0$. (iii) (a_n) is decreasing and bounded below, hence tends to a limit, say a . Then $a = \lim a_{n+1} = \lim (a_n^2 - 2a_n + 2) = a^2 - 2a + 2$, so $a = 1$ or 2 . (v) Since (a_n) is decreasing, $\forall n \in \mathbb{N}$, $a_n \leq a_1 = \alpha < 2$, so $a \leq a_1 < 2$, hence $a \neq 2$, so $a = 1$.

9. $a_{n+1} = (a_n - 1)^2/2 + 1$ can be used to show by induction that if $1 \leq \alpha < 3$ then for all n $1 \leq a_n < 3$ and if $\alpha > 3$ then for all n $a_n > 3$. Then use $a_{n+1} - a_n = (a_n - 1)(a_n - 3)/3$ to tell whether (a_n) is increasing or decreasing. For $1 \leq \alpha < 3$ we have (a_n) decreasing and bounded below, hence it tends to a limit, say a . The limit satisfies $a = (a^2 - 2a + 3)/2$ so $a = 1$ or 3 . If $\alpha > 3$ the sequence is increasing, and cannot be bounded above, for if it were it would tend to a limit a satisfying $a = (a^2 - 2a + 3)/2$, i.e. 1 or 3 ; but $\forall n \in \mathbb{N}$ $a_n \geq a_1 = \alpha > 3$, so $a \geq \alpha > 3$, a contradiction. For the last part notice that $a_2 = (\alpha - 1)^2/2 + 1 \geq 1$ and the sequence thereafter depends only on a_2 so the cases above determine that $a_n \to 1$ as $n \to \infty$ if $1 \leq a_2 < 3$, $a_n \to \infty$ as $n \to \infty$ if $a_2 > 3$. If $a_2 = 3$ then $\forall n \geq 2$ $a_n = 3$ and $a_n \to 3$ as $n \to \infty$. Finally $-1 < \alpha < 1 \Rightarrow 1 < a_2 < 3$, $\alpha = -1 \Rightarrow a_2 = 3$ and $\alpha < -1 \Rightarrow a_2 > 3$.

10. For $0 \leq \alpha < 1$ the sequence decreases and tends to 0 . For $\alpha = 1$ all terms are 1 . For $\alpha > 1$ the sequence increases and tends to ∞ . For $-1 \leq \alpha < 0$ we have $\forall n \in \mathbb{N}$ $-1/8 \leq a_n \leq 0$ and a_n increases and tends to 0 . For $\alpha < -1$ we have $a_2 > 1$ and the above cases recur, depending on where a_2 is. The sequence tends to 0 if $-2 < \alpha < -1$, tends to 1 if $\alpha = -2$ and to ∞ if $\alpha < -2$.

11. Clearly $a_n > 0 \Rightarrow a_{n+1} > 0$, so if $\alpha > 0$ $\forall n \in \mathbb{N}$ $a_n > 0$. Also $a_n - 1 = (a_n - 1)/(a_n + 1)$, so if $\alpha \in (0, 1]$ or $(1, \infty)$ all the terms a_n are in the same interval, and the sequence is increasing or decreasing respectively. For the last part, notice that the sequence is not properly defined if there is an N with $a_N = -1$. If this does not happen then a_n decreases until $a_n < -1$ then the next term is positive and the cases above apply.

12. $a_{n+1} = a_n^2(2 - a_n)$ so that $0 \leq a_n < 1 \Rightarrow a_{n+1} \geq 0$. Also $a_{n+1} - a_n = -a_n(a_n - 1)^2$ so $0 \leq a_n \Rightarrow a_{n+1} \leq a_n$. In particular $0 \leq a_n < 1 \Rightarrow a_{n+1} \leq a_n < 1$. Therefore $0 \leq \alpha < 1$ shows that for all n $0 \leq a_n < 1$ and a_n decreases. $a_{n+1} - 1 = -(a_n - 1)(a_n - (1+\sqrt{5})/2)(a_n - (1 - \sqrt{5})/2)$ so $1 \leq a_n \leq (1 + \sqrt{5})/2 \Rightarrow a_{n+1} - 1 \geq 0$ and by the first line above $0 \leq a_n \Rightarrow a_{n+1} \leq a_n$. so $1 \leq a_n \leq (1 + \sqrt{5})/2 \Rightarrow 1 \leq a_{n+1} \leq (1 + \sqrt{5})/2$. In this case the sequence decreases and tends to 1 .

13. If $a_n \to a$ as $n \to \infty$ then $a = 0, 1/2$ or $-1/2$. Also $a_n > 0 \Rightarrow a_{n+1} > 0$ so $\alpha > 0 \Rightarrow \forall n$ $a_n > 0$. Since $a_{n+1} - 1/2 = (a_n - 1/2)/(1 + 2|a_n|)$ we see that if α is in one of the intervals $(0, 1/2], (1/2, \infty)$ then all other terms are in this interval and looking at $a_{n+1} - a_n$ shows that a_n is respectively increasing or

decreasing. $a_n \to 1/2$ as $n \to \infty$ if $\alpha > 0$, $a_n \to 0$ if $\alpha = 0$ and $a_n \to -1/2$ if $\alpha < 0$.

15. Let $a_n = (-1)^n n$.

16. Use $|b_n - a| \le |b_n - a_n| + |a_n - a|$. Let $\varepsilon > 0$. Then (because $a_n \to a$ we can choose N_1 so that $\forall n \ge N_1$ $|a_n - a| < \varepsilon/2$, and $\forall n \ge 2/\varepsilon$, $1/n \le \varepsilon/2$ so that $\forall n \ge N = \max(N_1, 2/\varepsilon)$ $|b_n - a| \le |b_n - a_n| + |a_n - a| < 1/n + \varepsilon/2 \le \varepsilon$.

18.* If $\forall n \ge N$ $|a_n - a| < \varepsilon$ then $\forall n \ge 2N$ $|b_n - a| < \varepsilon$ and $\forall n \ge (1/2)N(N-1)$ $|c_n - a| < \varepsilon$

19.* $\forall n \in \mathbb{N}$ let $a_n = 1/n$ and $b_n = 2/n$.

20.* Prove (i) directly, choosing "ε" carefully, and (ii) using $a_n^2 = |a_n^2|$. (v) If $a = 0$ use part (i). If $a_n^2 \to a^2 \ne 0$ and $a_{n+1} - a_n \to 0$ then $|a_n| \to |a|$ so $\exists N$ s.t. $\forall n \ge N$ $|a_n| > |a|/2$ and $|a_{n+1} - a_n| < |a|/4$, whence a_{n+1} and a_n have the same sign. So either $\forall n \ge N$ $a_n = |a_n|$ or $\forall n \ge N$ $a_n = -|a_n|$.

Chapter 8

2. $0 \le a_n \le (3/4)^n$ and $\sum(3/4)^n$ converges.

3. The first two diverge, the last two converge.

4. Since $\sum a_n$ converges $a_n \to 0$ as $n \to \infty$, so $\exists N$ s.t. $\forall n \ge N$ $0 < a_n < 1$. Then $\forall n \ge N$ $0 \le a_n^2 \le a_n$, so $\sum a_n^2$ converges by comparison. Try $b_n = 1/n$.

6. The first three converge (ratio test), the fourth converges (root test). The fifth and sixth converge and seventh diverges (ratio test), the last uses Theorem 8.10.

8. With $a_n = x^n/(1 + x^{2n})$, $a_{n+1}/a_n \to x$ as $n \to \infty$ if $|x| < 1$, and $a_{n+1}/a_n \to 1/x$ if $|x| > 1$. If $x = \pm 1$ the nth term does not tend to 0. Converges for all $x \ne \pm 1$.

9. Radii: 1, 1/2, 1, 1, 1, $\sqrt{2}$, 2, 1, 0, ∞, 1/4, 4/27, ∞.

11. $\sum_{n=k}^{\infty} 1/n! \le \sum_{n=k}^{\infty} \dfrac{1}{(k+1)^{n-k} k!} = \dfrac{1}{k!} \dfrac{1}{1 - \frac{1}{k+1}} = \dfrac{k+1}{k\,k!}$.

13. $\forall n \ge N$ $(L - \varepsilon)^{1-N/n} |a_N| < |a_n|^{1/n} < (L + \varepsilon)^{1-N/n} |a_N|^{1/n}$. Use the fact that $c^{1/n} \to 1$ as $n \to \infty$, where c is constant, to deduce that $(L + \varepsilon)^{1-N/n} \to (L + \varepsilon)$ as $n \to \infty$ so $\forall n \ge N_1$ $L - 2\varepsilon < |a_n|^{1/n} < L + 2\varepsilon$.

16. Assume $\sum(a_n + b_n)$ convergent and use contradiction: $b_n = (a_n + b_n) - a_n$.

17. $\sum b_n$ and $\sum c_n$ cannot both converge or $\sum |a_n|$ would. By Q16 if one of them converges, so does the other, since $b_n = a_n - c_n$.

19. Let $s_n = a_1 + \dots + a_n$, $t_n = b_1 + \dots + b_n$, $u_n = c_1 + \dots + c_n$. $c_{2n} = a_n + b_n$ and $c_{2n-1} = a_n + b_{n-1}$.

21. $\binom{n}{k} \dfrac{1}{n^k} = \dfrac{n(n-1)\dots(n-k+1)}{n^k\, k!} = \dfrac{n}{n}\dfrac{n-1}{n}\dots\dfrac{n-k+1}{n}\dfrac{1}{k!} \le \dfrac{1}{k!}$.

Chapter 9

1. $[1, \infty)$, $[1/2, \infty)$, $[4, \infty)$, $[4, \infty)$, $[1, \infty)$, $[1/2, \infty)$, $[1/2, \infty)$, $[1, \infty)$, 0.

2. 2 is an upper bound, $\sup A$ is the *smallest* upper bound, so $\sup A \le 2$. Let $A = (1, 2)$ and $A = (0, 1)$ respectively.

4. Let $r \in \mathbb{R} \backslash \mathbb{Q}$ and $q \in \mathbb{Q} \backslash \{0\}$; here $A \backslash B$ means $\{x : x \in A \text{ and } x \notin B\}$. Then if $rq \in \mathbb{Q}$, $r = (1/q)rq \in \mathbb{Q}$, a contradiction. If $x < y$ then $x\sqrt{2} < y\sqrt{2}$. By Lemma 9.6 $\exists q \in \mathbb{Q}$ with $x\sqrt{2} < q < y\sqrt{2}$, whence $x/\sqrt{2}$ is an irrational number between x and y.

6. By Lemma 9.6 we can find $a_1 \in \mathbb{Q}$ with $\sqrt{2} - 1 < a_1 < \sqrt{2}$. Suppose we have found $a_1, a_2, ..., a_n$. Then $\max(a_n, \sqrt{2} - 1/(n+1)) < \sqrt{2}$ so there is a rational number a_{n+1} with $\max(a_n, \sqrt{2} - 1/(n+1)) < a_{n+1} < \sqrt{2}$. This defines (a_n) by induction.

7. Show that $(\sup A) - 1$ is an upper bound for B (so B is bounded above) and either $\sup B + 1$ is an upper bound for A or that if $b < \sup A - 1$ then b is not an upper bound for B. (Hint: $b + 1 < \sup A$ so $\exists x \in A$ with $b + 1 < x$.)

9. $x \in A \cup B \Rightarrow x \in A$ or $x \in B \Rightarrow x \leq \sup A$ or $x \leq \sup B \Rightarrow x \leq \max(\sup A, \sup B)$. Also $c < \max(\sup A, \sup B) \Rightarrow c < \sup A$ or $c < \sup B \Rightarrow \exists x \in A$ with $c < x$ or $\exists x \in B$ with $c < x \Rightarrow \exists x \in A \cup B$ with $c < x$ so c is not an upper bound for $A \cup B$.
It is easy to show that $\inf A + \inf B$ is a lower bound for $\{x + y : x \in A, y \in B\}$. Let $c > \inf A + \inf B$. Then $c - \inf B > \inf A$, so $c - \inf B$ is not a lower bound for A and $\exists x \in A$ with $c - \inf B > x$. For this x, $c - x > \inf B$ so $c - x$ is not a lower bound for B and $\exists y \in B$ with $c - x > y$. Then $x + y < c$ so c is not a lower bound for $\{x + y : x \in A, y \in B\}$, and $\inf A + \inf B$ is the greatest lower bound.

10. Let $a = \sup A$ and $b = \sup B$. Then $z \in C \Rightarrow z = xy$ where $x \leq a$ and $y \leq b$, so ab is an upper bound. Let $c < ab$. Then $c/b < a = \sup A$ so $\exists x \in A$ with $c/b < x$. Then (noting that $x \in A \Rightarrow x > 0$) $c/x < b$ so $\exists y \in B$ with $c/x < y$. Finally $xy \in C$ and $xy > c$ so c is not an upper bound for C.

Chapter 10

2. $|f(x) - 1| \leq \max(|x|, x^2)$. Let $\varepsilon > 0$. Then $|x - 0| < \min(\delta, \sqrt{\delta}) \Rightarrow |f(x) - 1| < \varepsilon$.

4. Let $\varepsilon > 0$. Then $\exists \delta_1 > 0$ and $\delta_2 > 0$ such that $|x - a| < \delta_1 \Rightarrow f(a) - \varepsilon < f(x) < f(a) + \varepsilon$ and $|x - a| < \delta_2 \Rightarrow h(a) - \varepsilon < h(x) < h(a) + \varepsilon$. So $|x - a| < \min(\delta_1, \delta_2) \Rightarrow g(a) - \varepsilon < f(x) \leq g(x) \leq h(x) < g(a) + \varepsilon$ (using $f(a) = g(a) = h(a)$).

6. $f(a) > 0$ so we may use it in place of "ε". Since f is continuous at a, $\exists \delta > 0$ with $|x - a| < \delta \Rightarrow |f(x) - f(a)| < f(a) \Rightarrow 0 < f(x) < 2f(a) \Rightarrow f(x) > 0$.

7. Let $\varepsilon > 0$. Then $x \geq 1/\varepsilon + 1 \Rightarrow x > 1/\varepsilon \Rightarrow |1/x - 0| < \varepsilon$. This shows that $1/x \to 0$ as $x \to \infty$. Again with $\varepsilon > 0$ let $x > \max(3|a_2|/(\varepsilon), \sqrt{(3|a_1|/\varepsilon)}, (3|a_0|/\varepsilon)^{1/3})$, then $|f(x) - 1| < \varepsilon$. Therefore, putting "ε" $= 1/2$, we see that $\exists R_1$ such that $\forall x \geq R_1$ $f(x) > 1/2 > 0$.

8 $h(0) = f(0) - 0 \geq 0 - 0 = 0$ and $h(1) = f(1) - 1 \leq 1 - 1 \leq 0$. If $h(0) \neq 0$ and $h(1) \neq 0$ by the Intermediate Value Theorem $\exists \xi \in (0, 1)$ with $h(\xi) = 0$.

9. Suppose $g(c) \neq 0$ (so $c \in \mathbb{R} \backslash \mathbb{Q}$). Let $\varepsilon = |g(c)| > 0$. Then by continuity $\exists \delta > 0$ such that $|x - c| < \delta \Rightarrow |g(x) - g(c)| < |g(c)| \Rightarrow |g(x)| \neq 0$. This is a contradiction for the interval $(c - \delta, c + \delta)$ must contain a rational number.

10. By Theorems 10.5 and 10.6 $f([a, b])$ is bounded and its sup and inf are of the form $f(x_0)$ and $f(x_1)$, with $x_0, x_1 \in [a, b]$, so $f([a, b]) \subset [f(x_0), f(x_1)]$. Let $f(x_0) < y < f(x_1)$, then by the Intermediate Value Theorem $\exists x \in [a, b]$ with $y = f(x)$ so that $[f(x_0), f(x_1)] \subset f([a, b])$.

11.* Since $f(x) \to L$ as $x \to \infty$ $\exists X$ such that $\forall x \geq X$ $L - \varepsilon < f(x) < L + \varepsilon$. As f is continuous on $[0, X]$ Theorem 10.5 shows that $\exists m, M$ such that $0 \leq x \leq X \Rightarrow m \leq f(x) \leq M$. So $\forall x \in [0, \infty)$ $\min(m, L - \varepsilon) \leq f(x) \leq \max(M, L + \varepsilon)$.

12.* Let $y_0 \in \mathbb{R}$ and $\varepsilon > 0$. Since f is continuous at y_0 $\exists \delta > 0$ for which

$x \in (y_0 - \delta, y_0 + \delta) \Rightarrow f(y_0) - \varepsilon < f(x) < f(y_0) + \varepsilon$. Now $f(y_0) \le g(y_0)$, and if $f(y_0) = g(y_0)$ then $\forall x \in (y_0 - \delta, y_0 + \delta)$ $g(y_0) - \varepsilon < f(x) < g(y_0) + \varepsilon$ so $g(x) \ge f(x) > g(y_0) - \varepsilon$. Also for $y_0 - \delta < x \le y_0$ $g(x) \le g(y_0)$ and for $y_0 < x < y_0 + \delta$ $g(x) = \max(g(y_0), \sup\{g(t): y_0 < t < x\}) \le g(y_0) + \varepsilon$. Therefore $\forall x \in (y_0 - \delta, y_0 + \delta)$ $g(y_0) - \varepsilon < g(x) < g(y_0) + \varepsilon$. If $f(y_0) < g(y_0)$ then $\exists \delta > 0$ such that $g(y_0) \in (y_0 - \delta, y_0 + \delta) \Rightarrow f(x) < g(y_0) \Rightarrow g(x) = g(y_0)$.

Chapter 11

2. $h'(c) = f_1'(c)f_2(c) + f_1(c)f_2'(c)$; divide by $h(c)$ for the result. $k'(c) = f_1'(c)f_2(c)...f_n(c) + f_1(c)f_2'(c)...f_n(c) + ... + f_1(c)f_2(c)...f_n'(c)$.

3. $2f'(2x)$, $2xf''(x)$, $f'(g(x)h(x))(g'(x)h(x) + g(x)h'(x))$, $f'(g(h(x)))g'(h(x))h'(x)$.

5. Let $x, y \in \mathbb{R}$. Then by the Mean Value Theorem (on the interval $[x, y]$ or $[y, x]$) there is a ξ between x and y for which $|f(y) - f(x)| = |y - x||f'(\xi)| \le M|y - x|$. As x and y were arbitrary, this holds for all $x, y \in \mathbb{R}$.

6. Fix $y \in \mathbb{R}$. Then $0 \le |g(x) - g(y)|/|x - y| \le |x - y| \to 0$ as $x \to y$, so $g'(y) = 0$. This holds $\forall y \in \mathbb{R}$, so g is constant by Q4.

7. Apply the Mean Value Theorem to $[y, z]$, so $\exists \xi \in (y, z)$ with $f(z) - f(y) = (z - y)f'(\xi) < 0$. Doing this on $[a, c]$, we see that f is strictly decreasing on $[a, c]$ and similarly on $[c, b]$, so if $y < z$ and both lie in $[a, c]$ or both lie in $[c, b]$, then $f(z) < f(y)$. Also if $y \in [a, c]$ and $z \in [c, b]$ $f(z) < f(c) < f(y)$, giving the reamining case for showing f is strictly decreasing on $[a, b]$.

9. Inflexion, minimum, minimum, minimum, inflexion.

10. By Rolle's Theorem there are points $\xi_1 \in (a, b)$ and $\xi_2 \in (b, c)$ for which $f'(\xi_i) = 0$ $(i = 1, 2)$. Then since $\xi_1 < b < \xi_2$ apply Rolle's Theorem to f' on $[\xi_1, \xi_2]$ so $\exists \xi_3 \in (\xi_1, \xi_2)$ with $f''(\xi_3) = 0$.

11. Example: let $f(x) = x^3$.

12. (i) $\phi(x) - \phi(b)/(x - b) \to \phi'(b) > 0$ as $x \to b$, so by Ex.10.5, $\exists \delta > 0$ such that $0 < |x - b| < \delta \Rightarrow \phi(x) - \phi(b)/(x - b) > 0 \Rightarrow \phi(x) - \phi(b)$ and $x - b$ have the same sign. (ii) Applying (i) to f' we see that $\exists \delta > 0$ such that $0 < |x - b| < \delta \Rightarrow f'(x) - f'(b)$ and $x - b$ have the same sign, hence $f'(x) > 0$ for $x \in (b, b + \delta)$. By Ex. 11.2 then, $f(x) > f(b)$ for $x \in (b, b + \delta)$. Hence f does not have a local maximum at b.

13. Suppose that y is not identically 0, so either its maximum is non-zero or its minimum is. In the first case, let the maximum occur at ξ, so $\xi \neq 0, 1$ hence $y'(\xi) = 0$ and $y''(\xi) = g(\xi)y(\xi) > 0$. This contradicts Q12(ii). If the minimum is negative at occurs at ξ, $y''(\xi) < 0$, another contradiction.

14. -1, -2, 1, 1/2.

15. $a_n > - \Rightarrow a_{n+1} > 0$ so $\forall n \in \mathbb{N}$ $a_n > 0$. Let $f(x) = 1 + 1/(1 + x)$ so $f(\sqrt{2}) = \sqrt{2}$. Then $|a_{n+1} - \sqrt{2}| = |f(a_n) - f(\sqrt{2})| = |f'(\xi)||a_n - \sqrt{2}| = (1/(1 + \xi)^2)|a_n - \sqrt{2}|$ for some ξ between a_n and $\sqrt{2}$, hence $\xi \ge 1$ so $1/(1 + \xi)^2 \le 1/4$. Therefore $|a_{n+1} - \sqrt{2}| \le (1/4)|a_n - \sqrt{2}| \le ... \le (1/4)^n |a_1 - \sqrt{2}| \to 0$ as $n \to \infty$.

16. f is differentiable and $f'(x) < 0$ on $\mathbb{R}\backslash\{0\}$, so in particular f is decreasing on $[3/2, 2]$. Therefore $3/2 \le x \le 2 \Rightarrow 5/3 = f(3/2) \ge f(x) \ge f(2) = 3/2 \Rightarrow 2 \ge f(x) \ge 3/2$. Since $a_{n+1} = f(a_n)$, this shows that $\forall n \in \mathbb{N}$ $a_n \in [3/2, 2]$. Now if a_n tends to a limit a then $a = f(a)$ so $a = (1 \pm \sqrt{5})/2$, and since $a \in$

[3/2, 2] $a = (1 + \sqrt{5})/2$. $|a_{n+1} - a| = |f(a_n) - f(a)| = |a_n - a||f'(\xi_n)|$, where $\xi_n \in$ [3/2, 2]. Then $|f'(\xi_n)| \le 4/9$, so $\forall n \in \mathbb{N}$ $|a_{n+1} - a| \le (4/9)^n |a_1 - a|$, hence $a_n \to a$.

17. By Taylor's Theorem, $f(n + 1) = f(n) + f'(n) + f''(\xi_n)/2!$ for some $\xi_n \in$ $(n, n + 1)$; $f'(n) = 1/(2\sqrt{n})$ and $f''(\xi_n) = -1/(4\xi_n^{3/2})$, giving $2\sqrt{(k + 1)} - 2\sqrt{k} = 1/\sqrt{k} - 1/(4\xi_k^{3/2})$. Summing this from $k = 1$ to n gives $2\sqrt{(n + 1)} - 2 = S_n - \sum_{k=1}^{n} \xi_k^{-3/2}/4$. As $\sum \xi_k^{-3/2}$ converges (compare with $\sum n^{-3/2}$) S_n tends to a limit as $n \to \infty$.

19.* Suppose $f(x) = f(y)$, then by Rolle's Theorem $\exists \xi \in [x, y]$ with $f'(\xi) = 0$, so if f' is nowhere zero in $[c, d]$ f is injective there, hence by Lemma 10.8 either strictly increasing or strictly decreasing. Let $c \ne d$, $f'(c) < 0 < f'(d)$. Then $\exists x > c$ with $f(x) < f(c)$ (for $(f(x) - f(c))/(x - c) \to f'(c) < 0$ as $x \to c+$) so f is not strictly increasing. Similarly $f'(d) > 0$ shows f is not strictly decreasing, which contradicts the assumption that f' is nowhere zero in $[c, d]$.

Chapter 12

2. If $x \ne 0$, the series is a geometric progression, common ratio $1/(1 + |x|) < 1$, so its sum to infinity is $x/(1 - 1/(1 + |x|)) = x(1 + |x|)/|x|$. Ths tends to 1 as $x \to 0+$ and to -1 as $x \to 0-$, so the sum to infinity is discontinuous.

4. $x^x = e^{x \log x} \to e^0 = 1$ as $x \to 0+$, since $\exp$ is continuous, and $x \log x \to 0$ as $x \to 0+$. $x^{1/x} = \exp((1/x)\log x)) \to e^0 = 1$ as $x \to \infty$.

6. $f'(x) = 1/x$, and in general $f^{(n)}(x) = (-1)^{n-1}(n-1)!/x^n$. The ratio test shows that the series has radius 1, and $g'(x) = \sum_{n=1}^{\infty}(-1)^{n-1}x^{n-1} = 1/(1 + x)$ by Theorem 12.5. Let $h(x) = (1 + x)e^{-g(x)}$; for all $x \in (-1, 1)$ $h'(x) = 0$ so h is constant, and putting $x = 0$ shows that constant is 1, so $g(x) = \log(1 + x)$.

8. Use Theorems 12.10 (with $\alpha = 1/2$) and 12.11 (with $N = 2$) for the first part. Then for $|x| \le 1/2$ $|\sqrt{(1 + x)} - \sqrt{(1 - x)} - x| \le |\sqrt{(1 + x)} - (1 + x/2) - \{\sqrt{(1 - x)} - (1 - x/2)\}| \le K|x|^2 + K|x|^2$. Divide by x and let x tend to 0.

9. Let $f(x) = x - \log(1 + x)$. $f(0) = 0$ and $\forall x > 0$ $f(x) > 0$ so by the Mean Value Theorem $f(x) = xf'(\xi) > 0$. Let $g(x) = x^2/2 - x + \log(1 + x)$; for the second inequality. For $k \in \mathbb{N}$ $0 \le 1/k - \log(1 + 1/k) \le 1/(2k^2)$; summing from $k = 1$ to n gives a_n. (a_n) is bounded above by $\sum_{k=1}^{\infty} 1/(2k^2)$

10. $a_0 = 0$, $a_n = (n - 1)!$ for $n \ge 1$. The series converges only for $x = 0$.

Chapter 13

2. Let $D_\delta = \{a, b, b+\delta, c\}$; and $S(D_\delta) = b - a + \delta$, $s(D_\delta) = b - a$ (Notice that $\sup\{f(x): b \le x \le b+\delta\} = 1$).. Then $\inf\{S(D): \text{all } D\} \le \inf\{S(D_\delta): \delta > 0\} = b - a = \sup\{s(D_\delta): \delta > 0\} \le \sup\{s(D): \text{all } D\} \le \inf\{S(D): \text{all } D\}$, the last inequality by Lemma 13.1. Therefore $\inf\{S(D)\} = \sup\{s(D)\} = b - a$.

3. Let $D = \{a, c - \delta/2, c + \delta/2, b\}$. As $\inf\{f(x): x \in (c - \delta/2, c + \delta/2)\} \ge \alpha$ and $f(x) \ge 0$ on the other intervals, $S(D) \ge \alpha\delta > 0$ so $\int_a^b f \ge S(D) > 0$.

5. Setting $\phi = f$ shows that $\int_a^b f^2 = 0$, then, since $f = |f|^2$, Q4 shows that f^2 and hence f is identically zero.

6. $|F(x) - F(y)| = |\int_x^y f| \le |x - y|M$ where $M = \sup\{|f(x)|: a \le x \le b\}$; this shows the continuity of F. Let $g(x) = \int_a^x f - \int_x^b f$. $g(a) = -g(b)$ so by the Intermediate Value Theorem, $\exists \xi \in [a, b]$ with $g(\xi) = 0$.

8. $\sin(x - t) = \sin x \cos t - \cos x \sin t$, Corollary 1 to Theorem 13.2 applies and
    $$y'(x) = \int_0^x \cos(x-t) f(t)\,dt + 0, \quad y''(x) = \int_0^x -\sin(x-t)f(t)\,dt + \cos 0 f(x).$$

10. $\int_1^n \log t\,dt = n\log n - n + 1$, and $\int_1^n \log t\,dt = \sum_{k=1}^{n-1}\{((\log k + \log(k+1))/2 + \varepsilon_k\}$
    $= \log n! - (\log n)/2 + \sum_{k=1}^{n-1}\varepsilon_k$. $\phi(n) = 1 - \sum_{k=1}^{n-1}\varepsilon_k$ and thends to a limit because $\Sigma\varepsilon_k$ converges. $n!/((n/e)^n\sqrt{n}) = \exp(\log n! - n\log n + n - (\log n)/2)$.

12. $\int_1^y \frac{\sin x}{x^\alpha}dx = \frac{\cos 1}{1} - \frac{\cos y}{y} - \int_1^y \alpha\frac{\cos x}{x^\alpha}dx$.   Since $\int_1^\infty \frac{1}{x^{\alpha+1}}dx$ exists and $|\frac{\cos x}{x^{\alpha+1}}| \le \frac{1}{x^{\alpha+1}}$ we see that $\int_1^y \frac{\sin x}{x^\alpha}dx$ tends to a limit as $y \to \infty$. $\int_a^1 \log x\,dx = -a\log a + a - 1 \to -1$ as $a \to 0+$. The rest are simple comparisons, but note that for $\int_0^1 \frac{1}{\sqrt{x(1-x)}}dx$ the integral should be split (e.g. at 1/2) and two comparisons made.

15. Let $\xi > 0$ and $y = \xi x$. $F(\xi) = F(\xi) = \int_0^\infty \frac{\sin \xi x}{\xi x}\xi\,dx = \int_0^\infty \frac{\sin y}{y}dy = F(1)$.

16.* $\int_{2(n-1)\pi}^{2n\pi} \frac{\sin x}{x}dx = \int_{2(n-1)\pi}^{(2n-1)\pi}\frac{\sin x}{x}dx + \int_{(2n-1)\pi}^{2n\pi}\frac{\sin x}{x}dx = \int_{2(n-1)\pi}^{2n\pi}\sin x(\frac{1}{x} - \frac{1}{x+\pi})dx > 0$.

17. The first does not exist (compare with $\int_0^1 \frac{1}{x-1}dx$). The second does: compare with $\int_0^1 \frac{1}{\sqrt{(1-x)}}dx$, $\int_1^2 \frac{1}{\sqrt{(x-1)}}dx$, and $\int_2^\infty \frac{1}{x^2}dx$. The third exists (direct calculation), while $\int_0^\infty x^\alpha\,dx$ exists for no real α.

18. Let $f(x) = 1/(x(\log x)^2)$ and use integral test. $\int_2^\infty \frac{1}{x(\log x)^2}dx = \left[-\frac{1}{\log x}\right]_2^\infty = \frac{1}{\log x}$.

19.* Let $\varepsilon > 0$. There is a constant M such that $\forall x \in [a, b]$ $|f(x)| \le M$. Choose x with $0 < x - a < \varepsilon/4M$. Let $D = \{c_0, ...c_n\}$ be a dissection of $[x, b]$ such that $S(D) - s(D) < \varepsilon/2$. Let $D' = \{a, c_0, ...c_n\}$. Then $(\sup_{[a,c_0]} f - \inf_{[a,c_0]} f)(c_0 - a) < \varepsilon/2$, so $S(D') - s(D') < \varepsilon$. Since ε was arbitrary $\int_a^b f$ exists.

Chapter 14

1. Only the first is continuous at $(0, 0)$. Consider $f(x, x)$ to see this for the second and third, and $f(x^2, \alpha x)$, with $\alpha \ne 1$, for the last.
3. Let $h(x) = (\sin x)/x$ $(x \ne 0)$ and $h(0) = 1$; $h: \mathbb{R} \to \mathbb{R}$ is continuous, so Q2 shows that f is continuous, since $f(x, y) = h(x + y)$. By the (one-variable)

Mean Value Theorem applied to the sin function, for $x \neq y$ $g(x, y) = \cos \xi$ for some ξ between x and y; by definition of g, this remains true if $x = y$. Let $\varepsilon > 0$ and choose $\delta > 0$ so that $|x| < \delta \Rightarrow |\cos x - 1| < \varepsilon$. Then $\|(x, y)\| < \delta$ $\Rightarrow |x| < \delta$ and $|y| < \delta$, so the ξ above satisfies $|\xi| < \delta$ and hence $|g(x, y) - 1|$ $= |\cos \xi - 1| < \varepsilon$.

5. The first two are closed (use the technique following Lemma 14.2), the third is not. (For $(-1 + 1/n, 0)$ belongs to the set if $n \in \mathbb{N}$ and tends to $(-1, 0)$ which is not in the set.)

6. Let (x_n) be Cauchy and $x_{n_k} \to x$ as $k \to \infty$. Let $\varepsilon > 0$. Then $\exists N$ such that $\forall m, n \geq N$ $|x_m - x_n| < \varepsilon$ and $\exists K$ such that $\forall k \geq K$ $|x_{n_k} - x| < \varepsilon$. Let $k = \max(N, K)$ so that $n_k \geq k \geq N$ and $\forall m \geq N$ $|x_m - x| \leq |x_m - x_{n_k}| + |x_{n_k} - x| < 2\varepsilon$.

7. Let $F(t) = f(\mathbf{b} + t(\mathbf{c} - \mathbf{b}))$ which is continuous, because $\mathbf{b} + t(\mathbf{c} - \mathbf{b}) = (1 - t)\mathbf{b} + t\mathbf{c} \in A$ for all $t \in [0, 1]$.

8. For all $(\partial f / \partial y)(0, 0) = 0$, $(\partial f / \partial x)(0, 0)$ does not exist for the third, but is zero for the others. The second and third functions are not differentiable at $(0, 0)$.

10. $\frac{\partial f}{\partial x}(0, 0) = \frac{\partial f}{\partial y}(0, 0) = 0$ and for $y \neq 0$ $\frac{\partial f}{\partial x}(0, y) = -y$. Also

$$\frac{\partial^2 f}{\partial y \partial x}(0, 0) = -1 \text{ and } \frac{\partial^2 f}{\partial x \partial y}(0, 0) = 1$$

12. $F'(x) = \frac{\partial f}{\partial x}(x, y, z) + \frac{\partial f}{\partial y}(x, y, z)y'(x) + \frac{\partial f}{\partial z}(x, y, z)z'(x)$.

14. Follow the notation of Example 14.7, with $F(x, y) = f(y, x)$, sao that in this case $F = f$. The Example shows the first part. Also

$$\frac{\partial^2 F}{\partial x^2}(a, b) = \frac{\partial}{\partial x}(\frac{\partial f}{\partial y}(g_1(x, y), g_2(x, y)))$$

$$= \frac{\partial^2 f}{\partial x \partial y}(g_1(x, y), g_2(x, y))\frac{\partial g_1}{\partial x}(x, y) + \frac{\partial^2 f}{\partial y^2}(g_1(x, y), g_2(x, y))\frac{\partial g_2}{\partial x}(x, y)$$

$$= \frac{\partial^2 f}{\partial y^2}(y, x) \text{ and then use } F = f.$$

15. If f attains a minimum at $\mathbf{a}$ with $\|\mathbf{a}\| < 1$ then the n functions obtained by fixing all but one of the variables $a_1, ..., a_n$ also attain their minimum, so their derivative with respect to the remaining varaible is zero. This shows that ther partial derivatives are all 0 at $\mathbf{a}$. Since A is closed an bounded f attains its maximum and minimum at points $\mathbf{x}_0$ and $\mathbf{x}_1 \in A$ respectively (Th. 14.8). If $\|\mathbf{x}_0\| < 1$ or $\|\mathbf{x}_1\| < 1$ then for $i = 1, ... m$, $\frac{\partial f}{\partial x_i} = 0$ at that point. If $\|\mathbf{x}_0\| = \|\mathbf{x}_1\| = 1$, then f is constant, since is it constant on $\{\mathbf{x}: \|\mathbf{x}_0\| = 1\}$.

16. Let $\varepsilon > 0$. Then $\exists \delta > 0$ and M for which $\forall \mathbf{x}, \mathbf{x}' \in [a, b] \times [c, d]$ $|f(\mathbf{x})| \leq M$ and $|f(\mathbf{x}) - f(\mathbf{x}')| < \varepsilon$. Then if $\|(x, y) - (x', y')\| < \delta$, $|F(x, y) - F(x', y')| \leq$ $|\int_{y'}^{y} f(x, t)dt| + \int_{c}^{y'} |f(x, t) - f(x', t)|dt \leq M\delta + (d - c)\varepsilon$. Let $\delta' = \min(\delta, \varepsilon)$.

18. $F'(z) = \int_{a}^{z} f(z, y)dy$, $G'(z) = \int_{a}^{z} \frac{\partial}{\partial z}(\int_{y}^{z} f(x, y)dx)dy + \int_{z}^{z} f(x, z)dx = \int_{a}^{z} f(z, y)dy$

19. Differentiate with respect to a and b and add the results.

21.* By Rolle's Theorem there is $\xi \in (0, 1)$ with $F_1'(\xi) = 0$; $\mathbf{x}_1 = \mathbf{a} + \xi(\mathbf{b} - \mathbf{a})$.

Applying Rolle's Theorem to $F_2(t) = f_2(\mathbf{a} + t(\mathbf{b} - \mathbf{a}))$ gives a point $\mathbf{x}_2 \in A$ with $(b_1 - a_1)\frac{\partial f_2}{\partial x_1}(\mathbf{x}_2) + (b_2 - a_2)\frac{\partial f_2}{\partial x_2}(\mathbf{x}_2) = 0$. Regarding these as simultaneous equations for the non-zero numbers $b_1 - a_1$, $b_2 - a_2$, or by elimination, we see that $G(\mathbf{x}_1, \mathbf{x}_2) = 0$. Since $G(\mathbf{x}, \mathbf{y})$ is continuous, because the partial derivatives of f_1 and f_2 are, if $G(\mathbf{0}, \mathbf{0}) \neq 0$ then $\exists \delta > 0$ such that for $\|\mathbf{x}\| < \delta$ and $\|\mathbf{y}\| < \delta$ $G(\mathbf{x}, \mathbf{y}) \neq 0$. If $r < \delta$ this contradicts the existence of distinct points $\mathbf{a}$, $\mathbf{b} \in A$ above with $f_1(\mathbf{a}) = f_1(\mathbf{b})$ and $f_2(\mathbf{a}) = f_2(\mathbf{b})$.

22.* (i) $y_1 - \lambda y_2$ and the zero function both satisfy (*) with $y(a) = y'(a) = 0$.

(ii) If $y_1 = \lambda y_2$, then $\alpha_1 \beta_2 = \alpha_2 \beta_1$ by (i).

(iii) Let y satisfy (*) and *define* α and β by $y(a) = \alpha$, $y'(a) = \beta$. Then there are constants λ and μ with $\lambda \alpha_1 + \mu \alpha_2 = \alpha$ and $\lambda \beta_1 + \mu \beta_2 = \beta$. The function $y_3 = \lambda y_1 + \mu y_2$ satisfies (*) and has $y_3(a) = \alpha$ and $y_3'(a) = \beta$ so by the uniqueness part of Theorem 14.17 $y = y_3$.

Notation Index

A stroke through a symbol denotes 'not', so $a \notin A$ means that a does not belong to A. The same applies to $\nless$, $\ngtr$, $\nleq$, $\ngeq$, $a_n \nrightarrow a$, etc.

Subject Index

Albion books on Mathematics and its Applications

FUNDAMENTALS OF UNIVERSITY MATHEMATICS
COLIN McGREGOR, JOHN NIMMO and WILSON W. STOTHERS, Department of Mathematics, University of Glasgow

A unified course for first year mathematics, bridging the school/university gap, suitable for pure and applied mathematics courses, and those leading to degrees in physics, chemical physics, computing science, or statistics.

SIGNAL PROCESSING IN ELECTRONIC COMMUNICATIONS
for engineers and mathematicians
MICHAEL J. CHAPMAN, DAVID P. GOODALL, and NIGEL C. STEELE, School of Mathematics and Information Sciences, University of Coventry

This text for advanced undergraduate and postgraduate courses in electronic engineering, applied mathematics, and computer science, deals with signal processing as an important aspect of electronic communications in its role of transmitting information, and the language of its expression. It develops the required mathematics in an interesting and informative way, leading to confidence on the part of the reader.

CALCULUS
Introduction to Theory and Applications in Physical and Life Science
R.M.JOHNSON, Senior Lecturer, Department of Mathematics and Statistics, University of Paisley, Paisley, Scotland.

A lucid text conveying clear understanding of the fundamentals and applications for first year undergraduates in applied mathematics, computing, physics, electrical, mechanical and civil engineering, chemical science, biology and life science.

LINEAR DIFFERENTIAL AND DIFFERENCE EQUATIONS
a systems approach for mathematicians and engineers
R.M.JOHNSON, Senior Lecturer, Department of Mathematics and Statistics, University of Paisley, Paisley, Scotland.

An advanced text for senior undergraduates and graduates, and professional workers in applied mathematics, and electrical and mechanical engineering.

"Should find wide application by undergraduate students in engineering and computer science ... the author is to be congratulated on the importance that he attaches to conveying the parallelism of continuous and discrete systems" - *Institute of Electrical Engineers (IEE) Proceedings*

TEACHING AND LEARNING MATHEMATICAL MODELLING
Editors: S.K. HOUSTON, University of Ulster, Northern Ireland; W.BLUM, University of Kassel, Germany; IAN HUNTLEY, University of Bristol;. N.T.NEILL, University of Ulster, Northern Ireland

Mathematicians from 10 countries contribute to mathematical modelling. Interdisci plinary topics reflect applications in mechanics and engineering, computing science, traffic control, business studies, and mathematics (fractals and analysis).

Full details available from Ellis Horwood, MBE, Albion Publishing Limited, Coll House, Westergate, Chichester, West Sussex PO20 6QL